Springer Proceedings in Mathematics

Volume 4

For further volumes:
http://www.springer.com/series/8806

Springer Proceedings in Mathematics

The book series will feature volumes of selected contributions from workshops and conferences in all areas of current research activity in mathematics. Besides an overall evaluation, at the hands of the publisher, of the interest, scientific quality, and timeliness of each proposal, every individual contribution will be refereed to standards comparable to those of leading mathematics journals. It is hoped that this series will thus propose to the research community well-edited and authoritative reports on newest developments in the most interesting and promising areas of mathematical research today.

Jaroslav Fořt • Jiří Fürst • Jan Halama
Raphaèle Herbin • Florence Hubert
Editors

Finite Volumes for Complex Applications VI
Problems & Perspectives

FVCA 6, International Symposium,
Prague, June 6-10, 2011, Volume 2

Springer

Editors
Jaroslav Fořt
Czech Technical University
Faculty of Mechanical Engineering
Karlovo náměstí 13
Prague
Czech Republic
Jaroslav.Fort@fs.cvut.cz

Jiří Fürst
Czech Technical University
Faculty of Mechanical Engineering
Karlovo náměstí 13
Prague
Czech Republic
Jiri.Furst@fs.cvut.cz

Jan Halama
Czech Technical University
Faculty of Mechanical Engineering
Karlovo náměstí 13
Prague
Czech Republic
Jan.Halama@fs.cvut.cz

Raphaèle Herbin
Université Aix-Marseille
LATP
Laboratoire d'Analyse
Probabilités et T
rue Joliot Curie 39
13453 Marseille
France
Raphaele.Herbin@latp.univ-mrs.fr

Florence Hubert
Université Aix-Marseille
LATP
Laboratoire d'Analyse
Probabilités et T
rue Joliot Curie 39
13453 Marseille
France
Florence.Hubert@latp.univ-mrs.fr

ISSN 2190-5614 e-ISSN 2190-5622
ISBN 978-3-642-20670-2 e-ISBN 978-3-642-20671-9
DOI 10.1007/978-3-642-20671-9
Springer Heidelberg Dordrecht London New York

Library of Congress Control Number: 2011930263

© Springer-Verlag Berlin Heidelberg 2011

This work is subject to copyright. All rights are reserved, whether the whole or part of the material is concerned, specifically the rights of translation, reprinting, reuse of illustrations, recitation, broadcasting, reproduction on microfilm or in any other way, and storage in data banks. Duplication of this publication or parts thereof is permitted only under the provisions of the German Copyright Law of September 9, 1965, in its current version, and permission for use must always be obtained from Springer. Violations are liable to prosecution under the German Copyright Law.

The use of general descriptive names, registered names, trademarks, etc. in this publication does not imply, even in the absence of a specific statement, that such names are exempt from the relevant protective laws and regulations and therefore free for general use.

Cover design: deblik, Berlin

Printed on acid-free paper

Springer is part of Springer Science+Business Media (www.springer.com)

Editors Preface

The sixth International Symposium on Finite Volumes for Complex Applications, held in Prague (Czech Republic, June 2011) follows the series of symposiums held successively in Rouen (France, 1996), Duisburg (Germany, 1999), Porquerolles (France, 2002), Marrakech (Morocco, 2005), Aussois (France, 2008).

The sixth symposium, similarly to the previous ones, gives the opportunity of a large and critical discussion about the various aspects of finite volumes and related methods: mathematical results, numerical techniques, but also validations via industrial applications and comparisons with experimental test results.

This book tries to assemble the recent advances in both the finite volume method itself (theoretical aspects of the methods, new or improved algorithms, numerical implementation problems, benchmark problems and efficient solvers) as well as its application in complex problems in industry, environmental sciences, medicine and other fields of technology, so as to bring together the academic world and the industrial world. The topics of the proceedings reflect this wide range of perspectives and include: advanced schemes and methods (complex grid topology, higher order methods, efficient implementation), convergence and stability analysis, global error analysis, limits of methods, purely multidimensional difficulties, non homogeneous systems with stiff source terms, complex geometries and adaptivity, complexity, efficiency and large computations, chaotic problems (turbulence, ignition, mixing, ...), new fields of application, comparisons with experimental results. The proceedings also include the results to a benchmark on three–dimensional anisotropic and heterogeneous diffusion problems, which was designed to test some 16 different schemes, among which finite volume methods, finite element methods, discontinuous Galerkin methods, mimetic methods and discrete gradient schemes. A new feature of this benchmark is the comparison of various iterative solvers on the matrices resulting from the different schemes.

Of course, the success of the symposium crucially depends on the quality of the contributions. Therefore we would like to express many thank all the authors of regular papers, who provided high quality papers on the above mentioned wide range of subjects, or contributed to the 3D anisotropic diffusion benchmark. The level the contributions was ensured by the Scientific Committee members, who organized the reviewing process of each paper. We express our gratitude to members of the Scientific Committee as well as to many other reviewers.

The symposium could not have been organized without the local support of Czech Technical University, Faculty of Mechanical Engineering and financial support of French contributors: CMLA ENS Cachan, IFP Energies nouvelles, IRSN, LATP Université Aix Marseille I, MOMAS group, Université Paris XIII, Université Paris Est Marne la Vallée, Université Pierre et Marie Curie.

Finally we would like to thank Springer Verlag Editor's team for their cooperation in the proceedings preparation, conference secretary T. Němcová and all others, who ensured logistic and communication before and during the conference.

Jaroslav Fořt
Jiří Fürst
Jan Halama
Raphaèle Herbin
Florence Hubert

Organization

Committees

Organizing Committee:

Jaroslav Fořt
Jiří Fürst
Jan Halama
Rémi Abgrall
Fayssal Benkhaldoun

Robert Eymard
Jean-Michel Ghidaglia
Jean-Marc Hérard
Raphaèle Herbin
Martin Vohralík

Scientific Committee:

Rémi Abgrall
Brahim Amaziane
Fayssal Benkhaldoun
Vít Dolejší
François Dubois
Denys Dutykh
Robert Eymard
Jaroslav Fořt
Jürgen Fuhrmann
Jiří Fürst
Thierry Gallouet
Jean-Michel Ghidaglia

Hervé Guillard
Jan Halama
Khaled Hassouni
Jean-Marc Hérard
Raphaèle Herbin
Florence Hubert
Raytcho Lazarov
Karol Mikula
Mario Ohlberger
Frédéric Pascal
Martin Vohralík

Contents

Vol. 1

Part I Regular Papers

Volume-Agglomeration Coarse Grid In Schwarz Algorithm 3
H. Alcin, O. Allain, and A. Dervieux

A comparison between the meshless and the finite volume
methods for shallow water flows... 13
Yasser Alhuri, Fayssal Benkhaldoun, Imad Elmahi, Driss Ouazar,
Mohammed Seaïd, and Ahmed Taik

Time Compactness Tools for Discretized Evolution Equations
and Applications to Degenerate Parabolic PDEs 21
Boris Andreianov

Penalty Methods for the Hyperbolic System Modelling
the Wall-Plasma Interaction in a Tokamak 31
Philippe Angot, Thomas Auphan, and Olivier Guès

A Spectacular Vector Penalty-Projection Method for Darcy
and Navier-Stokes Problems 39
Philippe Angot, Jean-Paul Caltagirone, and Pierre Fabrie

Numerical Front Propagation Using Kinematical Conservation
Laws 49
K.R. Arun, M. Lukáčová-Medviďová, and P. Prasad

Preservation of the Discrete Geostrophic Equilibrium
in Shallow Water Flows... 59
E. Audusse, R. Klein, D.D. Nguyen, and S. Vater

Arbitrary order nodal mimetic discretizations of elliptic problems on polygonal meshes 69
Lourenço Beirão da Veiga, Konstantin Lipnikov, and Gianmarco Manzini

Adaptive cell-centered finite volume method for non-homogeneous diffusion problems: Application to transport in porous media 79
Fayssal Benkhaldoun, Amadou Mahamane, and Mohammed Seaïd

A Generalized Rusanov method for Saint-Venant Equations with Variable Horizontal Density 89
Fayssal Benkhaldoun, Kamel Mohamed, and Mohammed Seaïd

Hydrostatic Upwind Schemes for Shallow–Water Equations 97
Christophe Berthon and Françoise Foucher

Finite Volumes Asymptotic Preserving Schemes for Systems of Conservation Laws with Stiff Source Terms 107
C. Berthon and R. Turpault

Development of DDFV Methods for the Euler Equations 117
Christophe Berthon, Yves Coudière, and Vivien Desveaux

Comparison of Explicit and Implicit Time Advancing in the Simulation of a 2D Sediment Transport Problem 125
M. Bilanceri, F. Beux, I. Elmahi, H. Guillard, and M.V. Salvetti

Numerical Simulation of the Flow in a Turbopump Inducer in Non-Cavitating and Cavitating Conditions 135
M. Bilanceri, F. Beux, and M.V. Salvetti

On Some High Resolution Schemes for Stably Stratified Fluid Flows ... 145
Tomáš Bodnár and Luděk Beneš

Convergence Analysis of the Upwind Finite Volume Scheme for General Transport Problems 155
Franck Boyer

A Low Degree Non–Conforming Approximation of the Steady Stokes Problem with an Eddy Viscosity 165
F. Boyer, F. Dardalhon, C. Lapuerta, and J.-C. Latché

Some Abstract Error Estimates of a Finite Volume Scheme for the Wave Equation on General Nonconforming Multidimensional Spatial Meshes 175
Abdallah Bradji

A Convergent Finite Volume Scheme for Two-Phase Flows in Porous Media with Discontinuous Capillary Pressure Field 185
K. Brenner, C. Cancès, and D. Hilhorst

Contents

Uncertainty Quantification for a Clarifier–Thickener Model with Random Feed 195
Raimund Bürger, Ilja Kröker, and Christian Rohde

Asymptotic preserving schemes in the quasi-neutral limit for the drift-diffusion system 205
Chainais-Hillairet Claire and Vignal Marie-Hélène

A Posteriori Error Estimates for Unsteady Convection–Diffusion–Reaction Problems and the Finite Volume Method 215
Nancy Chalhoub, Alexandre Ern, Tony Sayah, and Martin Vohralík

Large Time-Step Numerical Scheme for the Seven-Equation Model of Compressible Two-Phase Flows 225
Christophe Chalons, Frédéric Coquel, Samuel Kokh, and Nicole Spillane

Asymptotic Behavior of the Scharfetter–Gummel Scheme for the Drift-Diffusion Model 235
Marianne CHATARD

A Finite Volume Solver for Radiation Hydrodynamics in the Non Equilibrium Diffusion Limit 245
D. Chauveheid, J.-M. Ghidaglia, and M. Peybernes

An Extension of the MAC Scheme to some Unstructured Meshes 253
Eric Chénier, Robert Eymard, and Raphaèle Herbin

Multi-dimensional Optimal Order Detection (MOOD) — a Very High-Order Finite Volume Scheme for Conservation Laws on Unstructured Meshes 263
S. Clain, S. Diot, and R. Loubère

A Relaxation Approach for Simulating Fluid Flows in a Nozzle 273
Frédéric Coquel, Khaled Saleh, and Nicolas Seguin

A CeVeFE DDFV scheme for discontinuous anisotropic permeability tensors 283
Yves Coudière, Florence Hubert, and Gianmarco Manzini

Multi-Water-Bag Model And Method Of Moments For The Vlasov Equation 293
Anaïs Crestetto and Philippe Helluy

Comparison of Upwind and Centered Schemes for Low Mach Number Flows 303
Thu–Huyen DAO, Michael NDJINGA, and Frédéric MAGOULES

On the Godunov Scheme Applied to the Variable Cross-Section Linear Wave Equation 313
Stéphane Dellacherie and Pascal Omnes

Towards stabilization of cell-centered Lagrangian methods for compressible gas dynamics.. 323
Bruno Després and Emmanuel Labourasse

Hybrid Finite Volume Discretization of Linear Elasticity Models on General Meshes.. 331
Daniele A. Di Pietro, Robert Eymard, Simon Lemaire, and Roland Masson

An A Posteriori Error Estimator for a Finite Volume Discretization of the Two-phase Flow... 341
Daniele A. Di Pietro, Martin Vohralík, and Carole Widmer

A Two-Dimensional Relaxation Scheme for the Hybrid Modelling of Two-Phase Flows 351
Kateryna Dorogan, Jean-Marc Hérard, and Jean-Pierre Minier

Finite Volume Method for Well-Driven Groundwater Flow............... 361
Milan Dotlić, Dragan Vidović, Milan Dimkić, Milenko Pušić, and Jovana Radanović

Adaptive Reduced Basis Methods for Nonlinear Convection–Diffusion Equations 369
Martin Drohmann, Bernard Haasdonk, and Mario Ohlberger

Adaptive Time-Space Algorithms for the Simulation of Multi-scale Reaction Waves.. 379
Max Duarte, Marc Massot, Stéphane Descombes, and Thierry Dumont

Dispersive wave runup on non-uniform shores 389
Denys Dutykh, Theodoros Katsaounis, and Dimitrios Mitsotakis

MAC Schemes on Triangular Meshes...................................... 399
Robert Eymard, Jürgen Fuhrmann, and Alexander Linke

Multiphase Flow in Porous Media Using the VAG Scheme................ 409
Robert Eymard, Cindy Guichard, Raphaèle Herbin, and Roland Masson

Grid Orientation Effect and MultiPoint Flux Approximation 419
Robert Eymard, Cindy Guichard, and Roland Masson

Gradient Schemes for Image Processing 429
Robert Eymard, Angela Handlovičová, Raphaèle Herbin, Karol Mikula, and Olga Stašová

Gradient Scheme Approximations for Diffusion Problems................ 439
Robert Eymard and Raphaèle Herbin

Contents

Cartesian Grid Method for the Compressible Euler Equations 449
M. Asif Farooq and B. Müller

Compressible Stokes Problem with General EOS 457
A. Fettah and T. Gallouët

**Asymptotic Preserving Finite Volumes Discretization
For Non-Linear Moment Model On Unstructured Meshes** 467
Emmanuel Franck, Christophe Buet, and Bruno Després

**Mass Conservative Coupling Between Fluid Flow and Solute
Transport** 475
Jürgen Fuhrmann, Alexander Linke, and Hartmut Langmach

Large Eddy Simulation of the Stable Boundary Layer 485
Vladimír Fuka and Josef Brechler

3D Unsteady Flow Simulation with the Use of the ALE Method 495
Petr Furmánek, Jiří Fürst, and Karel Kozel

FVM-FEM Coupling and its Application to Turbomachinery 505
J. Fořt, J. Fürst, J. Halama, K. Kozel, P. Louda, P. Sváček,
Z. Šimka, P. Pánek, and M. Hajsman

Charge Transport in Semiconductors and a Finite Volume Scheme 513
Klaus Gärtner

Playing with Burgers's Equation 523
T. Gallouët, R. Herbin, J.-C. Latché, and T.T. Nguyen

**On Discrete Sobolev–Poincaré Inequalities
for Voronoi Finite Volume Approximations** 533
Annegret Glitzky and Jens A. Griepentrog

A Simple Second Order Cartesian Scheme for Compressible Flows 543
Y. Gorsse, A. Iollo, and L. Weynans

**Efficient Implementation of High Order Reconstruction
in Finite Volume Methods** 553
Florian Haider, Pierre Brenner, Bernard Courbet,
and Jean-Pierre Croisille

**A Well-Balanced Scheme For Two-Fluid Flows In Variable
Cross-Section ducts** 561
Philippe Helluy and Jonathan Jung

Discretization of the viscous dissipation term with the MAC scheme 571
F. Babik, R. Herbin, W. Kheriji, and J.-C. Latché

A Sharp Contact Discontinuity Scheme for Multimaterial Models 581
Angelo Iollo, Thomas Milcent, and Haysam Telib

Numerical Simulation of Viscous and Viscoelastic Fluids Flow by Finite Volume Method 589
Radka Keslerová and Karel Kozel

An Aggregation Based Algebraic Multigrid Method Applied to Convection-Diffusion Operators 597
Sana Khelifi, Namane Méchitoua, Frank Hülsemann, and Frédéric Magoulès

Stabilized DDFV Schemes For The Incompressible Navier-Stokes Equations 605
Stella Krell

Higher-Order Reconstruction: From Finite Volumes to Discontinuous Galerkin 613
Václav Kučera

Flux-Based Approach for Conservative Remap of Multi-Material Quantities in 2D Arbitrary Lagrangian-Eulerian Simulations 623
Milan Kucharik and Mikhail Shashkov

Optimized Riemann Solver to Compute the Drift-Flux Model 633
Anela Kumbaro and Michaël Ndjinga

Finite Volume Schemes for Solving Nonlinear Partial Differential Equations in Financial Mathematics 643
Pavol Kútik and Karol Mikula

Monotonicity Conditions in the Mimetic Finite Difference Method 653
Konstantin Lipnikov, Gianmarco Manzini, and Daniil Svyatskiy

Discrete Duality Finite Volume Method Applied to Linear Elasticity 663
Benjamin Martin and Frédéric Pascal

Model Adaptation for Hyperbolic Systems with Relaxation 673
Hélène Mathis and Nicolas Seguin

Inflow-Implicit/Outflow-Explicit Scheme for Solving Advection Equations 683
Karol Mikula and Mario Ohlberger

4D Numerical Schemes for Cell Image Segmentation and Tracking 693
K. Mikula, N. Peyriéras, M. Remešíková, and M. Smíšek

Rhie-Chow interpolation for low Mach number flow computation allowing small time steps 703
Yann Moguen, Tarik Kousksou, Pascal Bruel, Jan Vierendeels, and Erik Dick

Study and Approximation of IMPES Stability: the CFL Criteria .... C. Preux and F. McKee	713
Numerical Solution of 2D and 3D Atmospheric Boundary Layer Stratified Flows .... Jiří Šimonek, Karel Kozel, and Zbyněk Jaňour	723
On The Numerical Validation Study of Stratified Flow Over 2D–Hill Test Case .... Sládek Ivo, Kozel Karel, and Janour Zbynek	731
A Multipoint Flux Approximation Finite Volume Scheme for Solving Anisotropic Reaction–Diffusion Systems in 3D .... Pavel Strachota and Michal Beneš	741
Higher Order Chimera Grid Interface for Transonic Turbomachinery Applications .... Petr Straka	751
Application of Nonlinear Monotone Finite Volume Schemes to Advection-Diffusion Problems............................ Yuri Vassilevski, Alexander Danilov, Ivan Kapyrin, and Kirill Nikitin	761
Scale-selective Time Integration for Long-Wave Linear Acoustics .... Stefan Vater, Rupert Klein, and Omar M. Knio	771
Nonlocal Second Order Vehicular Traffic Flow Models And Lagrange-Remap Finite Volumes........................ Florian De Vuyst, Valeria Ricci, and Francesco Salvarani	781
Unsteady Numerical Simulation of the Turbulent Flow around an Exhaust Valve .... M. Žaloudek, H. Deconinck, and J. Fořt	791

Vol. 2

Part II Invited Papers

Lowest order methods for diffusive problems on general meshes: A unified approach to definition and implementation .... Daniele A. Di Pietro and Jean-Marc Gratien	803
A Unified Framework for a posteriori Error Estimation in Elliptic and Parabolic Problems with Application to Finite Volumes .... Alexandre Ern and Martin Vohralík	821
Staggered discretizations, pressure correction schemes and all speed barotropic flows .... L. Gastaldo, R. Herbin, W. Kheriji, C. Lapuerta, and J.-C. Latché	839

ALE Method for Simulations of Laser-Produced Plasmas 857
Liska R., Kuchařík M., Limpouch J., Renner O., Váchal P.,
Bednárik L., and Velechovský J.

A two-dimensional finite volume solution of dam-break hydraulics over erodible sediment beds 875
Fayssal Benkhaldoun, Imad Elmahi, Saïda Sari, and Mohammed Seaïd

Part III   Benchmark Papers

3D Benchmark on Discretization Schemes for Anisotropic Diffusion Problems on General Grids 895
Robert Eymard, Gérard Henry, Raphaèle Herbin, Florence Hubert, Robert Klöfkorn, and Gianmarco Manzini

Benchmark 3D: a linear finite element solver 931
Hanen Amor, Marc Bourgeois, and Gregory Mathieu

Benchmark 3D: a version of the DDFV scheme with cell/vertex unknowns on general meshes 937
Boris Andreianov, Florence Hubert, and Stella Krell

Benchmark 3D: Symmetric Weighted Interior Penalty Discontinuous Galerkin Scheme 949
Peter Bastian

Benchmark 3D: A Mimetic Finite Difference Method 961
Peter Bastian, Olaf Ippisch, and Sven Marnach

Benchmark 3D: A Composite Hexahedral Mixed Finite Element 969
Ibtihel Ben Gharbia, Jérôme Jaffré, N. Suresh Kumar, and Jean E. Roberts

Benchmark 3D: CeVeFE-DDFV, a discrete duality scheme with cell/vertex/face+edge unknowns 977
Yves Coudière, Florence Hubert, and Gianmarco Manzini

Benchmark 3D: The Cell-Centered Finite Volume Method Using Least Squares Vertex Reconstruction ("Diamond Scheme") 985
Yves Coudière and Gianmarco Manzini

Benchmark 3D: A Monotone Nonlinear Finite Volume Method for Diffusion Equations on Polyhedral Meshes 993
Alexander Danilov and Yuri Vassilevski

Benchmark 3D: the SUSHI Scheme 1005
Robert Eymard, Thierry Gallouët, and Raphaèle Herbin

Contents   xvii

Benchmark 3D: the VAG scheme.. 1013
Robert Eymard, Cindy Guichard, and Raphaèle Herbin

**Benchmark 3D: The Compact Discontinuous
Galerkin 2 Scheme** 1023
Robert Klöfkorn

**Benchmark 3D: Mimetic Finite Difference Method
for Generalized Polyhedral Meshes** 1035
Konstantin Lipnikov and Gianmarco Manzini

**Benchmark 3D: CeVe-DDFV, a Discrete Duality Scheme
with Cell/Vertex Unknowns** 1043
Yves Coudière and Charles Pierre

**Benchmark 3D: A multipoint flux mixed finite element method
on general hexahedra** 1055
Mary F. Wheeler, Guangri Xue, and Ivan Yotov

Part II
Invited Papers

Lowest order methods for diffusive problems on general meshes: A unified approach to definition and implementation

Daniele A. Di Pietro and Jean-Marc Gratien

Abstract In this work we propose an original point of view on lowest order methods for diffusive problems which lays the pillars of a `C++` multi-physics, `FreeFEM`-like platform. The key idea is to regard lowest order methods as (Petrov)-Galerkin methods based on possibly incomplete, broken polynomial spaces defined from a gradient reconstruction. After presenting some examples of methods entering the framework, we show how implementation strategies common in the finite element context can be extended relying on the above definition. Several examples are provided throughout the presentation, and programming details are often omitted to help the reader unfamiliar with advanced `C++` programming techniques.

Keywords Lowest-order methods, Domain specific embedded language, Petrov-Galerkin methods, cell centered Galerkin methods, hybrid finite volume methods
MSC2010: 65Y99, 65N08, 65N30

1 Introduction

An increasing amount of attention has recently been given to the discretization of diffusive problems on general meshes. Lowest order methods possibly featuring conservation of physical quantities are traditionally employed in industrial applications where computational cost is a crucial issue. In this context, the main interest of handling general meshes is to reduce the number of elements required to represent complicate domains. In sedimentary basin modeling, non-standard elements may also appear due to the erosion of geological layers. Different ways to adapt finite volume and finite element methods to general, possibly non-conforming

---

Daniele A. Di Pietro and Jean-Marc Gratien
IFP Energies nouvelles, e-mail: dipietrd@ifpenergiesnouvelles.fr, j-marc.gratien@ifpenergiesnouvelles.fr

polyhedral meshes have been proposed. In the context of cell centered finite volume methods, we recall, in particular, the classical works of Aavatsmark, Barkve, Bøe and Mannseth [1] and Edwards and Rogers [15] on multipoint fluxes. More recently, two ways of extending the mixed finite element philosophy to general meshes have been proposed independently by Brezzi, Lipnikov, Shashkov and Simoncini [5, 6] (mimetic finite difference methods) and by Droniou and Eymard [13] (mixed/hybrid finite volume methods). Yet another perspective is considered by Eymard, Gallouët and Herbin [17], who show, in particular, that face unknowns can be selectively used as Lagrange multipliers for the flux continuity constraint or be eliminated using a consistent interpolator (SUSHI scheme). The strong link between the strategies above has been highlighted by Droniou, Eymard, Gallouët and Herbin [14]. A slightly different approach based on the analogy between lowest order methods in variational formulation and discontinuous Galerkin methods has been proposed by the author in [8–10] (cell centered Galerkin methods). The key advantage of this approach is that it largely benefits from the well-established theory for discontinuous Galerkin methods applied to diffusive problems [11]. All of the methods above have been (or can be) extended to several classical problems for which the discretization of second order diffusive terms is central.

In this work we present a unified implementation covering a wide range of lowest order methods and applications based on similar experiences in the context of finite element methods. Finite element libraries have nowadays reached a good level of maturity, and user-friendly front-ends are provided in several cases. Just to mention a few, we recall Feel++ [20] (formerly known as Life), FEniCS [19], FreeFEM++ [7]. Our goal is to show that similar tools can be conceived and implemented for lowest order methods. The starting point is to reformulate the method at hand as a (Petrov)-Galerkin scheme based on possibly incomplete broken affine spaces. This new unified perspective, drawing on the lines of [9], allows, in particular, to recycle many ideas originally developed for finite elements. A major difference, however, is that the lowest order methods considered herein are often based on reconstructions of first order differential operators which may depend on problem data such as the diffusion coefficient or the boundary condition. As a consequence, the classical approach based on a table of degrees of freedom computed from a mesh and a finite element (see, e.g., [16, Chapters 7–8]) is no longer adequate. This issue is solved by introducing the programming counterpart of tensor-valued linear combinations of (globally numbered) degrees of freedom. This concept allows, in particular, to reproduce a finite element-like matrix assembly with local contributions stemming from integrals over mesh elements and faces. A further layer of abstraction is added by defining a domain-specific language (DSL) for variational formulations. The DSL is closely inspired by that of Feel++, the most noticeable differences being the type-based identification of test and trial functions and the possibility to store the expressions defining linear and bilinear forms independently of their algebraic representation. Another novelty is the introduction of tensor-like notation for systems of PDEs. Domain-specific languages and generative programming are an established tool to break down the complexity of industrial applications by distinguishing the actors that tackle different aspects of the problem, and providing each of them with means of expression as close as possible to his/her technical jargon. An important advantage of the DSL is that it potentially allows to combine lowest order methods with more standard discretizations techniques in a seamless way. In the presentation we try to avoid all technicalities and to pinpoint the main difficulties as well as the proposed solutions. Although the language of choice is C++, the listings are rather to be intended as pseudo-code since simplifications are often made to improve readability. The actual implementation is based on the Arcane framework [18], a proprietary platform conjointly developed at CEA-DAM and IFP Energies nouvelles which takes care of technical aspects such as memory management, parallelism and post-processing.

The material is organized as follows. In §2 we propose a unified perspective and show how several lowest order methods can fit in there for a simple diffusion problem. In §3 we discuss the implementation. More specifically, we first discuss the solutions to the issues that arise when trying to mimic the finite element approach and then present a DSL which allows to conceal the related technicalities.

2 Definition

2.1 Discrete setting

Let $\Omega \subset \mathbb{R}^d$, $d \geq 2$, denote a bounded connected polyhedral domain. The first ingredient in the definition of lowest order methods is a suitable discretization of Ω. We denote by $\mathcal{T}_h$ a finite collection of nonempty, disjoint open polyhedra $\mathcal{T}_h = \{T\}$ forming a partition of Ω such that $h = \max_{T \in \mathcal{T}_h} h_T$ and h_T denotes the diameter of the element $T \in \mathcal{T}_h$. Admissible meshes include general polyhedral discretizations with possibly nonconforming interfaces; see Fig. 1. Mesh nodes are collected in the set $\mathcal{N}_h$ and, for all $T \in \mathcal{T}_h$, $\mathcal{N}_T$ contains the nodes that lie on the boundary of T. We say that a hyperplanar closed subset F of $\overline{\Omega}$ is a mesh face if it has positive $(d-1)$-dimensional measure and if either there exist $T_1, T_2 \in \mathcal{T}_h$ such that $F \subset \partial T_1 \cap \partial T_2$ (and F is called an *interface*) or there exists $T \in \mathcal{T}_h$ such that $F \subset \partial T \cap \partial \Omega$ (and F is called a *boundary face*). Interfaces are collected in the set $\mathcal{F}_h^i$, boundary faces in $\mathcal{F}_h^b$ and we let $\mathcal{F}_h := \mathcal{F}_h^i \cup \mathcal{F}_h^b$. For all $T \in \mathcal{T}_h$ we set

$$\mathcal{F}_T := \{F \in \mathcal{F}_h \mid F \subset \partial T\}. \tag{1}$$

Symmetrically, for all $F \in \mathcal{F}_h$, we define

$$\mathcal{T}_F := \{T \in \mathcal{T}_h \mid F \subset \partial T\}.$$

The set $\mathcal{T}_F$ consists of exactly two mesh elements if $F \in \mathcal{F}_h^i$ and of one if $F \in \mathcal{F}_h^b$. For all mesh nodes $P \in \mathcal{N}_h$, $\mathcal{F}_P$ denotes the set of mesh faces sharing P, i.e.

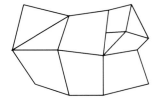 

Fig. 1 *Left.* Mesh $\mathscr{T}_h$ *Right.* Pyramidal submesh $\mathscr{P}_h$

$$\mathscr{F}_P := \{F \in \mathscr{F}_h \mid P \in F\}. \qquad (2)$$

The diameter of a face $F \in \mathscr{F}_h$ is denoted by h_F. For every interface $F \in \mathscr{F}_h^i$ we introduce an arbitrary but fixed ordering of the elements in $\mathscr{T}_F$ and let $\mathbf{n}_F = \mathbf{n}_{T_1,F} = -\mathbf{n}_{T_2,F}$, where $\mathbf{n}_{T_i,F}$, $i \in \{1,2\}$, denotes the unit normal to F pointing out of $T_i \in \mathscr{T}_F$. On a boundary face $F \in \mathscr{F}_h^b$, $\mathbf{n}_F$ denotes the unit normal pointing out of Ω. The barycenter of a face $F \in \mathscr{F}_h$ is denoted by $\overline{\mathbf{x}}_F := \int_F \mathbf{x}/|F|_{d-1}$. For each $T \in \mathscr{T}_h$ we identify a point $\mathbf{x}_T \in T$ (the *cell center*) such that T is star-shaped with respect to $\mathbf{x}_T$. For all $F \in \mathscr{F}_T$ we let

$$d_{T,F} := \mathrm{dist}(\mathbf{x}_T, F).$$

It is assumed that, for all $T \in \mathscr{T}_h$ and all $F \in \mathscr{F}_T$, $d_{T,F} > 0$ is comparable to h_T. Starting from cell centers we can define a pyramidal submesh of $\mathscr{T}_h$ as follows:

$$\mathscr{P}_h := \{\mathscr{P}_{T,F}\}_{T \in \mathscr{T}_h, F \in \mathscr{F}_T},$$

where, for all $T \in \mathscr{T}_h$ and all $F \in \mathscr{F}_T$, $\mathscr{P}_{T,F}$ denotes the open pyramid of apex $\mathbf{x}_T$ and base F, i.e.,

$$\mathscr{P}_{T,F} := \{\mathbf{x} \in T \mid \exists \mathbf{y} \in F \setminus \partial F, \exists \theta \in (0,1) \mid \mathbf{x} = \theta \mathbf{y} + (1-\theta)\mathbf{x}_T\}.$$

The pyramids $\{\mathscr{P}_{T,F}\}_{T \in \mathscr{T}_h, F \in \mathscr{F}_T}$ are nondegenerate by assumption. Let $\mathscr{S}_h$ be such that

$$\mathscr{S}_h = \mathscr{T}_h \text{ or } \mathscr{S}_h = \mathscr{P}_h. \qquad (3)$$

For all $k \geq 0$, we define the broken polynomial spaces of total degree $\leq k$ on $\mathscr{S}_h$,

$$\mathbb{P}_d^k(\mathscr{S}_h) := \{v \in L^2(\Omega) \mid \forall S \in \mathscr{S}_h, v_{|S} \in \mathbb{P}_d^k(S)\},$$

with $\mathbb{P}_d^k(S)$ given by the restriction to $S \in \mathscr{S}_h$ of the functions in $\mathbb{P}_d^k$.

Remark 1 (Admissible mesh sequence). In the context of *a priori* convergence analysis for vanishing mesh size h it is necessary to bound some quantities uniformly with respect to h. This leads to the concept of *admissible mesh sequence*. This topic is not addressed in detail herein since our focus is mainly on implementation. For a comprehensive discussion we refer to [5,6,9,13,17]; see also [11, Chapter 1].

We close this section by introducing trace operators which are of common use in the context of nonconforming finite element methods. Let v be a scalar-valued function defined on Ω smooth enough to admit on all $F \in \mathscr{F}_h$ a possibly two-valued trace. To any interface $F \subset \partial T_1 \cap \partial T_2$ we assign two nonnegtive real numbers $\omega_{T_1,F}$ and $\omega_{T_2,F}$ such that

$$\omega_{T_1,F} + \omega_{T_2,F} = 1,$$

and define the jump and weighted average of v at F for a.e. $\mathbf{x} \in F$ as

$$[\![v]\!]_F(\mathbf{x}) := v_{|T_1} - v_{|T_2}, \qquad \{\!\{v\}\!\}_{\omega,F}(\mathbf{x}) := \omega_{T_1,F} v_{|T_1}(\mathbf{x}) + \omega_{T_2,F} v_{|T_2}(\mathbf{x}). \qquad (4)$$

If $F \in \mathscr{F}_h^b$ with $F = \partial T \cap \partial \Omega$, we conventionally set $\{\!\{v\}\!\}_{\omega,F}(\mathbf{x}) = [\![v]\!]_F(\mathbf{x}) = v_{|T}(\mathbf{x})$. The subscript ω is omitted from the average operator when $\omega_{T_1,F} = \omega_{T_2,F} = \frac{1}{2}$. The dependence on $\mathbf{x}$ and on the face F is also omitted if no ambiguity arises.

2.2 An abstract perspective

The key idea to gain a unifying perspective is to regard lowest order methods as nonconforming methods based on incomplete broken affine spaces that are defined starting from the space of degrees of freedom (DOFs) $\mathbb{V}_h$. More precisely, we let

$$\mathbb{T}_h := \mathbb{R}^{\mathscr{T}_h}, \qquad \mathbb{F}_h := \mathbb{R}^{\mathscr{F}_h},$$

and consider the following choices:

$$\mathbb{V}_h = \mathbb{T}_h \text{ or } \mathbb{V}_h = \mathbb{T}_h \times \mathbb{F}_h.$$

In every case the elements of $\mathbb{V}_h$ are indexed with respect to the mesh entity they belong to. Other choices for $\mathbb{V}_h$ are possible but are not considered herein for the sake of conciseness. To fix the ideas, one can assume that the choice $\mathbb{V}_h = \mathbb{T}_h$ corresponds to cell centered finite volume (CCFV) and cell centered Galerkin (CCG) methods, while the choice $\mathbb{V}_h = \mathbb{T}_h \times \mathbb{F}_h$ leads to mimetic finite difference (MFD) and mixed/hybrid finite volume (MHFV) methods.

The key ingredient in the definition of the broken affine space is a piecewise constant linear gradient reconstruction $\mathfrak{G}_h : \mathbb{V}_h \to [\mathbb{P}_d^0(\mathscr{S}_h)]^d$ (the linearity of $\mathfrak{G}_h$ is a founding assumption for the implementation discussed in §3). Starting from $\mathfrak{G}_h$, we can define the linear operator $\mathfrak{R}_h : \mathbb{V}_h \to \mathbb{P}_d^1(\mathscr{S}_h)$ such that, for all $\mathbf{v}_h \in \mathbb{V}_h$,

$$\forall S \in \mathscr{S}_h, \ S \subset T_S \in \mathscr{T}_h, \ \forall \mathbf{x} \in S, \quad \mathfrak{R}_h(\mathbf{v}_h)_{|S} = v_{T_S} + \mathfrak{G}_h(\mathbf{v}_h)_{|S} \cdot (\mathbf{x} - \mathbf{x}_{T_S}) \in \mathbb{P}_d^1(\mathscr{S}_h). \tag{5}$$

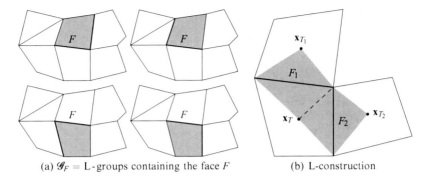

(a) $\mathscr{G}_F = $ L-groups containing the face F (b) L-construction

Fig. 2 L-construction

The operator $\mathfrak{R}_h$ maps every vector of DOFs onto a piecewise affine function belonging to $\mathbb{P}^1_d(\mathscr{S}_h)$. Hence, we can define a broken affine space as follows:

$$V_h = \mathfrak{R}_h(\mathbb{V}_h) \subset \mathbb{P}^1_d(\mathscr{S}_h). \qquad (6)$$

The operator $\mathfrak{R}_h$ is additionally assumed to be injective, so that a bijective operator can be obtained by restricting its codomain. The next section presents some examples covering the methods listed above.

2.3 Examples

In this section we focus on the model problem

$$-\nabla \cdot (\kappa \nabla u) = f, \qquad u = 0, \qquad (7)$$

where $f \in L^2(\Omega)$ and $\kappa \in [\mathbb{P}^0_d(\mathscr{T}_h)]^d$ is a piecewise constant, uniformly elliptic tensor field (possibly resulting from a homogeneization process). Problem (7) provides the paradigm to illustrate how selected lowest order methods can be recast in the framework of §2.2.

The G-method As a first example we consider the special instance of CCFV methods analyzed in [3]. A preliminary step consists in presenting the so-called L-construction introduced in [2]. The key idea of the L-construction is to use d cell and boundary face values (provided, in this case, by the homogeneous boundary condition) to express a continuous piecewise affine function with continuous diffusive fluxes. The values are selected using d neighboring faces belonging to a cell and sharing a common vertex. More precisely, we define the set of L-groups (see Fig. 2) as follows:

$$\mathscr{G} := \{\mathfrak{g} \subset \mathscr{F}_T \cap \mathscr{F}_P, T \in \mathscr{T}_h, P \in \mathscr{N}_T \mid card(\mathfrak{g}) = d\},$$

Lowest order methods for diffusive problems

with $\mathscr{F}_T$ and $\mathscr{F}_P$ given by (1) and (2) respectively. It is useful to introduce a symbol for the set of cells concurring in the L-construction: For all $\mathfrak{g} \in \mathscr{G}$, we let

$$\mathscr{T}_{\mathfrak{g}} := \{T \in \mathscr{T}_h \mid T \in \mathscr{T}_F,\ F \in \mathfrak{g}\}.$$

Let now $\mathfrak{g} \in \mathscr{G}$ and denote by $T_\mathfrak{g}$ an element $T_\mathfrak{g}$ such that $\mathfrak{g} \subset \mathscr{F}_{T_\mathfrak{g}}$ (this element may not be unique). For all $\mathbf{v}_h \in \mathbb{V}_h$ we construct the function $\xi_{\mathbf{v}_h}^\mathfrak{g}$ piecewise affine on the family of pyramids $\{\mathscr{P}_{T,F}\}_{F \in \mathfrak{g},\, T \in \mathscr{T}_\mathfrak{g}}$ such that: (i) $\xi_{\mathbf{v}_h}^\mathfrak{g}(\mathbf{x}_T) = v_T$ for all $T \in \mathscr{T}_\mathfrak{g}$ and $\xi_{\mathbf{v}_h}^\mathfrak{g}(\overline{\mathbf{x}}_F) = 0$ for all $F \in \mathfrak{g} \cap \mathscr{F}_h^\mathrm{b}$; (ii) $\xi_{\mathbf{v}_h}^\mathfrak{g}$ is affine inside $T_\mathfrak{g}$ and is continuous across every interface in the group: For all $F \in \mathfrak{g} \cap \mathscr{F}_h^\mathrm{i}$ such that $F \subset \partial T_1 \cap \partial T_2$,

$$\forall \mathbf{x} \in F, \qquad \xi_{\mathbf{v}_h\, |T_1}^\mathfrak{g}(\mathbf{x}) = \xi_{\mathbf{v}_h\, |T_2}^\mathfrak{g}(\mathbf{x});$$

(iii) $\xi_{\mathbf{v}_h}^\mathfrak{g}$ has continuous diffusive flux across every interface in the group: For all $F \in \mathfrak{g} \cap \mathscr{F}_h^\mathrm{i}$ such that $F \subset \partial T_1 \cap \partial T_2$,

$$(\kappa \nabla \xi_{\mathbf{v}_h}^\mathfrak{g})_{|T_1} \cdot \mathbf{n}_F = (\kappa \nabla \xi_{\mathbf{v}_h}^\mathfrak{g})_{|T_2} \cdot \mathbf{n}_F.$$

For further details on the L-construction including an explicit formula for $\xi_{\mathbf{v}_h}^\mathfrak{g}$ we refer to [3]. For every face $F \in \mathscr{F}_h$ we define the set $\mathscr{G}_F$ of L-groups containing F,

$$\mathscr{G}_F := \{\mathfrak{g} \in \mathscr{G} \mid F \in \mathfrak{g}\}, \qquad (8)$$

and introduce the set of nonnegative weights $\{\varsigma_{\mathfrak{g},F}\}_{\mathfrak{g} \in \mathscr{G}_F}$ such that $\sum_{\mathfrak{g} \in \mathscr{G}_F} \varsigma_{\mathfrak{g},F} = 1$. The trial space for the G-method is obtained as follows: (i) let $\mathscr{S}_h = \mathscr{P}_h$ and $\mathbb{V}_h = \mathbb{T}_h$; (ii) let $\mathfrak{G}_h = \mathfrak{G}_h^\mathrm{g}$ with $\mathfrak{G}_h^\mathrm{g}$ such that

$$\forall \mathbf{v}_h \in \mathbb{T}_h,\ \forall T \in \mathscr{T}_h,\ \forall F \in \mathscr{F}_T, \qquad \mathfrak{G}_h^\mathrm{g}(\mathbf{v}_h)_{|\mathscr{P}_{T,F}} = \sum_{\mathfrak{g} \in \mathscr{G}_F} \varsigma_{\mathfrak{g},F} \nabla \xi_{\mathbf{v}_h}^\mathfrak{g}{}_{|\mathscr{P}_{T,F}}.$$

We denote by $\mathfrak{R}_h^\mathrm{g}$ the reconstruction operator defined as in (5) with $\mathfrak{G}_h = \mathfrak{G}_h^\mathrm{g}$ and let $V_h^\mathrm{g} := \mathfrak{R}_h^\mathrm{g}(\mathbb{V}_h)$. The G-method of [3] is then equivalent to the following Petrov-Galerkin method:

$$\text{Find } u_h \in V_h^\mathrm{g} \text{ s.t. } a_h^\mathrm{g}(u_h, v_h) = \int_\Omega f v_h \text{ for all } v_h \in \mathbb{P}_d^0(\mathscr{T}_h),$$

where $a_h^\mathrm{g}(u_h, v_h) := -\sum_{F \in \mathscr{F}_h} \int_F \{\kappa \nabla_h u_h\} \cdot \mathbf{n}_F [\![v_h]\!]$ with ∇_h broken gradient on $\mathscr{S}_h$.

Remark 2 (An unconditionally stable method). The main drawback of the G-method is that stability can only be proven under quite stringent conditions; see, e.g., [3, Lemma 3.4]. A possible way to circumvent this difficulty has recently been proposed by one of the authors [10] in the context of CCG methods. The key idea is to use V_h^g both as a trial and test space, and modify the discrete bilinear form to recover both consistency and stability. Since the discrete functions in V_h^g are discontinuous across the lateral faces of the pyramids in $\mathscr{P}_h$, least-square penalization of the jumps is required to assert stability in terms of coercivity. The resulting method also enters the present framework, but is not detailed here for the sake of conciseness.

A cell centered Galerkin method The L-construction is used to define a trace reconstruction in the CCG method of [8, 10]. More specifically, for all $F \in \mathscr{F}_h^i$, we select one group $\mathfrak{g}_F \in \mathscr{G}_F$ with $\mathscr{G}_F$ defined by (8) and introduce the linear trace operator $\mathbf{T}_h^g : \mathbb{T}_h \to \mathbb{F}_h$ which maps every vector of cell centered DOFs $\mathbf{v}_h \in \mathbb{T}_h$ onto a vector $(v_F)_{F \in \mathscr{F}_h} \in \mathbb{F}_h$ such that

$$v_F = \begin{cases} \xi_{\mathbf{v}_h}^{\mathfrak{g}_F}(\overline{\mathbf{x}}_F) & \text{if } F \in \mathscr{F}_h^i, \\ 0 & \text{if } F \in \mathscr{F}_h^b. \end{cases} \qquad (9)$$

The trace operator $\mathbf{T}_h^g$ is then employed in a gradient reconstruction based on Green's formula and inspired from [17]. More precisely, we introduce the linear gradient operator $\mathfrak{G}_h^{\text{green}} : \mathbb{T}_h \times \mathbb{F}_h \to [\mathbb{P}_d^0(\mathscr{T}_h)]^d$ such that, for all $(\mathbf{v}^\mathscr{T}, \mathbf{v}^\mathscr{F}) \in \mathbb{T}_h \times \mathbb{F}_h$ and all $T \in \mathscr{T}_h$,

$$\mathfrak{G}_h^{\text{green}}(\mathbf{v}^\mathscr{T}, \mathbf{v}^\mathscr{F})_{|T} = \frac{1}{|T|_d} \sum_{F \in \mathscr{F}_T} |F|_{d-1}(v_F - v_T)\mathbf{n}_{T,F}. \qquad (10)$$

The discrete space for the CCG method under examination can then be obtained as follows: (i) let $\mathscr{S}_h = \mathscr{T}_h$ and $\mathbb{V}_h = \mathbb{T}_h$; (ii) let $\mathfrak{G}_h = \mathfrak{G}_h^{\text{ccg}}$ with $\mathfrak{G}_h^{\text{ccg}}$ such that

$$\forall \mathbf{v}_h \in \mathbb{V}_h, \qquad \mathfrak{G}_h^{\text{ccg}}(\mathbf{v}_h) = \mathfrak{G}_h^{\text{green}}(\mathbf{v}_h, \mathbf{T}_h^g(\mathbf{v}_h)). \qquad (11)$$

The reconstruction operator defined taking $\mathfrak{G}_h = \mathfrak{G}_h^{\text{ccg}}$ in (5) is denoted by $\mathfrak{R}_h^{\text{ccg}}$, and the corresponding discrete space by $V_h^{\text{ccg}} := \mathfrak{R}_h^{\text{ccg}}(\mathbb{T}_h)$. We define the weights in the average operator as follows: For all $F \in \mathscr{F}_h^i$ such that $F \subset \partial T_1 \cap \partial T_2$,

$$\omega_{T_1,F} = \frac{\lambda_{T_2,F}}{\lambda_{T_1,F} + \lambda_{T_2,F}}, \qquad \omega_{T_2,F} = \frac{\lambda_{T_1,F}}{\lambda_{T_1,F} + \lambda_{T_2,F}},$$

where $\lambda_{T_i,F} := \kappa_{|T_i} \mathbf{n}_F \cdot \mathbf{n}_F$ for $i \in \{1, 2\}$. Set, for all $(u_h, v_h) \in V_h^{\text{ccg}} \times V_h^{\text{ccg}}$,

$$a_h^{\text{ccg}}(u_h, v_h) := \int_\Omega \kappa \nabla_h u_h \cdot \nabla_h v_h - \sum_{F \in \mathscr{F}_h} \int_F [\{\kappa \nabla_h u_h\}_\omega \cdot \mathbf{n}_F [\![v_h]\!] + [\![u_h]\!] \{\kappa \nabla v_h\}_\omega \cdot \mathbf{n}_F]$$

$$+ \sum_{F \in \mathscr{F}_h} \eta \frac{\gamma_F}{h_F} \int_F [\![u_h]\!] [\![v_h]\!], \qquad (12)$$

with ∇_h broken gradient on $\mathscr{T}_h$, $\gamma_F = \frac{2\lambda_{T_1,F}\lambda_{T_2,F}}{\lambda_{T_1,F}+\lambda_{T_2,F}}$ on internal faces $F \subset \partial T_1 \cap \partial T_2$ and $\gamma_F = \kappa_{|T}\mathbf{n}_F \cdot \mathbf{n}_F$ on boundary faces $F \subset \partial T \cap \partial \Omega$. The user-dependent parameter η should be chosen large enough to ensure stability. The CCG method reads

$$\text{Find } u_h \in V_h^{\text{ccg}} \text{ s.t. } a_h^{\text{ccg}}(u_h, v_h) = \int_\Omega f v_h \text{ for all } v_h \in V_h^{\text{ccg}}. \tag{13}$$

The bilinear form a_h^{ccg} has been originally introduced by Di Pietro, Ern and Guermond [12] in the context of dG methods for degenerate advection-diffusion-reaction problems. For $\kappa = \mathbf{1}_d$, the bilinear form a_h^{ccg} becomes

$$a_h^{\text{sip}}(u_h, v_h) = \int_\Omega \nabla_h u_h \cdot \nabla_h v_h - \sum_{F \in \mathscr{F}_h} \int_F [\{\nabla_h u_h\} \cdot \mathbf{n}_F [\![v_h]\!] + [\![u_h]\!] \{\nabla_h v_h\} \cdot \mathbf{n}_F]$$
$$+ \sum_{F \in \mathscr{F}_h} \frac{\eta}{h_F} \int_F [\![u_h]\!][\![v_h]\!], \tag{14}$$

and a_h^{sip} is the bilinear form yielding the Symmetric Interior Penalty (SIP) method of Arnold [4]. For further details on the link between CCG and discontinuous Galerkin methods we refer to [8–10].

A hybrid finite volume method As a last example we consider a variant of the SUSHI scheme of [17]; see also [14] for a discussion on the link with the MFD methods of [5, 6]. This method is based on the gradient reconstruction (10), but stabilization is achieved in a rather different manner with respect to (12). More precisely, we define the linear residual operator $\mathfrak{r}_h : \mathbb{T}_h \times \mathbb{F}_h \to \mathbb{P}_d^0(\mathscr{P}_h)$ as follows: For all $T \in \mathscr{T}_h$ and all $F \in \mathscr{F}_T$,

$$\mathfrak{r}_h(\mathbf{v}_h^{\mathscr{T}}, \mathbf{v}_h^{\mathscr{F}})_{|\mathscr{P}_{T,F}} = \frac{d^{\frac{1}{2}}}{d_{T,F}} \left[v_F - v_T - \mathfrak{G}_h^{\text{green}}(\mathbf{v}_h^{\mathscr{T}}, \mathbf{v}_h^{\mathscr{F}})_{|T} \cdot (\mathbf{x}_F - \mathbf{x}_T) \right].$$

We observe, in passing, that the factor $d^{\frac{1}{2}}$ can in general be replaced by a user-defined stabilization parameter $\eta > 0$. The advantage of taking $\eta = d^{\frac{1}{2}}$ is that it yields the classical two-point method on κ-orthogonal meshes. The discrete space for SUSHI method with hybrid unknowns is obtained as follows: (i) let $\mathscr{S}_h = \mathscr{P}_h$ and $\mathbb{V}_h = \mathbb{T}_h \times \mathbb{F}_h$; (ii) let $\mathfrak{G}_h = \mathfrak{G}_h^{\text{hyb}}$ with $\mathfrak{G}_h^{\text{hyb}}$ such that, for all $(\mathbf{v}_h^{\mathscr{T}}, \mathbf{v}_h^{\mathscr{F}}) \in \mathbb{T}_h \times \mathbb{F}_h$, all $T \in \mathscr{T}_h$ and all $F \in \mathscr{F}_T$,

$$\mathfrak{G}_h^{\text{hyb}}(\mathbf{v}_h^{\mathscr{T}}, \mathbf{v}_h^{\mathscr{F}})_{|\mathscr{P}_{T,F}} = \mathfrak{G}_h^{\text{green}}(\mathbf{v}_h^{\mathscr{T}}, \mathbf{v}_h^{\mathscr{F}})_{|T} + \mathfrak{r}_h(\mathbf{v}_h^{\mathscr{T}}, \mathbf{v}_h^{\mathscr{F}})_{|\mathscr{P}_{T,F}} \mathbf{n}_{T,F}. \tag{15}$$

Denote by $\mathfrak{R}_h^{\text{hyb}}$ the reconstruction operator defined by (5) with $\mathfrak{G}_h = \mathfrak{G}_h^{\text{hyb}}$. The SUSHI method with hybrid unknowns reads

$$\text{Find } u_h \in V_h^{\text{hyb}} \text{ s.t. } a_h^{\text{sushi}}(u_h, v_h) = \int_\Omega f v_h \text{ for all } v_h \in V_h^{\text{hyb}},$$

with $a_h^{\text{sushi}}(u_h, v_h) := \int_\Omega \kappa \nabla_h u_h \cdot \nabla_h v_h$ and ∇_h broken gradient on $\mathscr{P}_h$. Alternatively, one can obtain a cell centered version by setting $\mathbb{V}_h = \mathbb{T}_h$ and replacing $\mathfrak{G}_h^{\text{hyb}}$ defined by (15) by $\mathfrak{G}_h = \mathfrak{G}_h^{\text{cc}}$ with $\mathfrak{G}_h^{\text{cc}}$ such that

$$\forall \mathbf{v}_h \in \mathbb{T}_h, \qquad \mathfrak{G}_h^{\text{cc}}(\mathbf{v}_h) = \mathfrak{G}_h^{\text{hyb}}(\mathbf{v}_h, \mathbf{T}_h^{\text{g}}(\mathbf{v}_h)), \qquad (16)$$

and trace operator $\mathbf{T}_h^{\text{g}}$ defined by (9). This variant coincides with the version proposed in [17] for homogeneous κ, but it has the advantage to reproduce piecewise affine solutions of (7) on $\mathscr{T}_h$ when κ is heterogeneous. The discrete space obtained taking $\mathfrak{G}_h = \mathfrak{G}_h^{\text{cc}}$ in (6) is labeled V_h^{cc}.

3 Implementation

The goal of this section is to lay the foundations for a DSL embedded in the C++ language which transposes the mathematical concepts of §2 into practical implementations. To illustrate the capabilities of the DSL in a nutshell, compare Listing 1 with the expression of the bilinear form a_h^{sip} (14). The material is organized as follows: §3.1 introduces the algebraic back-end aiming at replacing the table of DOFs in the context of a element-like assembly procedure; §3.2 deals with more abstract concepts that mimic function spaces, linear and bilinear forms to offer a functional front-end.

3.1 Algebraic back-end

In this section we focus on the elementary ingredients to build the terms appearing in the linear and bilinear forms of §2, which constitute the back-end of the DSL presented in §3.2.

Linear combination The point of view presented in §2 naturally leads to finite element-like assembly of local contributions stemming from integrals over elements or faces. However, a few major differences have to be taken into account: (i) the stencil of the local contributions may vary from term to term; (ii) the stencil may be data-dependent, as is the case for the methods of §2 based on the L-construction; (iii) the stencil may be non-local, as DOFs from neighboring elements may enter in local reconstructions. All of the above facts invalidate the classical approach based on a global table of DOFs inferred from a mesh and a finite element in the sense of Ciarlet. Our approach to meet the above requirements is to (i) drop the concept

Listing 1 Implementation of the bilinear form a_h^{sip} defined by (14) using the DSL of §3

```
// Define discrete spaces, test and trial functions; c.f. Table 1
typedef FunctionSpace<span<Polynomial<d, 1> >,
                     gradient<GreenFormula<LInterpolator> >
                     >::type CCGSpace;
CCGSpace Vh(𝒯ₕ);
Vh.gradientReconstruction().trace().set(DiffusionCoefficient, κ);
CCGSpace::TrialFunction uh(Vh, "uh");
CCGSpace::TestFunction  vh(Vh, "vh");
// Define the bilinear form
Form2 ah =
  integrate(All<Cell>::items(𝒯ₕ), dot(grad(uh),grad(vh)))
 -integrate(All<Face>::items(𝒯ₕ), dot(N(),avg(grad(uh)))*jump(vh))
                                 +dot(N(),avg(grad(vh)))*jump(uh))
 +integrate(All<Face>::items(𝒯ₕ), η/H()*jump(uh)*jump(vh));
// Evaluate the bilinear form
MatrixContext context(A);
evaluate(ah, context);
```

of local element, and to refer to DOFs by a unique global index; (ii) introduce the concept of LinearCombination (with template parameters to be specified in what follows), which realizes a linear application from $\mathbb{V}_h$ onto the space $\mathbb{T}_r$ of real tensors of order $r \leq 2$.

In practice, a LinearCombination is an efficient mapping of the DOFs in $\mathbb{V}_h$ onto the corresponding coefficients in $\mathbb{T}_r$. A LinearCombination l^r can indeed be thought of as a list of couples $(I, \tau_{1,I})_{I \in \mathbb{I}_1}$ where $\mathbb{I}_1 \subset \mathbb{V}_h$ is the stencil (i.e., a vector of global DOFs) and $\tau_{1,I} \in \mathbb{T}_r$, $I \in \mathbb{I}_1$, are the corresponding coefficients. To account for strongly enforced boundary conditions, LinearCombination also contains a constant coefficient $\tau_{1,0}$, so that the evaluation at $\mathbf{v}_h \in \mathbb{V}_h$ (obtained by calling the function LinearCombination.eval($\mathbf{v}_h$)) actually returns

$$l^r(\mathbf{v}_h) = \sum_{I \in \mathbb{I}_1} \tau_{1,I} v_I + \tau_{1,0} \in \mathbb{T}_r.$$

It is useful to define efficient operations such as the sum and subtraction of linear combinations, as well as different kinds of products by constants. This allows, e.g., to implement the gradient $\mathfrak{G}_h^{\text{green}}$ defined by (10) as described in line 6 of Listing 2. When needed, each DOF I can be represented as a LinearCombination containing only the couple $(I, 1)$. As a result, both the hybrid version with face unknowns (15) and the cell centered version (11) of the gradient reconstruction can be obtained from Listing 2 by simply changing the value returned by the trace interpolator $\mathbf{T}_h$.eval(F) in line 5. We also pinpoint that the tensor order is a template parameter of LinearCombination to reduce the need for dynamic allocation.

Listing 2 Implementation of the gradient reconstruction $\mathfrak{G}_h^{\mathrm{green}}$ (10) for an element $T \in \mathcal{T}_h$. The gradient $\mathfrak{G}_h^{\mathrm{ccg}}$ can be obtained by changing the value of the LinearCombination returned by $\mathbb{T}_h$ in line 5. Bufferization is used as a means to improve efficiency

```
LinearCombination<0> vT;
vT += LinearCombination<1>::Term(I_T,1.);
LinearCombination<1, Buffer> buffer;
for(F ∈ 𝓕_T) {
   const LinearCombination<0> & vF = T_h.eval(F);
   buffer += |F|_{d-1}/|T|_d (vF-vT) n_{T,F};
}
LinearCombination<1, Vector> GT;
buffer.compact(GT);
```

In the implementation, particular care must be devoted to expressions containing the sum or subtraction of two linear combinations 1_1^r and 1_2^r, since this involves computing the intersection of the corresponding set of DOFs, say $\mathbb{I}_{1_1}$ and $\mathbb{I}_{1_2}$ respectively. To overcome this difficulty, complicate expressions are computed in two steps: a first step in which duplicate DOF indices are allowed, followed by a compaction stage where the algebraic sums of the corresponding coefficients are performed. This is obtained by changing the value of the second template parameter of LinearCombination. Specifically, in Buffer-mode a LinearCombination efficiently supports adding terms and can appear in the left-hand side of an assignment operator, while Vector-mode (default) only allows to traverse its elements in a fixed order; c.f. lines 3 and 8 of Listing 2.

Linear and bilinear contributions Exploiting the concept of LinearCombination, it is possible to devise a unified treatment for local contributions stemming from integrals over elements or faces. We illustrate the main ideas using the an example: For a given $T \in \mathcal{T}_h$ and for $u_h, v_h \in V_h^{\mathrm{ccg}}$, we consider the local contribution $\mathbf{A}_{\mathrm{loc}}$ associated to the term

$$\int_T \kappa \nabla_h u_h \cdot \nabla_h v_h.$$

For the sake of simplicity we focus on the case when the constant coefficient $\tau_{1,0}$ is zero (in the example, this corresponds to the homogeneous Dirichlet boundary condition in problem (7)). The key remark is that both $(\kappa \nabla_h u_h)_{|T} = \kappa_{|T} \nabla(u_{h|T})$ and $(\nabla_h v_h)_{|T} = \nabla(v_{h|T})$ can be represented as objects of type LinearCombination<1>, say $1_u^1 = (J, \tau_{1_u,J})_{J \in \mathbb{J}}$ and $1_v^1 = (I, \tau_{1_v,I})_{I \in \mathbb{I}}$. The associated local contribution reads

$$\mathbf{A}_{\mathrm{loc}} = [|T|_d \tau_{1_v,I} \cdot \tau_{1_u,J}]_{I \in \mathbb{I}, J \in \mathbb{J}}. \tag{17}$$

Generalizing the above remark, one can implement local terms in matrix assembly as BilinearContributions which can be represented as triplets $(\mathbb{I}, \mathbb{J}, \mathbf{A}_{\mathrm{loc}})$

Lowest order methods for diffusive problems 815

Listing 3 Assembly of a bilinear and linear contribution (**A** represents here the global matrix **b** the global right-hand side vector)

```
LinearCombination<r> l_u^r, l_v^r;
// Assemble a bilinear contribution into the left-hand side
BilinearContribution<r> blc(γ, l_u^r, l_v^r);
A << blc;
// Assemble a linear contribution into the right-hand side
LinearContribution<r> lc(γ, l_v^r);
b << lc;
```

containing two vectors of DOF indices $\mathbb{I}$ and $\mathbb{J}$ and the local matrix $\mathbf{A}_{\mathrm{loc}}$. Observe, in particular, that $\mathbb{I}$ and $\mathbb{J}$ play the same role as the lines of the table of DOFs corresponding to test and trial functions supported in T in standard finite element implementations. As such, they are related to the lines and columns of the global matrix **A** to which $\mathbf{A}_{\mathrm{loc}}$ contributes,

$$\mathbf{A}(\mathbb{I},\mathbb{J}) \leftarrow \mathbf{A}(\mathbb{I},\mathbb{J}) + \mathbf{A}_{\mathrm{loc}}. \tag{18}$$

When the `LinearCombinations` concurring to a local term take values in $\mathbb{T}_r$, the vector inner product in (17) should be replaced by the appropriate tensor contraction. The additional argument γ appearing in Listing 3 serves as a multiplicative factor for the whole expression (in the above example, $\gamma = |T|_d$). More generally, γ can be a function of space and time, and may depend on discrete variables.

Similarly, `LinearContributions` serve to represent right-hand side contributions. `LinearContributions` are not detailed here for the sake of brevity. A typical assembly pattern is described in Listing 3. In particular, line 4 is the programming counterpart of (18).

3.2 Functional front-end

A further level of abstraction can be reached defining a DSL that allows to conceal all technical details and provide only the relevant components in a form as close as possible to the mathematical formulations of §2. We focus here, in particular, on the programming equivalent of discrete spaces and bilinear forms.

Function spaces Incomplete broken polynomial spaces defined by (6) are mapped onto C++ types conforming to the `FunctionSpace` concept detailed in Listing 4. The actual types are generated by a helper template class parametrized by a containing polynomial space, labeled **span**, and a piecewise constant gradient reconstruction, labeled **gradient** (labels for template arguments are here defined using the `boost::parameter` library). An example of usage is provided in lines 2–4 in Listing 1. The gradient reconstruction implicitly fixes both the

Listing 4 FunctionSpace concept

```
class FunctionSpace {
    // Types for trial and test functions
    class TrialFunction;
    class TestFunction;
    // Constructor
    FunctionSpace(const Mesh &);
    // Constant value of 𝔊_{h|S} for S ∈ 𝒮_h as a vector-valued linear combination of DOFs
    const LinearCombination<1> & Gh(S) const;
    // Value of ℜ_{h|S}(x) for x ∈ S and S ∈ 𝒮_h as a scalar-valued linear combination of DOFs
    const LinearCombination<0> & Rh(S, x) const;
};
```

Table 1 **span** and **gradient** template parameters for the class FunctionSpace corresponding to the discrete spaces of §2

Space	$\mathcal{S}_h$	**span**	**gradient**
$\mathbb{P}_d^0(\mathcal{T}_h)$	$\mathcal{T}_h$	Polynomial<d, 0>	Null
V_h^{g}	$\mathcal{P}_h$	Polynomial<d, 1>	GFormula
V_h^{ccg}	$\mathcal{T}_h$	Polynomial<d, 1>	GreenFormula<LInterpolator>
V_h^{hyb}	$\mathcal{P}_h$	Polynomial<d, 1>	SUSHIFormula<HybridUnknowns>
V_h^{cc}	$\mathcal{P}_h$	Polynomial<d, 1>	SUSHIFormula<LInterpolator>

choice (6) for the space of DOFs and the choice (3) for $\mathcal{S}_h$. The programming counterparts of the function spaces appearing in §2 are listed in Table 1.

The key role of a FunctionSpace is to bridge the gap between the algebraic representation of DOFs and the functional representation used in the methods of §2. This is achieved by the functions Gh and Rh, which are the C++ counterpart of the linear operators $\mathfrak{G}_h$ and $\mathfrak{R}_h$ respectively; see §2.1. More specifically, (i) for all $S \in \mathcal{S}_h$, Gh(S) returns a vector-valued linear combination corresponding to the (constant) restriction $\mathfrak{G}_{h|S}$; (ii) for all $S \in \mathcal{S}_h$ and all $\mathbf{x} \in S$, Rh(S, **x**) returns a scalar-valued linear combination corresponding to $\mathfrak{R}_{h|S}(\mathbf{x})$ defined according to (5). The linear combinations returned by Gh and Rh can be used to generate LinearContributions and BilinearContributions to build linear and bilinear terms as described above. A FunctionSpace also defines the types TestFunction and TrialFunction that correspond to the mathematical notions of test and trial functions in variational formulations. The main difference between a TestFunction and a TrialFunction is that the latter is associated to a vector of DOFs which is stored in memory. In addition, when used to define bilinear contributions, test and trial functions are associated to the lines and columns of the local matrix respectively. We conclude by observing that the choice of identifying test and trial functions by their type is in contrast with the approach of [20, §3.4], where special keywords accomplish this task.

Linear and bilinear forms Linear and bilinear forms are obtained as sums of linear and bilinear terms resulting from the composition of TestFunctions and

Lowest order methods for diffusive problems

Listing 5 CCG discretization of the Stokes problem

```
CCGSpace::VectorTrialFunction uh(d);
CCGSpace::VectorTestFunction vh(d);
P0Space::TrialFunction ph;
P0Space::TestFunction qh;
Range::Index i(Range(0,dim-1));
Form2 ah, bh, sh;
ah = integrate(All<Cell>::items(𝒯ₕ),
               sum(i)(dot(grad(uh(i)),grad(vh(i))) ))
   +integrate(Internal<Face>::items(𝒯ₕ),
               sum(i)(-dot(fn,avg(grad(uh(i))))*jump(vh(i))
                      -jump(uh(i))*dot(N(),avg(grad(vh(i))))
                      +η/H()*jump(uh(i))*jump(vh(i))));
bh =-integrate(Internal<Face>::items(𝒯ₕ),
               jump(ph)*dot(N(),avg(vh)));
sh = integrate(Internal<Face>::items(𝒯ₕ),H()*jump(ph)*jump(qh));
```

TrialFunctions (or unary modifications thereof) by suitable tensor contractions. Examples of tensor contractions in Listing 1 are the **dot** and * operators. Products by functions of space, time and possibly discrete variables are also allowed. In Listing 1 we also display examples of geometric operators such as **N()** and **H()**, which allow to access face normals and diameters respectively. Unary modifiers encountered in Listing 1 are **grad**, **avg** and **jump**, corresponding, respectively, to the broken gradient on $\mathscr{S}_h$ and to the average and jump operators defined by (4). When applied to a test or trial function, a unary modifier is an object capable of returning a LinearCombination at evaluation.

By default, linear and bilinear forms are represented by vectors and (sparse) matrices, but other representations are possible resulting, e.g., in matrix-free implementations. In contrast with [20], the expression corresponding to a linear (resp. bilinear) form is stored as a property of an object **Form1** (resp. **Form2**) instead of being evaluated on-the-fly. This allows, in particular, to change the operations actually performed at evaluation according to a context. Changing the representation of linear and bilinear forms thus amounts to changing the context of evaluation. An example of evaluation is provided in lines 16–17 of Listing 1, where the global matrix **A** is assembled according to the expression of ah and to the procedure defined in MatrixContext. During the evaluation, each term in the expression of ah generates a corresponding BilinearContribution, which is in turn assembled as described in §3.1.

To conclude, we present a more complicate example involving a system of PDEs. More specifically, we consider the Stokes problem:

$$-\Delta u + \nabla p = f \text{ in } \Omega, \quad \nabla \cdot u = 0 \text{ in } \Omega, \quad u = 0 \text{ on } \partial\Omega,$$

with $\langle p \rangle_\Omega = 0$ to ensure well-posedness. Let $X_h := [V_h^{\text{ccg}}]^d \times \mathbb{P}_d^0(\mathscr{T}_h)/\mathbb{R}$. In Listing 5 we present the implementation of the CCG method of [9, §3]: Find

$(u_h, p_h) \in X_h$ such that

$$a_h(u_h, v_h) + b_h(v_h, p_h) - b_h(u_h, q_h) + s_h(p_h, q_h) = \int_\Omega f \cdot v_h, \qquad \forall (v_h, q_h) \in X_h$$

where $a_h(u_h, v_h) := \sum_{i=1}^d a_h^{\text{sip}}(u_{h,i}, v_{h,i})$, $b_h(p_h, v_h) := -\sum_{F \in \mathscr{F}_h^i} \int_F [\![p_h]\!] \{\!\{v_h\}\!\} \cdot \mathbf{n}_F$ and $s_h(p_h, q_h) := \sum_{F \in \mathscr{F}_h^i} h_F \int_F [\![p_h]\!] [\![q_h]\!]$. Notice the use of the **sum** keyword.

Acknowledgements Fruitful discussions with Christophe Prud'homme (Laboratoire Jean Kuntzmann, University of Grenoble) are gratefully acknowledged. We also wish to thank all the contributors to the Arcane platform.

References

1. I. Aavatsmark, T. Barkve, Ø. Bøe, and T. Mannseth. Discretization on unstructured grids for inhomogeneous, anisotropic media, Part I: Derivation of the methods. *SIAM J. Sci. Comput.*, 19(5):1700–1716, 1998.
2. I. Aavatsmark, G. T. Eigestad, B. T. Mallison, and J. M. Nordbotten. A compact multipoint flux approximation method with improved robustness. *Numer. Methods Partial Differ. Eq.*, 24:1329–1360, 2008.
3. L. Agélas, D. A. Di Pietro, and J. Droniou. The G method for heterogeneous anisotropic diffusion on general meshes. *M2AN Math. Model. Numer. Anal.*, 44(4):597–625, 2010.
4. D. N. Arnold. An interior penalty finite element method with discontinuous elements. *SIAM J. Numer. Anal.*, 19:742–760, 1982.
5. F. Brezzi, K. Lipnikov, and M. Shashkov. Convergence of mimetic finite difference methods for diffusion problems on polyhedral meshes. *SIAM J. Numer. Anal.*, 43(5):1872–1896, 2005.
6. F. Brezzi, K. Lipnikov, and V. Simoncini. A family of mimetic finite difference methods on polygonal and polyhedral meshes. *M3AS*, 15:1533–1553, 2005.
7. I. Danaila, F. Hecht, and O. Pironneau. *Simulation numérique en C++*. Dunod, Paris, 2003. http://www.freefem.org.
8. D. A. Di Pietro. Cell centered Galerkin methods. *C. R. Acad. Sci. Paris, Ser. I*, 348:31–34, 2010.
9. D. A. Di Pietro. Cell centered Galerkin methods for diffusive problems. Submitted. Preprint available at http://hal.archives-ouvertes.fr/hal-00511125/en/, September 2010.
10. D. A. Di Pietro. A compact cell-centered Galerkin method with subgrid stabilization. *C. R. Acad. Sci. Paris, Ser. I.*, 348(1–2):93–98, 2011.
11. D. A. Di Pietro and A. Ern. *Mathematical aspects of discontinuous Galerkin methods*. Mathematics & Applications. Springer-Verlag, Berlin, 2010. To appear.
12. D. A. Di Pietro, A. Ern, and J.-L. Guermond. Discontinuous Galerkin methods for anisotropic semi-definite diffusion with advection. *SIAM J. Numer. Anal.*, 46(2):805–831, 2008.
13. J. Droniou and R. Eymard. A mixed finite volume scheme for anisotropic diffusion problems on any grid. *Num. Math.*, 105(1):35–71, 2006.
14. J. Droniou, R. Eymard, T. Gallouët, and R. Herbin. A unified approach to mimetic finite difference, hybrid finite volume and mixed finite volume methods. *M3AS, Math. Models Methods Appl. Sci.*, 20(2):265–295, 2010.
15. M. G. Edwards and C. F. Rogers. Finite volume discretization with imposed flux continuity for the general tensor pressure equation. *Comput. Geosci.*, 2:259–290, 1998.

16. A. Ern and J.-L. Guermond. *Theory and Practice of Finite Elements*, volume 159 of *Applied Mathematical Sciences*. Springer-Verlag, New York, NY, 2004.
17. R. Eymard, Th. Gallouët, and R. Herbin. Discretization of heterogeneous and anisotropic diffusion problems on general nonconforming meshes SUSHI: a scheme using stabilization and hybrid interfaces. *IMA J. Numer. Anal.*, 4(30):1009–1043, 2010.
18. G. Grospellier and B. Lelandais. The Arcane development framework. In *Proceedings of the 8th workshop on Parallel/High-Performance Object-Oriented Scientific Computing*, pages 4:1–4:11, New York, NY, USA, 2009. ACM.
19. A. Logg and G. N. Wells. DOLFIN: Automated finite element computing. *ACM TOMS*, 37, 2010.
20. C. Prud'homme. A domain specific embedded language in C++ for automatic differentiation, projection, integration and variational formulations. *Sci. Prog.*, 14(2):81–110, 2006.

The paper is in final form and no similar paper has been or is being submitted elsewhere.

A Unified Framework for a posteriori Error Estimation in Elliptic and Parabolic Problems with Application to Finite Volumes

Alexandre Ern and Martin Vohralík

Abstract We present a unified framework based on potential and flux reconstruction for guaranteed and efficient a posteriori error estimation. We consider as model problems the Laplace equation, the singularly perturbed convection-diffusion-reaction equation, and the heat equation. The analysis is performed for a wide class of space discretization schemes. Three simple conditions need to be verified, which we do for cell- and vertex-centered finite volumes for all model problems.

Keywords a posteriori error estimation, guaranteed upper bound, efficiency, robustness, elliptic and parabolic problems
MSC2010: 65M08, 65M15, 65M50, 65N08, 65N15, 65N50

1 Introduction

A posteriori error estimation is an important tool in practical computations for error control and computational efficiency by adapting the discretization parameters. In the context of finite element methods, residual-based a posteriori error estimation has been initiated by Babuška and Rheinboldt [2] over three decades ago. The application to finite volume (FV) schemes is more recent; we refer, among others, to Achdou, Bernardi, and Coquel [1], Nicaise [19], and Ohlberger [20, 21].

Alexandre Ern
Université Paris-Est, CERMICS, Ecole des Ponts, 77455 Marne la Vallée, France, e-mail: ern@cermics.enpc.fr

Martin Vohralík
UPMC Univ. Paris 06, UMR 7598, Laboratoire J.-L. Lions, 75005, Paris, France & CNRS, UMR 7598, Laboratoire J.-L. Lions, 75005, Paris, France, e-mail: vohralik@ann.jussieu.fr

The purpose of this work is to present some recent results (and extensions thereof) by the authors [9, 11, 28–30] in a general framework. The salient features of this framework can be summarized as follows. Firstly, the error upper bound is formulated in terms of a *potential* and a *flux reconstruction* which must comply with some basic physical properties related to the model problem at hand. This approach allows one to achieve *guaranteed* error upper bounds, that is, upper bounds *without undetermined constants*, which is a key feature in the context of error control. Flux-based a posteriori error estimation for elliptic problems hinges on the Prager–Synge equality [22] and was first developed, among others, by Ladevèze [18] and Haslinger and Hlaváček [14].

Next, the present approach does not rely on a specific discretization scheme (in space), that is, we bound the difference between the exact solution and an arbitrary approximate solution which is only required to be piecewise smooth. Owing to this generality, the approach encompasses a wide class of schemes including FVs and many other schemes (discontinuous Galerkin, mixed finite elements, etc.) in a *unified setting*. At this stage, quite *general meshes* (e.g., with polygonal elements and so-called hanging nodes) can be considered as well. Turning next to *local efficiency*, that is, to local lower bounds on the error, we still proceed generally without resorting to any specific discretization scheme under two additional assumptions. On the one hand, we suppose that the approximate solution, the potential and flux reconstructions, and the problem data are piecewise polynomials and that the meshes possess some regularity which we formulate by introducing a matching simplicial, shape-regular submesh. On the other hand, we assume that the potential and flux reconstructions satisfy some local approximation properties which are expressed in terms of suitable local residuals of the approximate solution (plus its jumps). Local lower bounds on the error then result from the combination of these two assumptions and the fact that the local residuals provide local lower bounds on the approximation error, as previously shown, e.g., in Verfürth [24].

This paper is organized as follows. In §2, we collect some useful notation and basic ingredients for the analysis. Then, we present our results on three model problems. In §3, we consider the Laplace equation. The aim is to present in detail the key ideas in the context of a simple model problem. In §4, we turn to the convection-diffusion-reaction equation. We focus on singularly perturbed regimes resulting from dominant convection or reaction and show how the present approach can achieve *robustness* with respect to physical parameters. In §5, we consider the heat equation and the backward Euler scheme to discretize in time. The purpose is to show how the present approach handles evolution problems including time-varying meshes. In all cases, we first derive upper and lower bounds on the approximation error in an abstract framework applicable to a wide class of discretization schemes in space. Then, we show how the framework can be applied to cell- and vertex-centered FV schemes. For the sake of simplicity, we only consider model problems with homogeneous Dirichlet boundary conditions. Inhomogeneous Dirichlet and Neumann boundary conditions can be taken into account following [29]. Finally, we observe that some interesting applications of a posteriori error estimates are not covered herein; we mention, in particular, the use of such estimates as adaptive stopping criteria for linear [15] and nonlinear [7] iterative solvers.

2 Basic ingredients

Let $\Omega \subset \mathbb{R}^d$, $d \geq 2$, be a polygonal (polyhedral) domain (open, bounded, and connected set). Let $\mathscr{T}_h$ be a partition of Ω into polygonal elements. The elements K can be *nonconvex* or *non star-shaped*. We denote by h_K the diameter of $K \in \mathscr{T}_h$ and by $\mathbf{n}_K$ its outward normal. The partition $\mathscr{T}_h$ can be *nonmatching*, that is, so-called hanging nodes are allowed. We only suppose later on (cf. Assumption 3 below) the existence of a simplicial matching and shape-regular submesh $\mathscr{S}_h$. We say that σ is a mesh side if σ has positive $(d-1)$-dimensional measure and if there are distinct $K, L \in \mathscr{T}_h$ such that $\sigma = \partial K \cap \partial L$ or if there is $K \in \mathscr{T}_h$ such that $\sigma = \partial K \cap \partial \Omega$. Mesh sides are collected in the set $\mathscr{E}_h$. We denote by h_σ the diameter of $\sigma \in \mathscr{E}_h$, we fix a unit normal to σ denoted by $\mathbf{n}_\sigma$, and define the jump across σ as the difference following the direction of $\mathbf{n}_\sigma$. Besides the usual Sobolev spaces $H^1(\Omega)$ and $H_0^1(\Omega)$, we consider the so-called broken Sobolev space $H^1(\mathscr{T}_h)$ spanned by those functions whose restriction to each element $K \in \mathscr{T}_h$ belongs to $H^1(K)$ and the so-called broken gradient operator ∇_h acting elementwise on functions in $H^1(\mathscr{T}_h)$. Additionally, we need the space $\mathbf{H}(\mathrm{div}, \Omega)$ spanned by those functions in $[L^2(\Omega)]^d$ with square-integrable weak divergence. The notation $\mathbb{P}_k(\mathscr{T}_h)$ stands for the space of piecewise polynomials of total degree $\leq k$ on $\mathscr{T}_h$, whereas, for $\mathscr{T}_h$ simplicial and matching, $\mathbf{RTN}(\mathscr{T}_h) \subset \mathbf{H}(\mathrm{div}, \Omega)$ stands for the (lowest-order) Raviart–Thomas–Nédélec finite element space [3]. For all $\mathbf{v}_h \in \mathbf{RTN}(\mathscr{T}_h)$, $\mathbf{v}_h \cdot \mathbf{n}_\sigma$ is constant on all sides $\sigma \in \mathscr{E}_h$, the univalued side fluxes $\langle \mathbf{v}_h \cdot \mathbf{n}_\sigma, 1 \rangle_\sigma$ representing the degrees of freedom.

Let $D \subset \Omega$ be a polygon or polyhedron. The Poincaré inequality states that

$$\|\varphi - \varphi_D\|_D^2 \leq C_{\mathrm{P},D} h_D^2 \|\nabla \varphi\|_D^2 \qquad \forall \varphi \in H^1(D), \tag{1}$$

where φ_D is the mean of φ over D given by $\varphi_D := (\varphi, 1)_D / |D|$. When D is convex, the constant $C_{\mathrm{P},D}$ can be evaluated as $1/\pi^2$. The constant $C_{\mathrm{P},D}$ can also be evaluated for nonconvex D, cf. [12, Lemma 10.2] or [5, §2]. Let now $K \subset \Omega$ be a simplex and let σ be one of its sides. The trace inequality states that

$$\|\varphi\|_\sigma^2 \leq C_{\mathrm{t},K,\sigma}(h_K^{-1}\|\varphi\|_K^2 + \|\varphi\|_K \|\nabla \varphi\|_K) \qquad \forall \varphi \in H^1(K). \tag{2}$$

It follows from [23, Lemma 3.12] that the constant $C_{\mathrm{t},K,\sigma}$ can be evaluated as $|\sigma| h_K / |K|$, see also [5, Theorem 4.1] for $d = 2$.

3 Laplace equation

We consider the second-order elliptic problem

$$-\Delta p = f \quad \text{in } \Omega, \tag{3a}$$

$$p = 0 \quad \text{on } \partial\Omega, \tag{3b}$$

with $f \in L^2(\Omega)$. The weak formulation consists in finding $p \in H_0^1(\Omega)$ such that

$$(\nabla p, \nabla \varphi) = (f, \varphi) \quad \forall \varphi \in H_0^1(\Omega). \tag{4}$$

The scalar-valued function $p \in H_0^1(\Omega)$ is called the *potential* and the vector-valued function $\mathbf{t} := -\nabla p \in \mathbf{H}(\text{div}, \Omega)$ the (diffusive) *flux*.

3.1 Abstract framework

The purpose of this section is to present a unified abstract framework for a posteriori error estimation in problem (3a)–(3b). In order to proceed generally, without the specification of the numerical scheme at hand, we merely suppose that we are given a function $p_h \in H^1(\mathcal{T}_h)$ (which will represent the discrete solution later on). We define the energy (semi-)norm as $|||v||| := \|\nabla_h v\|$ for all $v \in H^1(\mathcal{T}_h)$. The a posteriori estimate for the energy error $|||p - p_h|||$ is formulated in terms of a *potential reconstruction* s_h and a *flux reconstruction* $\mathbf{t}_h$. These reconstructions must comply with the following assumption.

Assumption 1 (Potential and flux reconstruction for (3a)–(3b)**)** *There holds* $s_h \in H_0^1(\Omega)$, $\mathbf{t}_h \in \mathbf{H}(\text{div}, \Omega)$, *and*

$$(\nabla \cdot \mathbf{t}_h, 1)_K = (f, 1)_K \quad \forall K \in \mathcal{T}_h. \tag{5}$$

Remark 1 (Assumption 1). Assumption 1 is concerned with basic physical *constraints* and *local conservation*. For the exact solution, $p \in H_0^1(\Omega)$ and $\mathbf{t} \in \mathbf{H}(\text{div}, \Omega)$ (physical constraints); moreover, $\nabla \cdot \mathbf{t} = f$ (conservation). The potential and flux reconstructions mimic these continuous properties.

We can now state and prove our main result concerning the error upper bound, see [27, Theorem 4.2] and [30, Theorem 4.5].

Theorem 2 (A posteriori estimate for (3a)–(3b)**).** *Let p be the solution of* (4) *and let $p_h \in H^1(\mathcal{T}_h)$ be arbitrary. Let Assumption 1 be satisfied. Then,*

A Unified Framework for a posteriori Error Estimation

$$|||p - p_h||| \leq \left\{ \sum_{K \in \mathcal{T}_h} \eta_{\text{NC},K}^2 + (\eta_{\text{R},K} + \eta_{\text{DF},K})^2 \right\}^{1/2},$$

where, for all $K \in \mathcal{T}_h$, *the diffusive flux estimator, the* nonconformity estimator, *and the* residual estimator *are respectively given by*

$$\eta_{\text{DF},K} := \|\nabla p_h + \mathbf{t}_h\|_K, \tag{6a}$$

$$\eta_{\text{NC},K} := \|\nabla(p_h - s_h)\|_K, \tag{6b}$$

$$\eta_{\text{R},K} := C_{\text{P},K}^{1/2} h_K \|f - \nabla \cdot \mathbf{t}_h\|_K. \tag{6c}$$

Proof. Following [17, Lemma 4.4], we obtain using $s_h \in H_0^1(\Omega)$,

$$|||p - p_h|||^2 \leq |||p_h - s_h|||^2 + \left\{ \sup_{\varphi \in H_0^1(\Omega), |||\varphi|||=1} (\nabla_h(p - p_h), \nabla \varphi) \right\}^2.$$

The first term equals the Hilbertian sum of the nonconformity estimators, and we are thus left with bounding the second term. Using (4) and $\mathbf{t}_h \in \mathbf{H}(\text{div}, \Omega)$, we obtain

$$(\nabla_h(p - p_h), \nabla \varphi) = (f, \varphi) - (\nabla_h p_h, \nabla \varphi) = (f, \varphi) - (\nabla_h p_h + \mathbf{t}_h, \nabla \varphi) + (\mathbf{t}_h, \nabla \varphi)$$
$$= (f - \nabla \cdot \mathbf{t}_h, \varphi) - (\nabla_h p_h + \mathbf{t}_h, \nabla \varphi).$$

We now bound the two above terms separately. For all $K \in \mathcal{T}_h$, let φ_K be the mean value of φ over K. Then, using (5), the Poincaré inequality (1), and the Cauchy–Schwarz inequality, we infer

$$|(f - \nabla \cdot \mathbf{t}_h, \varphi)_K| = |(f - \nabla \cdot \mathbf{t}_h, \varphi - \varphi_K)_K| \leq \eta_{\text{R},K} |||\varphi|||_K.$$

Moreover, bounding $|(\nabla p_h + \mathbf{t}_h, \nabla \varphi)_K| \leq \eta_{\text{DF},K} |||\varphi|||_K$ is immediate using the Cauchy–Schwarz inequality. The conclusion is straightforward. □

We now address local efficiency and we still proceed generally, without any notion of a particular numerical scheme. We make two more assumptions.

Assumption 3 (Local efficiency) *We suppose that*

- *there exists a shape-regular matching simplicial submesh $\mathcal{S}_h$ of $\mathcal{T}_h$ such that, for each $K \in \mathcal{T}_h$, the number of subelements $L \subset K$, $L \in \mathcal{S}_h$, is uniformly bounded;*
- *for a fixed integer $k \geq 1$, the approximate solution p_h and the datum f are in $\mathbb{P}_k(\mathcal{T}_h)$, and the flux reconstruction $\mathbf{t}_h$ is in $[\mathbb{P}_k(\mathcal{S}_h)]^d$;*

Henceforth, we use $A \lesssim B$ when there exists a positive constant C, that can only depend on the space dimension d, the shape-regularity parameter of the mesh $\mathcal{S}_h$, and the polynomial degree k, such that $A \leq CB$. For all $K \in \mathscr{T}_h$, let $\mathfrak{T}_K$ denote all the elements in $\mathscr{T}_h$ having a nonempty intersection with K, $\mathfrak{E}_K$ all the sides in $\mathscr{E}_h$ having a nonempty intersection with K, and $\mathfrak{E}_K^{\text{int}}$ the subset of $\mathfrak{E}_K$ collecting those sides lying in the interior of Ω. We introduce the *classical residual estimators* for problem (3a)–(3b) (cf. [24] for conforming methods and [1,6] for nonconforming methods) given by

$$\eta_{\text{res},K} := h_K \|f + \Delta p_h\|_{\mathfrak{T}_K} + h_K^{1/2} \|[\![\nabla_h p_h \cdot \mathbf{n}]\!]\|_{\mathfrak{E}_K^{\text{int}}}, \tag{7a}$$

$$|p_h|_{\text{J},K} := h_K^{-1/2} \|[\![p_h]\!]\|_{\mathfrak{E}_K}. \tag{7b}$$

Assumption 4 (Approximation property for (3a)–(3b)**)** *We assume that, for all $K \in \mathscr{T}_h$,*

$$\|\nabla(p_h - s_h)\|_K + \|\nabla p_h + \mathbf{t}_h\|_K \lesssim \eta_{\text{res},K} + |p_h|_{\text{J},K}. \tag{8}$$

We can now state and prove our main result concerning efficiency.

Theorem 5 (Efficiency of the estimate of Theorem 2). *Let p be the solution of* (4) *and let Assumptions 3 and 4 be satisfied. Then, for all $K \in \mathscr{T}_h$,*

$$\eta_{\text{NC},K} + \eta_{\text{R},K} + \eta_{\text{DF},K} \lesssim |||p - p_h|||_{\mathfrak{T}_K} + |p_h|_{\text{J},K}.$$

Proof. Our first step is to observe that $\eta_{\text{NC},K} + \eta_{\text{R},K} + \eta_{\text{DF},K} \lesssim \eta_{\text{res},K} + |p_h|_{\text{J},K}$. This bound is immediate for $\eta_{\text{NC},K}$ and $\eta_{\text{DF},K}$ owing to Assumption 4, while for $\eta_{\text{R},K}$, the triangle and inverse inequalities yield $\eta_{\text{R},K} \lesssim h_K \|f + \Delta p_h\|_K + \|\nabla p_h + \mathbf{t}_h\|_K \lesssim \eta_{\text{res},K} + |p_h|_{\text{J},K}$, owing to Assumptions 3 and 4. Our second step is to observe that $\eta_{\text{res},K} \lesssim |||p - p_h|||_{\mathfrak{T}_K}$, as can be derived using suitable bubble functions [24]. □

Remark 2 (Equivalence result). If p_h is in $H_0^1(\Omega)$, the jump seminorm $|p_h|_{\text{J},K}$ vanishes. If the jumps of p_h have zero mean on each side, proceeding as in [1, Theorem 10] yields $|p_h|_{\text{J},K} \lesssim |||p - p_h|||_{\mathfrak{T}_K}$. Finally, in the general case, an equivalence result is achieved by adding the jump seminorm $|p - p_h|_{\text{J},K} = |p_h|_{\text{J},K}$ to both the error measure and the nonconformity estimator.

3.2 Application to finite volumes

We apply here the framework of §3.1 to cell- and vertex-centered finite volume schemes, i.e., we specify s_h and $\mathbf{t}_h$, and we verify Assumptions 1, 3, and 4.

3.2.1 Cell-centered finite volumes

Definition 1 (Cell-centered FVs for (3a)–(3b)**).** A cell-centered FV scheme for discretizing (3a)–(3b), cf. [12], reads: find $\bar{p}_h \in \mathbb{P}_0(\mathscr{T}_h)$ such that

$$\sum_{\sigma \in \mathcal{E}_K} F_{K,\sigma} = (f, 1)_K \quad \forall K \in \mathcal{T}_h. \tag{9}$$

Here, $\mathcal{E}_K$ collects the sides of K and $F_{K,\sigma}$ is the diffusive flux through the side σ, which depends on $\bar{p}_h$. A simple example is the so-called "two-point" scheme. In what follows, we do not need the specific form of $F_{K,\sigma}$, but only the conservation property $F_{K,\sigma} = -F_{L,\sigma}$ for all interior sides σ shared by the elements K and L.

Let us first suppose that $\mathcal{T}_h$ is simplicial and matching. Following [13], let $\mathbf{t}_h \in \mathbf{RTN}(\mathcal{T}_h)$ be prescribed on all $K \in \mathcal{T}_h$ by the fluxes $F_{K,\sigma}$ as

$$(\mathbf{t}_h|_K \cdot \mathbf{n}_K)|_\sigma := F_{K,\sigma}/|\sigma|. \tag{10}$$

Since $\bar{p}_h$ is piecewise constant, the energy error $|||p - \bar{p}_h||| = \|\nabla p\|$ is not relevant. Instead, following [28, §3.2], we first postprocess $\bar{p}_h$ locally into $p_h \in \mathbb{P}_2(\mathcal{T}_h)$ such that, for all $K \in \mathcal{T}_h$,

$$-\nabla p_h|_K = \mathbf{t}_h|_K, \quad (p_h, 1)_K/|K| = \bar{p}_h|_K. \tag{11}$$

The potential s_h is constructed by applying an averaging operator $\mathcal{I}_{\text{av}} : \mathbb{P}_k(\mathcal{T}_h) \to \mathbb{P}_k(\mathcal{T}_h) \cap H_0^1(\Omega)$ to p_h. This operator sets the Lagrangian degrees of freedom inside Ω to the average of the values and sets 0 on $\partial\Omega$. Theorem 2 can now used to bound the error $|||p - p_h|||$ observing that (5) in Assumption 1 results from $(\nabla \cdot \mathbf{t}_h, 1)_K = \langle \mathbf{t}_h \cdot \mathbf{n}_K, 1 \rangle_{\partial K} = \sum_{\sigma \in \mathcal{E}_K} F_{K,\sigma} = (f, 1)_K$. Note that $\eta_{\text{DF},K} = 0$ from (11), which is typical for cell-centered finite volumes. To apply Theorem 5, we verify Assumptions 3 and 4. Assumption 3 is straightforward with $\mathcal{S}_h = \mathcal{T}_h$, whereas Assumption 4 is trivial for $\mathbf{t}_h$ since $\|\nabla p_h + \mathbf{t}_h\|_K = 0$, while the bound on $\|\nabla(p_h - \mathcal{I}_{\text{av}}(p_h))\|_K$ results from [1, 4, 16].

When $\mathcal{T}_h$ is not simplicial or is nonmatching, the submesh $\mathcal{S}_h$ needs to be introduced. We can then proceed as in [28, §5] and [10]. The averaging operator for potential reconstruction maps into $\mathbb{P}_k(\mathcal{S}_h) \cap H_0^1(\Omega)$, while the flux is reconstructed in $\mathbf{RTN}(\mathcal{S}_h)$ either by direct prescription of its degrees of freedom or by solving local Neumann problems.

3.2.2 Vertex-centered finite volumes

We suppose here that $\mathcal{T}_h$ is simplicial and matching. Let $\mathcal{D}_h$ be a dual mesh with dual volumes D associated with the vertices of $\mathcal{T}_h$. We refer to Fig. 1, left, for an illustration. We decompose $\mathcal{D}_h$ into $\mathcal{D}_h^{\text{int}}$ and $\mathcal{D}_h^{\text{ext}}$, with $\mathcal{D}_h^{\text{int}}$ associated with interior vertices and $\mathcal{D}_h^{\text{ext}}$ with boundary ones.

Definition 2 (Vertex-centered FVs for (3a)–(3b)). A vertex-centered FV scheme for discretizing (3a)–(3b), cf. [12], reads: find $p_h \in \mathbb{P}_1(\mathcal{T}_h) \cap H_0^1(\Omega)$ such that

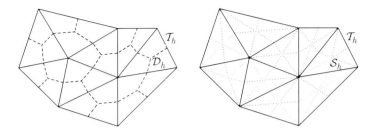

Fig. 1 Simplicial mesh $\mathcal{T}_h$ and the dual mesh $\mathcal{D}_h$ (left); simplicial submesh $\mathcal{S}_h$ (right)

$$-\langle \nabla p_h \cdot \mathbf{n}_D, 1 \rangle_{\partial D} = (f, 1)_D \qquad \forall D \in \mathcal{D}_h^{\text{int}}. \tag{12}$$

To apply the framework of §3.1, we first note that, since $p_h \in H_0^1(\Omega)$, we can set $s_h = p_h$. Consequently, $\eta_{\text{NC},K} = 0$ in Theorem 2, which is typical for vertex-centered finite volumes. To construct the flux $\mathbf{t}_h$, we introduce a matching simplicial submesh $\mathcal{S}_h$, cf. Fig. 1, right. Such $\mathcal{S}_h$ is a refinement of both $\mathcal{T}_h$ and $\mathcal{D}_h$. The flux $\mathbf{t}_h$ is reconstructed in $\mathbf{RTN}(\mathcal{S}_h)$ such that, at all interior sides σ of $\mathcal{S}_h$ which lie on the boundary of some $D \in \mathcal{D}_h$, $\mathbf{t}_h \cdot \mathbf{n}_\sigma := -\nabla p_h \cdot \mathbf{n}_\sigma$. Owing to the Green theorem, $(\nabla \cdot \mathbf{t}_h, 1)_D = (f, 1)_D$ for all $D \in \mathcal{D}_h^{\text{int}}$. There are various ways of prescribing the remaining degrees of freedom of $\mathbf{t}_h$. We can merely prescribe them directly, but better computational results are obtained if a local Neumann or Neumann/Dirichlet problem is solved using mixed finite elements in each $D \in \mathcal{D}_h$ [30, §4.3]. Verifying Assumptions 1 and 3 is immediate, while Assumption 4 is proven as in [30, §5].

4 Convection-diffusion-reaction equation

We consider the convection-diffusion-reaction equation

$$-\nabla \cdot (\varepsilon \nabla p - \mathbf{w} p) + rp = f \quad \text{in } \Omega, \tag{13a}$$

$$p = 0 \quad \text{on } \partial\Omega, \tag{13b}$$

with $\varepsilon > 0$, $r \in L^\infty(\Omega)$, $\mathbf{w} \in [W^{1,\infty}(\Omega)]^d$, and $f \in L^2(\Omega)$. We assume that $\mathbf{w}$ is divergence-free with piecewise polynomial components and that r is piecewise constant taking nonnegative values. We introduce the bilinear form $\mathcal{B} := \mathcal{B}_S + \mathcal{B}_A$ on $H_0^1(\Omega) \times H_0^1(\Omega)$ such that

$$\mathcal{B}_S(p, \varphi) := \varepsilon(\nabla p, \nabla \varphi) + (rp, \varphi), \tag{14a}$$

$$\mathcal{B}_A(p, \varphi) := -(\mathbf{w} p, \nabla \varphi). \tag{14b}$$

The weak formulation consists in finding $p \in H_0^1(\Omega)$ such that

$$\mathscr{B}(p,\varphi) = (f,\varphi) \qquad \forall \varphi \in H_0^1(\Omega). \tag{15}$$

The vector-valued functions $\mathbf{t} := -\varepsilon \nabla p$ and $\mathbf{q} := \mathbf{w}p$ are in $\mathbf{H}(\text{div},\Omega)$ and are, respectively, called the *diffusive* and *convective flux*.

4.1 Abstract framework

We present here a unified abstract framework for a posteriori error estimation in problem (13a)–(13b). Extending the above bilinear forms to $H^1(\mathscr{T}_h) \times H^1(\mathscr{T}_h)$ using broken gradients, we now define the energy (semi-)norm as

$$|||v||| := \mathscr{B}_S(v,v)^{1/2} = \left(\|\varepsilon^{1/2}\nabla_h v\|^2 + \|r^{1/2}v\|^2 \right)^{1/2} \qquad \forall v \in H^1(\mathscr{T}_h). \tag{16}$$

To achieve robustness of the a posteriori error estimates in the singularly perturbed regime resulting from dominant convection, we introduce, following Verfürth [26], the augmented (semi-)norm defined as

$$|||v|||_\oplus := |||v||| + \sup_{\varphi \in H_0^1(\Omega),\,|||\varphi|||=1} \mathscr{B}_A(v,\varphi) \qquad \forall v \in H^1(\mathscr{T}_h). \tag{17}$$

The a posteriori error estimate for $|||p - p_h|||_\oplus$ is formulated in terms of a *potential reconstruction* s_h, a *diffusive flux reconstruction* $\mathbf{t}_h$, and a *convective flux reconstruction* $\mathbf{q}_h$. These reconstructions must comply with the following assumption.

Assumption 6 (Potential and flux reconstruction for (13a)–(13b)**)** *There holds* $s_h \in H_0^1(\Omega)$, $\mathbf{t}_h, \mathbf{q}_h \in \mathbf{H}(\text{div},\Omega)$, *and*

$$(\nabla \cdot \mathbf{t}_h + \nabla \cdot \mathbf{q}_h + r p_h, 1)_K = (f, 1)_K \qquad \forall K \in \mathscr{T}_h. \tag{18}$$

We can now state and prove our main result concerning the error upper bound. For simplicity, we assume that the mesh $\mathscr{T}_h$ is matching and simplicial so as to use the trace inequality (2). The general case can be treated by resorting to a matching simplicial submesh.

Theorem 7 (A posteriori estimate for (13a)–(13b)**).** *Let p be the solution of* (15) *and let $p_h \in H^1(\mathscr{T}_h)$ be arbitrary. Let Assumption 6 be satisfied. Assume that $\mathscr{T}_h$ is matching and simplicial. Then,*

$$|||p - p_h|||_\oplus \leq \eta := 2 \left\{ \sum_{K \in \mathscr{T}_h} \eta_{\text{NC},K}^2 \right\}^{1/2} + \left\{ \sum_{K \in \mathscr{T}_h} \widetilde{\eta}_{\text{NC},K}^2 \right\}^{1/2}$$

$$+ 3 \left\{ \sum_{K \in \mathscr{T}_h} (\eta_{\text{R},K} + \eta_{\text{CDF},K})^2 \right\}^{1/2}.$$

For all $K \in \mathcal{T}_h$, *the* convective-diffusive flux estimator *is given by*

$$\eta_{\text{CDF},K} := \min(\eta_{\text{CDF},1,K}, \eta_{\text{CDF},2,K}), \tag{19a}$$

$$\eta_{\text{CDF},1,K} := \varepsilon^{-1/2}\|\mathbf{a}_h\|_K, \tag{19b}$$

$$\eta_{\text{CDF},2,K} := m_K\|(I-\Pi_0)\nabla\cdot\mathbf{a}_h\|_K + \widetilde{m}_K^{1/2}\sum_{\sigma\in\mathcal{E}_K} C_{\text{t},K,\sigma}^{1/2}\|\mathbf{a}_h\cdot\mathbf{n}_\sigma\|_\sigma, \tag{19c}$$

with $\mathbf{a}_h := \mathbf{t}_h + \mathbf{q}_h + \varepsilon\nabla_h p_h - \mathbf{w}s_h$ *and* Π_0 *the L^2-orthogonal projector onto* $\mathbb{P}_0(\mathcal{T}_h)$, *the* nonconformity estimators *by*

$$\eta_{\text{NC},K} := \|\|p_h - s_h\|\|_K, \qquad \widetilde{\eta}_{\text{NC},K} := \min(\widetilde{\eta}_{\text{NC},1,K}, \widetilde{\eta}_{\text{NC},2,K}), \tag{20a}$$

$$\widetilde{\eta}_{\text{NC},1,K} := \varepsilon^{-1/2}\|\mathbf{b}_h\|_K, \tag{20b}$$

$$\widetilde{\eta}_{\text{NC},2,K} := m_K\|(I-\Pi_0)\nabla\cdot\mathbf{b}_h\|_K + \widetilde{m}_K^{1/2}\sum_{\sigma\in\mathcal{E}_K} C_{\text{t},K,\sigma}^{1/2}\|\mathbf{b}_h\cdot\mathbf{n}_\sigma\|_\sigma, \tag{20c}$$

with $\mathbf{b}_h := \mathbf{w}(p_h - s_h)$, *and the* residual estimator *by*

$$\eta_{\text{R},K} := m_K\|f - \nabla\cdot\mathbf{t}_h - \nabla\cdot\mathbf{q}_h - rp_h\|_K. \tag{21}$$

Here $m_K := \min(C_{\text{P},K}^{1/2}\varepsilon^{-1/2}h_K, r_K^{-1/2})$ *and* $\widetilde{m}_K := 2(1 + C_{\text{P},K}^{1/2})\varepsilon^{-1/2}m_K$.

Proof. Following [27, Lemma 7.1] and [8, Lemma 3.1], we infer

$$\|\|p - p_h\|\| \leq \|\|p_h - s_h\|\| + \sup_{\varphi\in H_0^1(\Omega), \|\|\varphi\|\|=1}\{\mathcal{B}(p - p_h, \varphi) + \mathcal{B}_{\text{A}}(p_h - s_h, \varphi)\},$$

and proceeding as in [9, Lemma 4.2] yields

$$\|\|p - p_h\|\|_\oplus \leq 2\|\|p_h - s_h\|\| + \sup_{\varphi\in H_0^1(\Omega), \|\|\varphi\|\|=1}\mathcal{B}_{\text{A}}(p_h - s_h, \varphi)$$
$$+ 3\sup_{\varphi\in H_0^1(\Omega), \|\|\varphi\|\|=1}\{\mathcal{B}(p - p_h, \varphi) + \mathcal{B}_{\text{A}}(p_h - s_h, \varphi)\}.$$

For the second term on the right-hand side, we obtain

$$\mathcal{B}_{\text{A}}(p_h - s_h, \varphi) = -(\mathbf{b}_h, \nabla\varphi) \leq \sum_{K\in\mathcal{T}_h}\widetilde{\eta}_{\text{NC},K}\|\|\varphi\|\|_K.$$

Indeed, for all $K \in \mathcal{T}_h$, the Cauchy–Schwarz inequality on the one hand yields $-(\mathbf{b}_h, \nabla\varphi)_K \leq \varepsilon^{-1/2}\|\mathbf{b}_h\|_K\|\|\varphi\|\|_K = \widetilde{\eta}_{\text{NC},1,K}\|\|\varphi\|\|_K$, while integrating by parts on K leads to

$$-(\mathbf{b}_h, \nabla\varphi)_K = ((I - \Pi_0)\nabla\cdot\mathbf{b}_h, \varphi - \varphi_K)_K - \sum_{\sigma \in \mathcal{E}_K}(\mathbf{b}_h\cdot\mathbf{n}_\sigma, \varphi - \varphi_K)_\sigma \leq \widetilde{\eta}_{\mathrm{NC},2,K}|||\varphi|||_K,$$

owing to the Poincaré inequality (1) and the trace inequality (2). For the third term on the right-hand side, we observe that

$$\mathscr{B}(p - p_h, \varphi) + \mathscr{B}_\mathrm{A}(p_h - s_h, \varphi) = (f - \nabla\cdot\mathbf{t}_h - \nabla\cdot\mathbf{q}_h - rp_h, \varphi) - (\mathbf{a}_h, \nabla\varphi)$$
$$\leq \sum_{K \in \mathcal{T}_h}(\eta_{\mathrm{R},K} + \eta_{\mathrm{CDF},K})|||\varphi|||_K,$$

using Assumption 6 for the residual term and proceeding for $\mathbf{a}_h$ as for $\mathbf{b}_h$. □

We now address the efficiency of the estimate of Theorem 7. In what follows, $\lesssim$ can include factors depending on the maximal ratio m_K/m_L for K, L having a nonempty intersection. We introduce the *classical residual estimators* for problem (13a)–(13b) given by

$$\eta_{\mathrm{res},K} := m_K\|f + \nabla\cdot(\varepsilon\nabla p_h - \mathbf{w}p_h) - rp_h\|_{\mathfrak{T}_K} + m_K^{1/2}\varepsilon^{-1/4}\|[\![\varepsilon\nabla_h p_h]\!]\cdot\mathbf{n}\|_{\mathcal{E}_K^{\mathrm{int}}}, \tag{22a}$$

$$|p_h|_{\mathrm{J},K} := (\varepsilon^{1/2}h_K^{-1/2} + m_K^{1/2}\varepsilon^{-1/4}\|\mathbf{w}\|_{[L^\infty(K)]^d} + r_K^{1/2}h_K^{1/2})\|[\![p_h]\!]\|_{\mathcal{E}_K}. \tag{22b}$$

We also set $|v|_\mathrm{J} := \left\{\sum_{K \in \mathcal{T}_h}|v|_{\mathrm{J},K}^2\right\}^{1/2}$ for all $v \in H^1(\mathcal{T}_h)$.

Assumption 8 (Approximation property for (13a)–(13b)) *We assume that, for all $K \in \mathcal{T}_h$, with $\mathbf{c}_h = \mathbf{a}_h$ or $\mathbf{b}_h$,*

$$m_K\|(I - \Pi_0)\nabla\cdot\mathbf{c}_h\|_K + m_K^{1/2}\varepsilon^{-1/4}\sum_{\sigma \in \mathcal{E}_K}\|\mathbf{c}_h\cdot\mathbf{n}_\sigma\|_\sigma \lesssim \eta_{\mathrm{res},K} + |p_h|_{\mathrm{J},K}.$$

Proceeding as in [9, Theorems 3.2 and 3.4] leads to the following lower bound, which is global in space owing to the use of a dual norm.

Theorem 9 (Efficiency of the estimate of Theorem 7). *Let p be the solution of (15) and let Assumption 8, and the second item of Assumption 3, be satisfied. Then,*

$$\eta \lesssim |||p - p_h|||_\oplus + |p - p_h|_\mathrm{J}. \tag{23}$$

Remark 3 (Fully robust equivalence result). Adding the jump seminorm $|\cdot|_\mathrm{J}$ to the error measure, a fully robust equivalence result is finally achieved in the form

$$|||p - p_h|||_\oplus + |p - p_h|_\mathrm{J} \leq \eta + |p_h|_\mathrm{J} \lesssim |||p - p_h|||_\oplus + |p - p_h|_\mathrm{J}. \tag{24}$$

4.2 Application to finite volumes

We apply here the framework of §4.1 to cell- and vertex-centered finite volume schemes, i.e., we specify s_h, $\mathbf{t}_h$, and $\mathbf{q}_h$, and we verify Assumption 6, and, at least in some cases, Assumption 8.

4.2.1 Cell-centered finite volumes

Definition 3 (**Cell-centered FVs for** (13a)–(13b)). A cell-centered FV scheme for discretizing (13a)–(13b), cf. [12], reads: find $\bar{p}_h \in \mathbb{P}_0(\mathcal{T}_h)$ such that

$$\sum_{\sigma \in \mathcal{E}_K} F_{K,\sigma} + \sum_{\sigma \in \mathcal{E}_K} W_{K,\sigma} + r_K \bar{p}_h|_K = (f, 1)_K \quad \forall K \in \mathcal{T}_h. \tag{25}$$

In addition to the diffusive fluxes $F_{K,\sigma}$, $W_{K,\sigma}$ are the convective fluxes, also depending on $\bar{p}_h$. We do not need the precise form of the fluxes, but only $F_{K,\sigma} = -F_{L,\sigma}$ and $W_{K,\sigma} = -W_{L,\sigma}$ for all interior sides σ shared by the elements K and L.

Following the ideas exposed in §3.2.1, we first define $\mathbf{t}_h$, $\mathbf{q}_h \in \mathbf{RTN}(\mathcal{T}_h)$ by

$$(\mathbf{t}_h|_K \cdot \mathbf{n}_K)|_\sigma := F_{K,\sigma}/|\sigma|, \qquad (\mathbf{q}_h|_K \cdot \mathbf{n}_K)|_\sigma := W_{K,\sigma}/|\sigma|. \tag{26}$$

Define p_h similarly to (11). It is immediate to see using the Green theorem that (26) and (25) yield (18). A reasonable condition on $W_{K,\sigma}$ in the context of upwind or centered convective fluxes is that

$$\|W_{K,\sigma}/|\sigma| - \mathbf{w} \cdot \mathbf{n}_K p_h|_K\|_\sigma \lesssim \|\mathbf{w}\|_{[L^\infty(K)]^d} \|[\![\bar{p}_h]\!]\|_\sigma. \tag{27}$$

Then, Assumption 8 holds, up to the oscillation terms $m_K \|(I - \Pi_0)\nabla \cdot (\mathbf{w} p_h)\|_K$, when additionally including $|\bar{p}_h|_{J,K}$ on the right-hand side, and the efficiency result (23) holds when additionally including $|p - \bar{p}_h|_J$ on the right-hand side.

4.2.2 Vertex-centered finite volumes

Definition 4 (**Vertex-centered FVs for** (13a)–(13b)). A vertex-centered FV scheme for discretizing (13a)–(13b), cf. [12], reads: find $p_h \in \mathbb{P}_1(\mathcal{T}_h) \cap H_0^1(\Omega)$ such that

$$-\langle \varepsilon \nabla p_h \cdot \mathbf{n}_D, 1 \rangle_{\partial D} + \langle \mathbf{w} \cdot \mathbf{n}_D \, p_h, 1 \rangle_{\partial D} + (r p_h, 1)_D = (f, 1)_D \quad \forall D \in \mathcal{D}_h^{\text{int}}. \tag{28}$$

Note that we only consider a centered convective flux.

As in §3.2.2, we set $s_h = p_h$ in Assumption 6. Consequently, $\eta_{NC,K} = \widetilde{\eta}_{NC,K} = 0$ in Theorem 7. For the convective flux reconstruction, we simply set $\mathbf{q}_h := \mathbf{w}p_h$. For the diffusive flux reconstruction, we introduce the mesh $\mathscr{S}_h$ (cf. Fig. 1, right) and we define $\mathbf{t}_h \in \mathbf{RTN}(\mathscr{S}_h)$ such that $\mathbf{t}_h \cdot \mathbf{n}_\sigma := -\varepsilon \nabla p_h \cdot \mathbf{n}_\sigma$ at all interior sides σ of $\mathscr{S}_h$ which lie on the boundary of some $D \in \mathscr{D}_h$. As in §3.2.2, local problems can be solved to fulfill Assumption 6. Assumption 8 can be verified as in §3.2.2 for the diffusive part, while the convective part is trivial owing to the choice of $\mathbf{q}_h$.

5 Heat equation

We consider the heat equation

$$\partial_t p - \Delta p = f \quad \text{in } \Omega \times (0,T), \tag{29a}$$

$$p = 0 \quad \text{on } \partial\Omega \times (0,T), \tag{29b}$$

$$p(\cdot, 0) = p_0 \quad \text{in } \Omega, \tag{29c}$$

with $f \in L^2(\Omega \times (0,T))$, initial condition $p_0 \in L^2(\Omega)$, and final time $T > 0$. The exact solution is in the space $Y := \{y \in X; \partial_t y \in X'\}$, with $X := L^2(0,T; H_0^1(\Omega))$ and $X' = L^2(0,T; H^{-1}(\Omega))$, satisfies (29c) in $L^2(\Omega)$, and is such that, for a.e. $t \in (0,T)$,

$$\langle \partial_t p, \varphi \rangle(t) + (\nabla p, \nabla \varphi)(t) = (f, \varphi)(t) \quad \forall \varphi \in H_0^1(\Omega). \tag{30}$$

The space-time energy norm is given by $\|y\|_X := \left\{\int_0^T \|\nabla y\|^2(t)\, dt\right\}^{1/2}$ for all $y \in X$. Following Verfürth [25], we augment the energy norm by a dual norm of the time derivative as $\|y\|_Y := \|y\|_X + \|\partial_t y\|_{X'}$ with $\|\partial_t y\|_{X'} := \left\{\int_0^T \|\partial_t y\|_{H^{-1}}^2(t)\, dt\right\}^{1/2}$.

5.1 Abstract framework

We consider an increasing sequence of discrete times $\{t^n\}_{0 \le n \le N}$ such that $t^0 = 0$ and $t^N = T$ and introduce the time intervals $I_n := (t^{n-1}, t^n]$ and the time steps $\tau^n := t^n - t^{n-1}$ for all $1 \le n \le N$. The meshes are allowed to vary in time; we denote by $\mathscr{T}_h^n$ the mesh used to march in time from t^{n-1} to t^n, for all $1 \le n \le N$, and by $\mathscr{T}_h^0$ the initial mesh. We suppose that the approximate solution on t^n, denoted by p_h^n, is in $H^1(\mathscr{T}_h^n)$, and we let $p_{h\tau}$ be the space-time approximate solution, given by $p_{h\tau}^n$ at each discrete time t^n and piecewise affine and continuous in time. We denote the space of such functions by $P_\tau^1(H^1(\mathscr{T}_h))$. We also denote by $P_\tau^1(H_0^1(\Omega))$ the space of functions that are piecewise affine and continuous in time and $H_0^1(\Omega)$ in space and by $P_\tau^0(\mathbf{H}(\text{div}, \Omega))$ the space of functions that are piecewise constant in time and $\mathbf{H}(\mathrm{div}, \Omega)$ in space. For all $1 \leq n \leq N$, we set $\widetilde{f}^n := \frac{1}{\tau^n} \int_{I_n} f(\cdot, t) \, dt$, and, for $\varphi_{h\tau} \in P_\tau^1(H^1(\mathcal{T}_h))$, $\partial_t p_{h\tau}^n := \frac{1}{\tau^n}(\varphi_{h\tau}^n - \varphi_{h\tau}^{n-1})$.

We aim at measuring the error $(p - p_{h\tau})$ in the $\|\cdot\|_Y$-norm using the broken gradient operator in the energy norm. The a posteriori error estimate is formulated in terms of a *space-time potential reconstruction* $s_{h\tau}$ and a *space-time flux reconstruction* $\mathbf{t}_{h\tau}$. These reconstructions must comply with the following assumption.

Assumption 10 (Potential and flux reconstruction for (29a)–(29c)) *There holds* $s_{h\tau} \in P_\tau^1(H_0^1(\Omega))$, $\mathbf{t}_{h\tau} \in P_\tau^0(\mathbf{H}(\mathrm{div}, \Omega))$, *and, for all* $1 \leq n \leq N$ *and for all* $K \in \mathcal{T}_h^n$,

$$(\partial_t s_{h\tau}^n, 1)_K = (\partial_t p_{h\tau}^n, 1)_K, \tag{31a}$$

$$(\widetilde{f}^n - \partial_t p_{h\tau}^n - \nabla \cdot \mathbf{t}_{h\tau}^n, 1)_K = 0. \tag{31b}$$

We can now state our main result concerning the error upper bound, see [11, Theorem 3.6] and also [11, Theorem 3.2] for a slightly sharper bound.

Theorem 11 (A posteriori estimate for (29a)–(29c)). *Let p be the solution of (30) and let $p_{h\tau} \in P_\tau^1(H^1(\mathcal{T}_h))$ be arbitrary. Let Assumption 10 be satisfied. Then,*

$$\|p - p_{h\tau}\|_Y \leq \left\{\sum_{n=1}^N (\eta_{\mathrm{sp}}^n)^2\right\}^{1/2} + \left\{\sum_{n=1}^N (\eta_{\mathrm{tm}}^n)^2\right\}^{1/2} + \eta_{\mathrm{IC}} + 3\|f - \widetilde{f}\|_{X'}, \tag{32}$$

with, for all $1 \leq n \leq N$, the space *and* time *error estimators given by*

$$(\eta_{\mathrm{sp}}^n)^2 := \sum_{K \in \mathcal{T}_h^n} 3 \left\{ \tau^n (9(\eta_{\mathrm{R},K}^n + \eta_{\mathrm{DF},K}^n)^2 + (\eta_{\mathrm{NC},2,K}^n)^2) + \int_{I_n} (\eta_{\mathrm{NC},1,K}^n)^2(t) \, dt \right\}, \tag{33a}$$

$$(\eta_{\mathrm{tm}}^n)^2 := \sum_{K \in \mathcal{T}_h^n} 3\tau^n \|\nabla(s_{h\tau}^n - s_{h\tau}^{n-1})\|_K^2. \tag{33b}$$

For all $K \in \mathcal{T}_h^n$, the residual estimator, *the* diffusive flux estimator, *and the* nonconformity estimators *are given by*

$$\eta_{\mathrm{R},K}^n := C_{\mathrm{P},K}^{1/2} h_K \|\widetilde{f}^n - \partial_t s_{h\tau}^n - \nabla \cdot \mathbf{t}_{h\tau}^n\|_K, \tag{34a}$$

$$\eta_{\mathrm{DF},K}^n := \|\nabla s_{h\tau}^n + \mathbf{t}_{h\tau}^n\|_K, \tag{34b}$$

$$\eta_{\mathrm{NC},1,K}^n(t) := \|\nabla_h^n (s_{h\tau} - p_{h\tau})(t)\|_K, \quad \forall t \in I_n, \tag{34c}$$

$$\eta_{\mathrm{NC},2,K}^n := C_{\mathrm{P},K}^{1/2} h_K \|\partial_t (s_{h\tau} - p_{h\tau})^n\|_K. \tag{34d}$$

Finally, the initial condition estimator *is given by* $\eta_{\mathrm{IC}} := 2^{1/2} \|s_{h\tau}^0 - p_0\|$.

We next turn to the efficiency of the estimate of Theorem 11. We introduce the *classical residual estimators* for problem (29a)–(29c) given by

$$\eta_{\text{res},K}^n := h_K \|\widetilde{f}^n - \partial_t p_{h\tau}^n + \Delta p_{h\tau}^n\|_{\mathcal{T}_K} + h_K^{1/2} \|[\![\nabla_h^n p_{h\tau}^n \cdot \mathbf{n}]\!]\|_{\mathcal{E}_K^{\text{int}}}, \tag{35a}$$

$$|p_{h\tau}^n|_{J,K} := h_K^{-1/2} \|[\![p_{h\tau}^n]\!]\|_{\mathcal{E}_K}. \tag{35b}$$

Assumption 12 (Approximation property for (29a)–(29c)**)** *We assume that for all* $1 \le n \le N$ *and for all* $K \in \mathcal{T}_h^n$,

$$\|\nabla_h^n(p_{h\tau}^n - s_{h\tau}^n)\|_K + \|\nabla_h^n p_{h\tau}^n + \mathbf{t}_{h\tau}^n\|_K \lesssim \eta_{\text{res},K}^n + |p_{h\tau}^n|_{J,K}. \tag{36}$$

We can now state our efficiency result, see [11, Theorem 3.9]. As in [25], the lower bound is local in time, but global in space.

Theorem 13 (Efficiency of the estimate of Theorem 11). *Let Assumption 12 hold, let Assumption 3 hold at all discrete times, let both the refinement and coarsening in time be not too abrupt, and let, for all* $1 \le n \le N$, $(h^n)^2 \lesssim \tau^n$. *Then, for all* $1 \le n \le N$,

$$\eta_{\text{sp}}^n + \eta_{\text{tm}}^n \lesssim \|p - p_{h\tau}\|_{Y(I_n)} + \mathscr{J}^n(p_{h\tau}) + \|f - \widetilde{f}\|_{X'(I_n)}, \tag{37}$$

where $\mathscr{J}^n(p_{h\tau}) := \left\{ \tau^n \sum_{K \in \mathcal{T}_h^{n-1}} |p_{h\tau}^{n-1}|_{J,K}^2 + \tau^n \sum_{K \in \mathcal{T}_h^n} |p_{h\tau}^n|_{J,K}^2 \right\}^{1/2}$.

Remark 4 (Equivalence result). We refer to [11, Remark 3.10] for bounding the jumps $\mathscr{J}^n(p_{h\tau})$, see also Remark 2.

5.2 Application to finite volumes

We apply here the framework of §5.1 to cell- and vertex-centered finite volume schemes, i.e., we specify $s_{h\tau}$ and $\mathbf{t}_{h\tau}$, and we verify Assumptions 10 and 12. For simplicity, we only discuss matching simplicial meshes.

5.2.1 Cell-centered finite volumes

Definition 5 (Cell-centered FVs for (29a)–(29c)**).** A cell-centered FV scheme for (29a)–(29c), cf. [12], reads: for all $1 \le n \le N$, find $\bar{p}_{h\tau}^n \in \mathbb{P}_0(\mathcal{T}_h^n)$ s. t.

$$\frac{1}{\tau^n}(\bar{p}_{h\tau}^n - p_{h\tau}^{n-1}, 1)_K + \sum_{\sigma \in \mathcal{E}_K} F_{K,\sigma}^n = (\widetilde{f}^n, 1)_K \qquad \forall K \in \mathcal{T}_h^n. \tag{38}$$

As in §3.2.1, the fluxes $\mathbf{t}_{h\tau}^n$ are constructed from the side fluxes $F_{K,\sigma}^n$ by an equivalent of (10). An elementwise postprocessing as (11) is applied to obtain $p_{h\tau}^n$ from $\bar{p}_{h\tau}^n$. The potential is reconstructed at each discrete time from a modification of the averaging operator of §3.1 where local bubble functions are used to satisfy (31a) (cf. [11]). Then, owing to the construction of $\mathbf{t}_{h\tau}^n$, (31b) is also satisfied, whence Assumption 10 follows. Finally, we set $\mathscr{S}_h^n = \mathscr{T}_h^n$; Assumption 12 is trivial for $\mathbf{t}_{h\tau}$ since $\|\nabla_h^n p_{h\tau}^n + \mathbf{t}_{h\tau}^n\|_K = 0$ and is proven for $s_{h\tau}^n$ in [11].

5.2.2 Vertex-centered finite volumes

Definition 6 (Vertex-centered FVs for (29a)–(29c)). A vertex-centered FV scheme for (29a)–(29c), cf. [12], reads: for all $1 \leq n \leq N$, find $p_{h\tau}^n \in \mathbb{P}_1(\mathscr{T}_h^n) \cap H_0^1(\Omega)$ s. t.

$$(\partial_t p_{h\tau}^n, 1)_D - \langle \nabla p_{h\tau}^n \cdot \mathbf{n}_D, 1 \rangle_{\partial D} = (\widetilde{f}^n, 1)_D \quad \forall D \in \mathscr{D}_h^{\text{int},n}. \tag{39}$$

As in §3.2.2, $p_{h\tau}^n \in H_0^1(\Omega)$ for all $1 \leq n \leq N$, so that we set $s_{h\tau}^n = p_{h\tau}^n$. Consequently, $\eta_{\text{NC},1,K}^n = \eta_{\text{NC},2,K}^n = 0$ in Theorem 11. The fluxes $\mathbf{t}_{h\tau}$ are constructed as in §3.2.2, using the simplicial submeshes $\mathscr{S}_h^n$. Assumptions 10 and 12 are then verified by proceeding as in §3.2.2.

Acknowledgements This work was partly supported by the Groupement MoMaS (PACEN/CNRS, ANDRA, BRGM, CEA, EdF, IRSN).

References

1. Achdou, Y., Bernardi, C., Coquel, F.: A priori and a posteriori analysis of finite volume discretizations of Darcy's equations. Numer. Math. **96**(1), 17–42 (2003)
2. Babuška, I., Rheinboldt, W.C.: Error estimates for adaptive finite element computations. SIAM J. Numer. Anal. **15**(4), 736–754 (1978)
3. Brezzi, F., Fortin, M.: Mixed and hybrid finite element methods, *Springer Series in Computational Mathematics*, vol. 15. Springer-Verlag, New York (1991)
4. Burman, E., Ern, A.: Continuous interior penalty hp-finite element methods for advection and advection-diffusion equations. Math. Comp. **76**(259), 1119–1140 (2007)
5. Carstensen, C., Funken, S.A.: Constants in Clément-interpolation error and residual based a posteriori error estimates in finite element methods. East-West J. Numer. Math. **8**(3), 153–175 (2000)
6. Dari, E., Durán, R., Padra, C., Vampa, V.: A posteriori error estimators for nonconforming finite element methods. RAIRO Modél. Math. Anal. Numér. **30**(4), 385–400 (1996)
7. El Alaoui, L., Ern, A., Vohralík, M.: Guaranteed and robust a posteriori error estimates and balancing discretization and linearization errors for monotone nonlinear problems. Comput. Methods Appl. Mech. Engrg. (2010). DOI 10.1016/j.cma.2010.03.024
8. Ern, A., Stephansen, A.F.: A posteriori energy-norm error estimates for advection-diffusion equations approximated by weighted interior penalty methods. J. Comp. Math. **26**(4), 488–510 (2008)

9. Ern, A., Stephansen, A.F., Vohralík, M.: Guaranteed and robust discontinuous Galerkin a posteriori error estimates for convection–diffusion–reaction problems. J. Comput. Appl. Math. **234**(1), 114–130 (2010)
10. Ern, A., Vohralík, M.: Flux reconstruction and a posteriori error estimation for discontinuous Galerkin methods on general nonmatching grids. C. R. Math. Acad. Sci. Paris **347**(7-8), 441–444 (2009)
11. Ern, A., Vohralík, M.: A posteriori error estimation based on potential and flux reconstruction for the heat equation. SIAM J. Numer. Anal. **48**(1), 198–223 (2010)
12. Eymard, R., Gallouët, T., Herbin, R.: Finite volume methods. In: Handbook of Numerical Analysis, Vol. VII, pp. 713–1020. North-Holland, Amsterdam (2000)
13. Eymard, R., Gallouët, T., Herbin, R.: Finite volume approximation of elliptic problems and convergence of an approximate gradient. Appl. Numer. Math. **37**(1-2), 31–53 (2001)
14. Haslinger, J., Hlaváček, I.: Convergence of a finite element method based on the dual variational formulation. Apl. Mat. **21**(1), 43–65 (1976)
15. Jiránek, P., Strakoš, Z., Vohralík, M.: A posteriori error estimates including algebraic error and stopping criteria for iterative solvers. SIAM J. Sci. Comput. **32**(3), 1567–1590 (2010)
16. Karakashian, O.A., Pascal, F.: A posteriori error estimates for a discontinuous Galerkin approximation of second-order elliptic problems. SIAM J. Numer. Anal. **41**(6), 2374–2399 (2003)
17. Kim, K.Y.: A posteriori error analysis for locally conservative mixed methods. Math. Comp. **76**(257), 43–66 (2007)
18. Ladevèze, P.: Comparaison de modèles de milieux continus. Ph.D. thesis, Université Pierre et Marie Curie (Paris 6) (1975)
19. Nicaise, S.: A posteriori error estimations of some cell-centered finite volume methods. SIAM J. Numer. Anal. **43**(4), 1481–1503 (2005)
20. Ohlberger, M.: A posteriori error estimate for finite volume approximations to singularly perturbed nonlinear convection–diffusion equations. Numer. Math. **87**(4), 737–761 (2001)
21. Ohlberger, M.: A posteriori error estimates for vertex centered finite volume approximations of convection–diffusion–reaction equations. M2AN Math. Model. Numer. Anal. **35**(2), 355–387 (2001)
22. Prager, W., Synge, J.L.: Approximations in elasticity based on the concept of function space. Quart. Appl. Math. **5**, 241–269 (1947)
23. Stephansen, A.F.: Méthodes de Galerkine discontinues et analyse d'erreur a posteriori pour les problèmes de diffusion hétérogène. Ph.D. thesis, Ecole Nationale des Ponts et Chaussées (2007)
24. Verfürth, R.: A review of a posteriori error estimation and adaptive mesh-refinement techniques. Teubner-Wiley, Stuttgart (1996)
25. Verfürth, R.: A posteriori error estimates for finite element discretizations of the heat equation. Calcolo **40**(3), 195–212 (2003)
26. Verfürth, R.: Robust a posteriori error estimates for stationary convection-diffusion equations. SIAM J. Numer. Anal. **43**(4), 1766–1782 (2005)
27. Vohralík, M.: A posteriori error estimates for lowest-order mixed finite element discretizations of convection-diffusion-reaction equations. SIAM J. Numer. Anal. **45**(4), 1570–1599 (2007)
28. Vohralík, M.: Residual flux-based a posteriori error estimates for finite volume and related locally conservative methods. Numer. Math. **111**(1), 121–158 (2008)
29. Vohralík, M.: Two types of guaranteed (and robust) a posteriori estimates for finite volume methods. In: Finite Volumes for Complex Applications V, pp. 649–656. ISTE and John Wiley & Sons, London, UK and Hoboken, USA (2008)
30. Vohralík, M.: Guaranteed and fully robust a posteriori error estimates for conforming discretizations of diffusion problems with discontinuous coefficients. J. Sci. Comput. **46**, 397–438 (2011)

The paper is in final form and has not been or is not being submitted elsewhere.

Staggered discretizations, pressure correction schemes and all speed barotropic flows

L. Gastaldo, R. Herbin, W. Kheriji, C. Lapuerta, and J.-C. Latché

Abstract We present in this paper a class of schemes for the solution of the barotropic Navier-Stokes equations. These schemes work on general meshes, preserve the stability properties of the continuous problem, irrespectively of the space and time steps, and boil down, when the Mach number vanishes, to discretizations which are standard (and stable) in the incompressible framework. Finally, we show that they are able to capture solutions with shocks to the Euler equations.

Keywords barotropic Navier-Stokes, staggered discretizations
MSC2010: 65M12

1 Introduction

The problem addressed in this paper is the system of the so-called barotropic compressible Navier-Stokes equations, which reads:

$$\partial_t \bar{\rho} + \mathrm{div}(\bar{\rho}\bar{u}) = 0, \tag{1a}$$

$$\partial_t(\bar{\rho}\bar{u}) + \mathrm{div}(\bar{\rho}\bar{u} \otimes \bar{u}) + \nabla \bar{p} - \mathrm{div}(\tau(\bar{u})) = 0, \tag{1b}$$

$$\bar{\rho} = \wp(\bar{p}), \tag{1c}$$

where t stands for the time, $\bar{\rho}$, $\bar{u}$ and $\bar{p}$ are the density, velocity and pressure in the flow, and $\tau(\bar{u})$ stands for the shear stress tensor. The function $\wp(\cdot)$ is the equation

---

L. Gastaldo, W. Kheriji, C. Lapuerta, and J.-C. Latché
Institut de Radioprotection et de Sûreté Nucléaire (IRSN)
e-mail: [laura.gastaldo,walid.kheriji, celine.lapuerta, jean-claude.latche]@irsn.fr

R. Herbin
Université de Provence
e-mail: herbin@cmi.univ-mrs.fr

of state used for the modelling of the particular flow at hand, which may be the actual equation of state of the fluid or may result from assumptions concerning the flow; typically, laws as $\wp(\bar p) = \bar p^{1/\gamma}$, where $\gamma > 1$ is a coefficient which is specific to the considered fluid, are obtained by making the assumption that the flow is isentropic. This system of equations is posed over $\Omega \times (0, T)$, where Ω is a domain of $\mathbb{R}^d$, $d \leq 3$ supposed to be polygonal ($d = 2$) or polyhedral ($d = 3$), and the final time T is finite. We suppose that the boundary of Ω is split into $\partial \Omega = \partial \Omega_D \cup \partial \Omega_N$, and we suppose that the velocity and density are prescribed on $\partial \Omega_D$, while Neumann boundary conditions are prescribed on $\partial \Omega_N$. The flow is assumed to enter the domain through $\partial \Omega_D$ and to leave it through Ω_N. This system must be supplemented by initial conditions for $\bar \rho$ and $\bar u$.

The objective of this paper is to present a class of schemes which enjoy three essential features. First, these schemes work on quite general two and three dimensional meshes, including locally refined non-conforming (*i.e.* with hanging nodes) discretizations. Second, they respect the (expected) stability properties of the continuous problem at hand, irrespectively of the space and time steps: positivity of the density, conservation of mass, energy inequality. Third, they boil down, for vanishing Mach numbers, to usual stable coupled or pressure correction schemes, which means that the discretization enjoys a discrete *inf-sup* condition. Even if this is beyond the scope of this paper, we remark that this latter property allows a control of the pressure to be obtained through a control of its gradient; this property is used as a central argument to obtain convergence results on model problems [5, 6, 8].

This paper is organized as follows. First, we describe the general form of the schemes (Sect. 2). Then we show how stability requirements are taken into account to design the discretization of the velocity convection term (Sect. 3). The final expression for the schemes is given in Sect. 4, and their stability properties are stated. Finally, we discuss their capability to capture solutions of the Euler equations with shocks (Sect. 5).

2 The schemes: general form

2.1 Meshes and unknowns

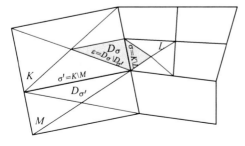

Fig. 1 Notations for primal and dual cells

A finite volume mesh of Ω is defined by a set $\mathcal{M}$ of non–empty convex open disjoint subsets K of Ω (the control volumes), such that $\bar \Omega = \bigcup_{K \in \mathcal{M}} \bar K$. We denote by $\mathcal{E}$ the set of edges (in 2D) or faces (in 3D), by $\mathcal{E}(K) \subset \mathcal{E}$ the set of faces of the cell $K \in \mathcal{M}$, by $\mathcal{E}_{\text{ext}}$ and $\mathcal{E}_{\text{int}}$ the set of boundary and interior faces, respectively. The set of external faces $\mathcal{E}_{\text{ext}}$ is split in $\mathcal{E}_N$ and $\mathcal{E}_D$, which stand for the set of the faces included in $\partial\Omega_N$ and $\partial\Omega_D$, respectively. Each internal face, denoted by $\sigma \in \mathcal{E}_{\text{int}}$, is supposed to have exactly two neighboring cells, say $K, L \in \mathcal{M}$, and $\bar{K} \cap \bar{L} = \bar{\sigma}$ which we denote by $\sigma = K|L$. By analogy, we write $\sigma = K|\text{ext}$ for an external face σ of K, even if this notation is somewhat incorrect, since K may have more than one external face. The mesh $\mathcal{M}$ will be referred to hereafter as the "primal mesh".

The outward normal vector to a face σ of K is denoted by $n_{K,\sigma}$. For $K \in \mathcal{M}$ and $\sigma \in \mathcal{E}$, we denote by $|K|$ the measure of K and by $|\sigma|$ the $(d-1)$-measure of the face σ.

Then, for $\sigma \in \mathcal{E}$ and $K \in \mathcal{M}$ such that $\sigma \in \mathcal{E}(K)$ (in fact, the only cell if $\sigma \in \mathcal{E}_{\text{ext}}$ and one among the two possible cells if $\sigma \in \mathcal{E}_{\text{int}}$), we denote by $D_{K,\sigma}$ a subvolume of K having σ as a face (see Fig. 1), and by $|D_{K,\sigma}|$ the measure of $D_{K,\sigma}$. For $\sigma \in \mathcal{E}_{\text{int}}$, $\sigma = K|L$, we set $D_\sigma = D_{K,\sigma} \cup D_{L,\sigma}$, so $|D_\sigma| = |D_{K,\sigma}| + |D_{L,\sigma}|$, and for $\sigma \in \mathcal{E}_{\text{ext}}$, $\sigma = K|\text{ext}$, $D_\sigma = D_{K,\sigma}$, so $|D_\sigma| = |D_{K,\sigma}|$. The set of faces of the dual cell D_σ is denoted by $\bar{\mathcal{E}}(D_\sigma)$, and the face separating two adjacent dual cells D_σ and $D_{\sigma'}$ is denoted by $\varepsilon = \sigma|\sigma'$.

For $1 \le i \le d$, the degree of freedom for the ith component of the velocity are assumed to be associated to a subset of $\mathcal{E}$, denoted by $\mathcal{E}^{(i)} \subset \mathcal{E}$, and are denoted by:

$$\{u_{\sigma,i}, \ \sigma \in \mathcal{E}^{(i)}\}.$$

The sets of internal, external, Neumann and Dirichlet faces associated to the component i are denoted by $\mathcal{E}^{(i)}_{\text{int}}$, $\mathcal{E}^{(i)}_{\text{ext}}$, $\mathcal{E}^{(i)}_N$ and $\mathcal{E}^{(i)}_D$ (so, for instance, $\mathcal{E}^{(i)}_{\text{int}} = \mathcal{E}_{\text{int}} \cap \mathcal{E}^{(i)}$). We consider the following assumption:

(H1) for $1 \le i \le d$, $\forall K \in \mathcal{M}$,

$$\cup_{\sigma \in \mathcal{E}^{(i)} \cap \mathcal{E}(K)} \overline{D}_{K,\sigma} = \overline{K} \quad \text{and} \quad \sum_{\sigma \in \mathcal{E}^{(i)} \cap \mathcal{E}(K)} |D_{K,\sigma}| = |K|,$$

which means that the volumes $D_{K,\sigma}$, $\sigma \in \mathcal{E}^{(i)}$, are disjoint, and that, for $1 \le i \le d$, $(D_\sigma)_{\sigma \in \mathcal{E}^{(i)}}$ is a partition of Ω. The sets of faces, internal faces and Neumann faces of this dual mesh are denoted by $\bar{\mathcal{E}}^{(i)}$, $\bar{\mathcal{E}}^{(i)}_{\text{int}}$ and $\bar{\mathcal{E}}^{(i)}_N$ respectively.

We suppose that the degrees of freedom for the pressure and the density are associated to the primal cells, so they read

$$\{p_K, \ K \in \mathcal{M}\}, \quad \{\rho_K, \ K \in \mathcal{M}\}.$$

We denote by V the approximation space for the velocity, by $V^{(i)}$, $1 \le i \le d$, the approximation spaces for the velocity components and by Q the approximation space for the pressure and the density, and we identify the discrete functions to their degrees of freedom:

$$\forall v \in V, \ v_i \in V^{(i)}, \ 1 \le i \le d \text{ and } v_i = (v_{\sigma,i})_{\sigma \in \mathcal{E}^{(i)}}; \quad \forall q \in Q, \ q = (q_K)_{K \in \mathcal{M}}.$$

For the velocity, since the concerned degrees of freedom are located on the boundary, the Dirichlet boundary conditions are enforced in the approximation space:

$$\text{for } 1 \leq i \leq d, \; \forall v_i \in V^{(i)}, \; \forall \sigma \in \mathcal{E}_D^{(i)}, \quad v_{\sigma,i} = \frac{1}{|\sigma|} \int_\sigma \bar{u}_{D,i} \, d\gamma,$$

where $\bar{u}_{D,i}$ stands for the ith component of the prescribed velocity.

2.2 The schemes

We now introduce the following notations and assumptions:

- for $K \in \mathcal{M}$ and $\sigma \in \mathcal{E}(K)$, we denote by $u \cdot n_{K,\sigma}$ an approximation of the normal velocity to the face σ outward K,
- for $v \in V$, $1 \leq i \leq d$ and $\sigma \in \mathcal{E}^{(i)}$, we denote by $(\text{div}\tau(v))_\sigma^{(i)}$ an approximation of the viscous term associated to σ and to the component i, and we suppose that the following assumption is satisfied:

$$\text{(H2)} \quad \sum_{i=1}^d \sum_{\sigma \in \mathcal{E}^{(i)}} |D_\sigma| \, (\text{div}\tau(v))_\sigma^{(i)} \, v_{\sigma,i} \geq 0.$$

- for $q \in Q$, $1 \leq i \leq d$ and $\sigma \in \mathcal{E}^{(i)}$, we denote by $(\nabla q)_\sigma^{(i)}$ the component i of the discrete gradient of q at the face σ, and we suppose that the following assumption is satisfied for any $q \in Q$ and $v \in V$:

$$\text{(H3)} \quad \sum_{i=1}^d \sum_{\sigma \in \mathcal{E}^{(i)}} |D_\sigma| \, (\nabla q)_\sigma^{(i)} \, v_{\sigma,i} = \sum_{K \in \mathcal{M}} q_K \sum_{\sigma \in \mathcal{E}(K)} |\sigma| \, v \cdot n_{K,\sigma}.$$

With these notations, we are able to write the general form of the implicit scheme:

$$\forall K \in \mathcal{M}, \quad \frac{|K|}{\delta t}(\rho_K - \rho_K^*) + \sum_{\sigma \in \mathcal{E}(K)} F_{K,\sigma} = 0. \tag{2a}$$

For $1 \leq i \leq d$, $\forall \sigma \in \mathcal{E}_{\text{int}}^{(i)} \cup \mathcal{E}_N^{(i)}$,

$$\frac{|D_\sigma|}{\delta t}(\rho_\sigma u_{\sigma,i} - \rho_\sigma^* u_{\sigma,i}^*) + \sum_{\varepsilon \in \bar{\mathcal{E}}(D_\sigma)} F_{\sigma,\varepsilon} u_{\varepsilon,i} \tag{2b}$$
$$+ |D_\sigma| \, (\nabla p)_\sigma^{(i)} + |D_\sigma| \, (\text{div}\tau(u))_\sigma^{(i)} = 0,$$

$$\forall K \in \mathcal{M}, \quad \rho_K = \wp(p_K). \tag{2c}$$

where the * superscript denotes the beginning-of-step quantities, $F_{K,\sigma}$ stands for the mass flux leaving K through σ, ρ_σ stands for an approximation of the density at the face, and $F_{\sigma,\varepsilon}$ is a mass flux leaving D_σ through ε. For the flux $F_{K,\sigma}$ at the internal face $\sigma = K|L$, we choose an upwind approximation of the density:

$$F_{K,\sigma} = |\sigma|\, u \cdot n_{K,\sigma}\, \rho_\sigma^{\mathrm{up}}, \quad \text{with } \rho_\sigma^{\mathrm{up}} = \rho_K \text{ if } F_{K,\sigma} \geq 0,\ \rho_\sigma^{\mathrm{up}} = \rho_L \text{ otherwise.} \tag{3}$$

On $\sigma \in \mathcal{E}_D$, the density $\rho_\sigma^{\mathrm{up}}$ is given by the boundary condition, and, on $\sigma \in \mathcal{E}_N$, $\sigma = K|\mathrm{ext}$, $\rho_\sigma^{\mathrm{up}} = \rho_K$, which is indeed an upwind choice, since the flow is supposed to enter the domain through $\partial\Omega_D$ and to leave it through $\partial\Omega_N$. For the velocity components at the dual faces, $u_{\varepsilon,i}$, we choose either the centred or upwind approximation on the internal faces, and the value at the face for the outflow faces.

A pressure correction scheme is obtained from (2) by splitting the resolution in two steps:

1- Velocity prediction step – Solve for $\tilde{u} \in V$ the momentum balance equation with the beginning-of-step pressure:

For $1 \leq i \leq d$, $\forall \sigma \in \mathcal{E}_{\mathrm{int}}^{(i)} \cup \mathcal{E}_N^{(i)}$,

$$\frac{|D_\sigma|}{\delta t}(\rho_\sigma \tilde{u}_{\sigma,i} - \rho_\sigma^* u_{\sigma,i}^*) + \sum_{\varepsilon \in \bar{\mathcal{E}}(D_\sigma)} F_{\sigma,\varepsilon} \tilde{u}_{\varepsilon,i} \tag{4}$$
$$+ |D_\sigma|\,(\nabla p^*)_\sigma^{(i)} + |D_\sigma|\,(\mathrm{div}\tau(\tilde{u}))_\sigma^{(i)} = 0,$$

2 - Correction step – Solve for $u \in V$ and $p \in Q$:

$$\forall K \in \mathcal{M}, \qquad \frac{|K|}{\delta t}(\rho_K - \rho_K^*) + \sum_{\sigma \in \mathcal{E}(K)} F_{K,\sigma} = 0. \tag{5a}$$

For $1 \leq i \leq d$, $\forall \sigma \in \mathcal{E}_{\mathrm{int}}^{(i)} \cup \mathcal{E}_N^{(i)}$,

$$\frac{|D_\sigma|}{\delta t}\rho_\sigma\,(u_{\sigma,i} - \tilde{u}_{\sigma,i}) + |D_\sigma|\,(\nabla(p - p^*))_\sigma^{(i)} = 0, \tag{5b}$$

$$\forall K \in \mathcal{M}, \qquad \rho_K = \wp(p_K). \tag{5c}$$

The equations of the correction step are combined to produce a nonlinear parabolic problem for the pressure, which reads, $\forall K \in \mathcal{M}$:

$$\frac{|K|}{\delta t}\bigl(\wp(p_K) - \rho_K^*\bigr) + \sum_{\sigma = K|L} \frac{\rho_\sigma^{\mathrm{up}}}{\rho_\sigma} \frac{|\sigma|^2}{|D_\sigma|}(\phi_K - \phi_L) + \sum_{\sigma \in \mathcal{E}(K) \cap \mathcal{E}_N} \frac{\rho_\sigma^{\mathrm{up}}}{\rho_\sigma} \frac{|\sigma|^2}{|D_\sigma|}\phi_K$$
$$= \frac{1}{\delta t} \sum_{\sigma \in \mathcal{E}(K)} |\sigma|\, \rho_\sigma^{\mathrm{up}}\, \tilde{u} \cdot n_{K,\sigma}, \tag{6}$$

where $\phi \in Q$ is defined by $\phi = p - p^*$. Note that the second and third terms at the left-hand side look like a finite volume discretization of a diffusion operator, with homogeneous Neumann boundary conditions on $\mathcal{E}_D$ and Dirichlet boundary conditions on $\mathcal{E}_N$ for the pressure increment, as usual in pressure correction schemes (see [4] for a discussion on the effects of these spurious boundary conditions).

The standard discretizations entering the present framework are either low-degree non-conforming finite elements, namely the Crouzeix-Raviart element [3] for simplicial meshes or the Rannacher-Turek element [23] for quadrangles and hexahedra, or, for structured cartesian grids, the MAC scheme [13, 14]. We describe here the construction of the diffusion and pressure gradient terms for the finite element schemes, supposing for short that the velocity obeys homogeneous Dirichlet boundary conditions on $\partial \Omega$. Let $\sigma \in \mathcal{E}_{\text{int}}$ and φ_σ be the finite element shape function associated to σ. In Rannacher-Turek or Crouzeix-Raviart elements, a degree of freedom for each component of the velocity is associated to each face, so $\mathcal{E}_{\text{int}}^{(i)} = \mathcal{E}_{\text{int}}$, for $1 \leq i \leq d$. Let $1 \leq i \leq d$ be given, let $e^{(i)}$ be the ith vector of the canonical basis of $\mathbb{R}^d$ and let us define $\varphi_\sigma^{(i)}$ by:

$$\varphi_\sigma^{(i)} = \varphi_\sigma \, e^{(i)}.$$

Then the usual finite element discretization of the diffusion term reads, for a constant viscosity Newtonian fluid (that is supposing $\text{div}\tau(u) = \mu \Delta u + (\mu/3)\nabla \text{div}(u)$, with μ the viscosity):

$$|D_\sigma| \, (\text{div}\tau(u))_\sigma^{(i)} = \sum_{K \in \mathcal{M}} \mu \int_K \nabla u : \nabla \varphi_\sigma^{(i)} \, dx + \frac{\mu}{3} \int_K \text{div} u \, \text{div} \varphi_\sigma^{(i)} \, dx.$$

The pressure gradient term at the internal face $\sigma = K|L$ reads:

$$|D_\sigma| \, (\nabla p)_\sigma^{(i)} = \sum_{K \in \mathcal{M}} \int_K p \, \text{div} \varphi_\sigma^{(i)} \, dx = |\sigma| \, (p_L - p_K) \, n_{K,\sigma} \cdot e^{(i)}.$$

3 The stability issue and consequences

3.1 A stability result for the convection

At the continuous level, let us assume that the mass balance $\partial_t \rho + \text{div}(\beta) = 0$ holds, with β a regular vector-valued function. Then, for all scalar regular functions u and v, we have:

$$\int_\Omega \left[\partial_t (\rho u) + \text{div}(u \beta) \right] v \, dx =$$

$$\int_\Omega \left[\partial_t (\rho u) - \frac{1}{2} (\partial_t \rho) u \right] v \, dx + s(u, v) + \frac{1}{2} \int_{\partial \Omega} u v \beta \cdot n \, d\gamma \quad (7)$$

where s is the following skew-symmetric bilinear form:

$$s(u,v) = \frac{1}{2}\int_\Omega v\beta \cdot \nabla u\,dx - \frac{1}{2}\int_\Omega u\beta \cdot \nabla v\,dx.$$

Taking $u = v = u_i$ and summing over i, the first term gives the time derivative of the kinetic energy, the second term vanishes and the last term corresponds to the kinetic energy flux through the boundary of the domain. The following Lemma, proven in [20], states a discrete counterpart of this computation in the case where the (possible) Dirichlet boundary conditions are homogeneous (see also [1] and [9] for a direct estimate of the kinetic energy, for an implicit and explicit scheme respectively).

Lemma 1. *Let us suppose that, for an index i, $1 \le i \le d$, the following discrete mass balance holds over the dual cells associated to the ith component of the velocity:*

$$\forall \sigma \in \mathcal{E}_{int}^{(i)} \cup \mathcal{E}_N^{(i)}, \qquad \frac{|D_\sigma|}{\delta t}(\rho_\sigma - \rho_\sigma^*) + \sum_{\varepsilon \in \bar{\mathcal{E}}(D_\sigma)} F_{\sigma,\varepsilon} = 0. \qquad (8)$$

Let $u, v \in V^{(i)}$, and let us suppose that these discrete functions obey homogeneous Dirichlet boundary. Then we have:

$$\sum_{\sigma \in \mathcal{E}_{int}^{(i)} \cup \mathcal{E}_N^{(i)}} v_\sigma \left[\frac{|D_\sigma|}{\delta t}(\rho_\sigma u_\sigma - \rho_\sigma^* u_\sigma^*) + \sum_{\varepsilon \in \bar{\mathcal{E}}(D_\sigma)} F_{\sigma,\varepsilon} u_\varepsilon\right]$$

$$\ge T_{\Omega,k}(u,v) + T_{\Omega,s}(u,v) + T_{\partial\Omega}(u,v), \qquad (9)$$

with:

$$T_{\Omega,k}(u,v) = \sum_{\sigma \in \mathcal{E}_{int}^{(i)} \cup \mathcal{E}_N^{(i)}} \frac{|D_\sigma|}{\delta t}(\rho_\sigma u_\sigma - \rho_\sigma^* u_\sigma^*)v_\sigma - \frac{1}{2}(\rho_\sigma - \rho_\sigma^*)u_\sigma v_\sigma,$$

$$T_{\Omega,s}(u,v) = S(u,v) - S(v,u), \qquad S(u,v) = \frac{1}{2}\sum_{\varepsilon \in \bar{\mathcal{E}}_{int}^{(i)},\, \varepsilon = D_\sigma|D_{\sigma'}} F_{\sigma,\varepsilon} v_\varepsilon (u_{\sigma'} - u_\sigma),$$

$$T_{\partial\Omega}(u,v) = \frac{1}{2}\sum_{\varepsilon \in \bar{\mathcal{E}}_N^{(i)},\, \sigma = D_\sigma|ext} F_{\sigma,\varepsilon} u_\varepsilon v_\varepsilon.$$

Inequality (9) becomes an equality for a centred choice of the discretization of the face values u_ε.

Of course, $T_{\Omega,s}(u,u) = 0$, and an easy computation shows that:

$$T_{\Omega,k}(u,u) \ge \frac{1}{2\delta t} \sum_{\sigma \in \mathcal{E}_{int}^{(i)} \cup \mathcal{E}_N^{(i)}} |D_\sigma|\left[\rho_\sigma u_\sigma^2 - \rho_\sigma^*(u_\sigma^*)^2\right].$$

Applying Lemma 1 to each component of the velocity, the obtained term is thus the discrete time-derivative of the kinetic energy, and may be used to obtain stability estimates for the scheme (see Sect. 4).

Remark 1 (Non-homogeneous Dirichlet boundary conditions). The limitation to homogeneous Dirichlet boundary conditions may be seen, from the proof, to stem from the fact that no balance equation is written on the dual cells associated to faces lying on $\partial \Omega_D$. The problem may thus be fixed by keeping these degrees of freedom and using a penalization technique.

Remark 2 (Artificial boundary conditions). Lemma 1 may be used to derive artificial boundary conditions allowing the flow to enter the domain through $\partial \Omega_N$, by first collecting the boundary terms in the variational form of the momentum balance equation (*i.e.* adding to $T_{\partial \Omega}(u, v)$ the terms issued from the diffusion and the pressure gradient) and then imposing that the result may be written as a linear form acting on the test function (see [2] for a similar development in the incompressible case). The so-built boundary condition is observed in practice to give quite good results when modelling external flows [20].

3.2 Discretization of the convection term

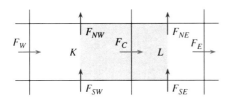

Fig. 2 Local notations for the definition of the mass fluxes at the dual edges with the MAC scheme

The problem to tackle is now the following one: on one side, the discrete mass balance over the dual cells (8) is necessary for the stability of the scheme; on the other side, the mass balance is only written by the scheme(s) for the primal cells (Equation (2a) or (5a)). We are thus lead to express the mass fluxes ($F_{\sigma,\varepsilon}$) through the dual faces as a function of the mass fluxes ($F_{K,\sigma}$) through the primal faces, in such a way that the discrete balance over the primal cells yields a discrete balance over the dual cells. We describe in this section how this may be done, first for the MAC (structured) mesh (see also [15]) and, second, for the Rannacher-Turek element on general quadrangles.

3.2.1 MAC scheme

For the MAC scheme, in two space dimensions and with the local notations introduced on Fig. 2, the mass balance on the primal cells reads:

$$K : \frac{|K|}{\delta t}(\rho_K - \rho_K^*) - F_W - F_{SW} + F_C + F_{NW} = 0,$$

$$L : \frac{|L|}{\delta t}(\rho_L - \rho_L^*) - F_C - F_{SE} + F_E + F_{NE} = 0.$$

Multiplying both equations by $1/2$ and summing them yields, for $\sigma = K|L$:

$$\frac{|D_\sigma|}{\delta t}(\rho_\sigma - \rho_\sigma^*)$$
$$-\frac{1}{2}[F_W + F_C] - \frac{1}{2}[F_{SW} + F_{SE}] + \frac{1}{2}[F_C + F_E] + \frac{1}{2}[F_{NW} + F_{NE}] = 0, \quad (10)$$

with the usual definition of the dual cell D_σ, which implies that $|D_{K,\sigma}| = |K|/2$ and $|D_{L,\sigma}| = |L|/2$, and with the following definition of the density on the face:

$$|D_\sigma|\rho_\sigma = |D_{K,\sigma}|\rho_K + |D_{L,\sigma}|\rho_L. \quad (11)$$

Equation (10) thus suggests the following definition for the mass fluxes at the dual faces:

left face: $F_{\sigma,\varepsilon} = -\frac{1}{2}[F_W + F_C]$; right face: $F_{\sigma,\varepsilon} = \frac{1}{2}[F_C + F_E]$;

bottom face: $F_{\sigma,\varepsilon} = -\frac{1}{2}[F_{SW} + F_{SE}]$; top face: $F_{\sigma,\varepsilon} = \frac{1}{2}[F_{NW} + F_{NE}]$.

Note that this definition is rather non-standard: for instance, the flux at the left face of D_σ, which is included in K, may involve densities of the neighbouring primal cells. The extension of the above construction to the three-dimensional case is straightforward.

3.2.2 Rannacher-Turek element

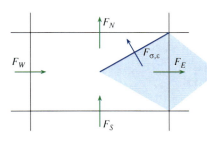

Fig. 3 Local notations for the definition of the mass fluxes at the dual edges with the Rannacher-Turek element

A construction similar to that of the MAC scheme may be performed for rectangular meshes. For K and L two neighbouring cells of $\mathcal{M}$, the half-diamond cell $D_{K,\sigma}$ (resp. $D_{L,\sigma}$) associated to the common face $\sigma = K|L$ is defined as the cone with vertex the mass center of K (resp. L) and with basis σ, the density ρ_σ is defined by the weighted average (11), and the dual mass fluxes are obtained by multiplying the mass balances over K and L by $1/4$ and summing. With the local notations of Fig. 3, this yields, for the dual mass flux $F_{\sigma,\varepsilon}$, an expression of the form:

$$F_{\sigma,\varepsilon} = -\frac{1}{8}F_W + \frac{3}{8}F_N - \frac{3}{8}F_E + \frac{1}{8}F_S. \quad (12)$$

We now explain how to extend this formulation to general meshes.

Let us suppose that, for any cell $K \in \mathcal{M}$, we are able to define the fluxes through the dual faces included in K in such a way that:

(A1) The mass balance over the half-diamond cells is proportional to the mass balance over K, in the following sense:

$$\forall \sigma \in \mathcal{E}(K), \quad F_{K,\sigma} + \sum_{\varepsilon \in \bar{\mathcal{E}}(D_\sigma),\, \varepsilon \subset K} F_{\sigma,\varepsilon} = \xi_K^\sigma \sum_{\sigma \in \mathcal{E}(K)} F_{K,\sigma},$$

with $\displaystyle\sum_{\sigma \in \mathcal{E}(K)} \xi_K^\sigma = 1$ and, for any $\sigma \in \mathcal{E}(K), \xi_K^\sigma \geq 0$.

(A2) The dual fluxes are conservative, i.e., for any $\varepsilon = D_\sigma | D'_\sigma$, $F_{\sigma,\varepsilon} = -F_{\sigma',\varepsilon}$.

(A3) The dual fluxes are bounded with respect to the $(F_{K,\sigma})_{\sigma \in \mathcal{E}(K)}$:

$$\forall \sigma \in \mathcal{E}(K),\ \forall \varepsilon \in \bar{\mathcal{E}}(D_\sigma) \subset K \quad |F_{\sigma,\varepsilon}| \leq C\, \max\left\{|F_{K,\sigma}|,\ \sigma \in \mathcal{E}(K)\right\}.$$

In addition, let us define $|D_{K,\sigma}|$ as:

$$|D_{K,\sigma}| = \xi_K^\sigma\, |K|, \quad (13)$$

and ρ_σ, once again, by the weighted average (11). Then the dual fluxes satisfy the required mass balance. Indeed, for $\sigma \in \mathcal{E}_{\text{int}}, \sigma = K|L$, we have:

$$\frac{|D_\sigma|}{\delta t}(\rho_\sigma - \rho_\sigma^*) + \sum_{\varepsilon \in \mathcal{E}(D_\sigma)} F_{\sigma,\varepsilon}$$

$$= \frac{|D_{K,\sigma}|}{\delta t}(\rho_K - \rho_K^*) + F_{K,\sigma} + \sum_{\varepsilon \in \bar{\mathcal{E}}(D_\sigma),\, \varepsilon \subset K} F_{\sigma,\varepsilon}$$

$$+ \frac{|D_{L,\sigma}|}{\delta t}(\rho_L - \rho_L^*) + F_{L,\sigma} + \sum_{\varepsilon \in \bar{\mathcal{E}}(D_\sigma),\, \varepsilon \subset L} F_{\sigma,\varepsilon}$$

$$= \xi_K^\sigma \left[\frac{|K|}{\delta t}(\rho_K - \rho_K^*) + \sum_{\sigma \in \mathcal{E}(K)} F_{K,\sigma}\right] + \xi_L^\sigma \left[\frac{|L|}{\delta t}(\rho_L - \rho_L^*) + \sum_{\sigma \in \mathcal{E}(L)} F_{L,\sigma}\right] = 0.$$

A similar computation leads to the same conclusion for the (half-)dual cells associated to the Neumann boundary faces.

The next issue is to check whether Assumptions (A1)-(A3) are sufficient for the consistency of the scheme. In this respect, the following lemma [16] brings a decisive argument.

Lemma 2. *Let Assumptions (A1)-(A3) hold. For $v \in V$ and $K \in \mathcal{M}$, let v_K be defined by $v_K = \sum_{\sigma \in \mathcal{E}(K)} \xi_K^\sigma v_\sigma$. Let $u \in V$, and $R(u, v)$ be the quantity defined by:*

$$R(u,v) = \sum_{\sigma \in \mathcal{E}_{int}} v_\sigma \sum_{\substack{\varepsilon \in \bar{\mathcal{E}}(D_\sigma), \\ \varepsilon = D_\sigma | D'_\sigma}} F_{\sigma,\varepsilon} \frac{u_\sigma + u_{\sigma'}}{2} - \sum_{K \in \mathcal{M}} v_K \sum_{\sigma \in \mathcal{E}(K)} F_{K,\sigma} u_\sigma.$$

Let us suppose that the primal fluxes are associated to a convection momentum field β, i.e. $\forall K \in \mathcal{M}$, $\forall \sigma \in \mathcal{E}(K)$, $F_{K,\sigma} = |\sigma| \, \beta_\sigma \cdot n_{K,\sigma}$. (For the schemes used here, of course, β depends the density and the velocity, see (3).) Then there exists C depending only on the regularity of the mesh such that:

$$|R(u,v)| \leq C \, h \, \|\beta\|_{l^\infty} \, \|u\|_1 \, \|v\|_1,$$

with $\|\beta\|_{l^\infty} = \max_{\sigma \in \mathcal{E}} |\beta_\sigma|$ and the discrete H^1-norm on the dual mesh is defined by:

$$\forall v \in V, \quad \|v\|_1 = \sum_{K \in \mathcal{M}} h_K^{d-2} \sum_{\sigma,\sigma' \in \mathcal{E}(K)} (v_\sigma - v_{\sigma'})^2.$$

The quantity $R(u, v)$ compares two discrete analogues to $\int_\Omega v \, \text{div}(u\beta) \, dx$; the first analogue is defined with the divergence taken over the dual meshes while the second analogue is defined with the divergence over the primal cells. Let us suppose that the discrete H^1-norm of the solution is controlled thanks to the diffusion term. Then, in a convergence or error analysis study in the linear case (*i.e.* with a given regular convection field β), Lemma 2 allows to replace the first discrete analogue by the second one, thus substituting well defined quantities to quantities only defined through (A1)-(A3). It is used in [16] to prove that the scheme is first-order for the stationary convection-diffusion equation. The convergence for the constant density Navier-Stokes equations (that is with $\beta = u$) was also proven, controlling now $\|u\|_{l^\infty}$ by $\|u\|_1$ thanks to an inverse inequality.

The last task is now to build fluxes satisfying (A1)-(A3); this is easily done by choosing $\xi_K^\sigma = 1/4$, and keeping for the expression of the dual fluxes as a function of the primal fluxes the same linear combination (12) as in the rectangular case. Note that this implicitly implies that the geometrical definition of the dual cells has been generalized, since it is not possible in general to split a quadrangle in four simplices of same measure (even if the quadrangle is convex). The extension to three dimensions only needs to deal with the rectangular parallelepipedic case, which is quite simple [1]. Finding directly a solution to (A1)-(A3) may also be an alternative route, to deal with more complex cases, as done in [16] to extend the scheme to locally refined non-conforming grids.

4 Schemes and stability estimates

In order to obtain the complete formulation of the considered schemes, we now have to fix the time-marching procedure. This is straightforward for the implicit scheme, and we concentrate here on the pressure correction scheme. The problem which we face in this case is that the mass balance is not yet solved when performing the prediction step. In our implementations in the ISIS computer code [18] developed at IRSN on the basis of the software component library PELICANS [22], it is circumvented by just shifting in time the density ρ_σ; the mass balance on the dual cells is recovered from the mass balance on the primal cells at the previous time step. This has essentially two drawbacks. First, the trick indeed works only if the time step is constant; for a variable time step, one has to choose between loosing stability or consistency (locally in time, so fortunately, without observed impact in practice). Second, the scheme is only first order in time.

In addition, stability seems to require an initial pressure renormalization step, which is an algebraic variant of the one introduced in [12]. It seems however that this step may be omitted in practice.

The algorithm (keeping in this presentation the pressure renormalization step) reads, assuming that u^n, p^n, ρ^n and the family $(F_{K,\sigma}^n)$ are known:

1- Pressure renormalization step – Let $(\lambda_\sigma)_{\sigma \in \mathcal{E}_{\text{int}}}$ be a family of positive real numbers, and let $-\text{div}(\lambda \nabla)_\mathcal{M}$ be the discrete elliptic operator from Q to Q defined by, $\forall K \in \mathcal{M}$ and $q \in Q$:

$$\left[-\text{div}(\lambda \nabla)_\mathcal{M}(q)\right]_K = \sum_{\sigma=K|L} \lambda_\sigma \frac{|\sigma|^2}{|D_\sigma|}(q_K - q_L) + \sum_{\sigma \in \mathcal{E}_N, \sigma=K|\text{ext}} \lambda_\sigma \frac{|\sigma|^2}{|D_\sigma|} q_K.$$

Then $\tilde{p}^{n+1} \in Q$ is given by:

$$-\text{div}(\frac{1}{\rho^n}\nabla)_\mathcal{M}(\tilde{p}^{n+1}) = -\text{div}(\frac{1}{[\rho^n \rho^{n-1}]^{1/2}}\nabla)_\mathcal{M}(p^n), \quad (14)$$

the weights $(\rho_\sigma^n)_{\sigma \in \mathcal{E}_{\text{int}} \cup \mathcal{E}_N}$ and $(\rho_\sigma^{n-1})_{\sigma \in \mathcal{E}_{\text{int}} \cup \mathcal{E}_N}$ being the densities involved in the time-derivative term of the momentum balance equation.

2- Velocity prediction step – Solve for $\tilde{u}^{n+1} \in V$, for $1 \le i \le d$ and $\forall \sigma \in \mathcal{E}_{\text{int}}^{(i)} \cup \mathcal{E}_N^{(i)}$:

$$\frac{|D_\sigma|}{\delta t}(\rho_\sigma^n \tilde{u}_{\sigma,i}^{n+1} - \rho_\sigma^{n-1} u_{\sigma,i}^n) + \sum_{\varepsilon \in \bar{\mathcal{E}}(D_\sigma)} F_{\sigma,\varepsilon}^n \tilde{u}_{\varepsilon,i}^{n+1}$$

$$+ |D_\sigma|(\nabla \tilde{p}^{n+1})_\sigma^{(i)} + |D_\sigma|(\text{div}\tau(\tilde{u}^{n+1}))_\sigma^{(i)} = 0, \quad (15)$$

where the $(F_{\sigma,\varepsilon}^n)_{\varepsilon \in \bar{\mathcal{E}}(D_\sigma)}$ are built as explained in the previous section, from the primal fluxes at time t^n.

3 - Correction step – Solve for $u^{n+1} \in V$ and $p^{n+1} \in Q$:

$$\forall K \in \mathcal{M}, \quad \frac{|K|}{\delta t}(\rho_K^{n+1} - \rho_K^n) + \sum_{\sigma \in \mathcal{E}(K)} F_{K,\sigma}^{n+1} = 0. \quad (16a)$$

For $1 \leq i \leq d$, $\forall \sigma \in \mathcal{E}_{\text{int}}^{(i)} \cup \mathcal{E}_N^{(i)}$,

$$\frac{|D_\sigma|}{\delta t} \rho_\sigma^n (u_{\sigma,i}^{n+1} - \tilde{u}_{\sigma,i}^{n+1}) + |D_\sigma| \left(\nabla(p^{n+1} - \tilde{p}^{n+1})\right)_\sigma^{(i)} = 0, \quad (16b)$$

$$\forall K \in \mathcal{M}, \quad \rho_K^{n+1} = \wp(p_K^{n+1}). \quad (16c)$$

The algorithm must be initialized by the data $u^0 \in V$, $\rho^{-1} \in Q$ and $\rho^0 \in Q$ satisfying the discrete mass balance equation, and with the corresponding mass fluxes $(F_{K,\sigma}^0)$. A possible way to obtain these quantities is to evaluate u^0 and ρ^{-1} from the initial conditions, and, as a preliminary step, to solve for ρ^0 the mass balance equation.

The upwinding in the discretization of the mass balance equation has for consequence that any density appearing in the algorithm is positive (provided that the initial density is positive). The existence and uniqueness of a solution to Steps 1 and 2 is then clear: these are linear problems with coercive operators (for Step 2, thanks to the stability of the convection term). The existence of a solution to Step 3 may be obtained by a Brouwer fixed point argument, using the fact that the conservativity of the mass balance yields an estimate for ρ, so for p, and finally for u (in any norm, since we work on finite dimensional spaces). The algorithm is thus well-posed.

Let us now turn to the energy estimate. At the continuous level, this relation is obtained for the barotropic Navier-Stokes equations by choosing the velocity u as a test function in the variational form of the momentum balance equation, writing the convection term as the time derivative of the kinetic energy, and setting the pressure work, namely $-\int_\Omega p \, \text{div}(u) \, dx$, under a convenient form. This latter step is done by the following formal computation. Let $b(\cdot)$ be a regular function from $(0, +\infty)$ to $\mathbb{R}$, and let us multiply the mass balance by $b'(\rho)$. Using:

$$b'(\rho)\text{div}(\rho u) = b'(\rho)[u \cdot \nabla \rho + \rho \text{div}(u)] = u \cdot \nabla b(\rho) + \rho b'(\rho)\text{div}(u)$$
$$= \text{div}(b(\rho)u) + [\rho b'(\rho) - b(\rho)]\text{div}(u),$$

we get:

$$\partial_t [b(\rho)] + \text{div}[b(\rho) u] + [\rho b'(\rho) - b(\rho)] \text{div}(u) = 0.$$

Choosing now the function $b(\cdot)$ in such a way that $\rho b'(\rho) - b(\rho) = \wp^{-1}(p)$, integrating over Ω and supposing homogeneous Dirichlet boundary conditions over $\partial \Omega$ yields:

$$-\int_\Omega p \operatorname{div}(u)\,dx = \frac{d}{dt}\int_\Omega b(\rho)\,dx.$$

The following lemma [7] states a discrete counterpart of this computation.

Lemma 3. *Let us suppose that the velocity field obeys homogeneous Dirichlet boundary conditions. Let $b(\cdot)$ be a regular convex function from $(0,+\infty)$ to $\mathbb{R}$, and $(\rho_K^*)_{K\in\mathcal{M}}$ be a positive family of real numbers. Then, with the upwind discretization (3) of the mass balance equation, the family $(\rho_K)_{K\in\mathcal{M}}$ is also positive, and we get:*

$$\sum_{K\in\mathcal{M}} b'(\rho_K)\left[\frac{|K|}{\delta t}(\rho_K - \rho_K^*) + \sum_{\sigma\in\mathcal{E}(K)} F_{K,\sigma}\right] \geq$$
$$\frac{1}{\delta t}\sum_{K\in\mathcal{M}}|K|\left[b(\rho_K) - b(\rho_K^*)\right] + \sum_{K\in\mathcal{M}}\left[\rho_K b'(\rho_K) - b(\rho_K)\right]\sum_{\sigma\in\mathcal{E}(K)}|\sigma|\,u_\sigma \cdot n_{K,\sigma}.$$

We are now in position to state the following stability result.

Theorem 1. *Let us suppose that the velocity field obeys homogeneous Dirichlet boundary conditions. The scheme (14)-(16) satisfies the following energy identity, for $1 \leq n \leq N$:*

$$\frac{1}{2}\sum_{i=1}^{d}\sum_{\sigma\in\mathcal{E}_{\mathrm{int}}^{(i)}}|D_\sigma|\,\rho_\sigma^{n-1}\,(u_{\sigma,i}^n)^2 + \delta t\sum_{k=1}^{n}\sum_{\sigma\in\mathcal{E}^{(i)}}|D_\sigma|\,(\operatorname{div}\tau(u^k))_\sigma^{(i)}\,u_{\sigma,i}^k$$
$$+\sum_{K\in\mathcal{M}}|K|\,b(\rho_K^n) \leq \frac{1}{2}\sum_{i=1}^{d}\sum_{\sigma\in\mathcal{E}_{\mathrm{int}}^{(i)}}|D_\sigma|\,\rho_\sigma^{(-1)}\,(u_{\sigma,i}^0)^2 + \sum_{K\in\mathcal{M}}|K|\,b(\rho_K^0).$$

The proof of this theorem is based on Lemma 1 and Lemma 3, and may be found, for the essential arguments, in [7].

Remark 3. Let us suppose that the equation of state reads $p = \rho^\gamma$, with $\gamma \in (1,+\infty)$. Then an easy computation yields $b(\rho) = \rho^\gamma/(\gamma-1) = p/(\gamma-1)$. Theorem 1 thus yields an estimate for the pressure in $L^\infty(0,T;L^1)$-norm. Note that this estimate is however not sufficient to ensure that a sequence of pressures obtained as discrete solutions converges to a function; in fact, in convergence studies of numerical schemes [5,6,8] as well as in mathematical analysis of the continuous problem [21], the pressure has to be controlled from estimates of its gradient.

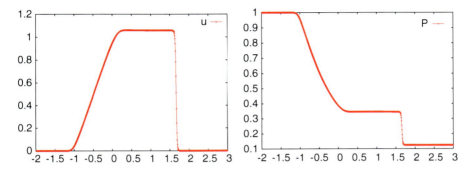

Fig. 4 Solution for the Sod shock-tube problem, obtained with a uniform mesh of 800 cells, with a residual viscosity – *left:* velocity, *right:* pressure

5 Euler equations and solutions with shocks

In this section we briefly discuss the capability of the considered numerical schemes to compute irregular (*i.e.* with discontinuities) solutions of inviscid flows.

The results obtained with the above described pressure correction scheme for the so-called one-dimensional Sod shock-tube problem are displayed on Fig. 4 (see [19] for a more detailed presentation). From numerical experiments, it seems that this scheme converges when the velocity space translates are controlled, either by upwinding the discretization of the velocity convection term, or by keeping a residual viscosity in the (discrete) momentum balance equation. Numerical experiments reported in [19] (addressing also an extension of this algorithm to the barotropic homogeneous two-phase flow model [11]) confirm the stability of the scheme, and show that the qualitative behaviour of the solution is captured up to very large values of the CFL number (typically, in the range of 50).

From the theoretical point of view, for Euler equations (*i.e.*, precisely speaking, with a diffusion vanishing with the space step), the control that we are able to prove on the solution of course does not yield (weak or strong) convergence in strong enough norms to pass to the limit in the scheme. We can however prove the following result: supposing convergence for the density in $L^p(\Omega)$, $p \in [1, +\infty)$ and for the velocity in $L^r(\Omega)$, $r \in [1, 3]$, it is possible to pass to the limit in the discrete equations, provided that the viscosity vanishes as h^α, $\alpha \in (0, 2)$ for both the implicit and the pressure correction scheme. In this case, the limit of a sequence of discrete solutions is proven to satisfy the weak form of the Euler equations, and so, in particular, the Rankine-Hugoniot conditions at the shocks.

6 Discussion and perspectives

The theoretical analysis of the schemes presented here has been undertaken for model stationary problems: in [5, 8], we prove the convergence for the Crouzeix-Raviart discretization of the Stokes equations (with the additional stabilization term needed for purely technical reasons); in [6], we prove the same result for the (standard) MAC scheme. An extension, still for the MAC discretization, to the stationary Navier-Stokes equations is underway.

From a practical point of view, a next step for the barotropic Navier-Stokes equations should be to derive an upwind explicit version of the scheme presented here; in this direction, an extension of Lemma 1 (stability of the velocity convection term) to the explicit case may be found in [9].

The main objective is however to deal with the full (*i.e.* non barotropic, therefore including an energy balance) Navier-Stokes equations. An unconditionally stable pressure correction scheme has been derived for this problem [17], but extensive tests of this scheme remain to be done. In particular, stability requires that the internal energy remains non-negative (in practice, positive); the way we obtained this property was to solve the internal energy balance, with a scheme able to preserve the sign of the unknown. However, it is commonly agreed that, for the scheme to converge toward the correct weak solution, a conservative discretization of the total energy balance should be used. The actual occurrence of this problem, and the possibility to circumvent it, possibly by adding stabilizing viscous terms, will deserve investigations in the near future; a preliminary step on this route may be found in [10].

References

1. G. Ansanay-Alex, F. Babik, J.-C. Latché, D. Vola: An L^2-stable approximation of the Navier-Stokes convection operator for low-order non-conforming finite elements. IJNMF, online (2010).
2. C.H. Bruneau, P. Fabrie: Effective downstream boundary conditions for incompressible Navier-Stokes equations. IJNMF, **99**, 693–705 (1994).
3. M. Crouzeix, P.-A. Raviart: Conforming and nonconforming finite element methods for solving the stationary Stokes equations I, Revue Française d'Automatique, Informatique et Recherche Opérationnelle (R.A.I.R.O.), **R-3**, 33–75 (1973).
4. F. Dardalhon, J.-C. Latché, S. Minjeaud: Analysis of a projection method for low-order nonconforming finite elements. Submitted (2011).
5. R. Eymard, T. Gallouët, R. Herbin, J.-C. Latché: A convergent Finite Element-Finite Volume scheme for the compressible Stokes problem. Part II: the isentropic case. Mathematics of Computation, **79**, 649–675 (2010).
6. R. Eymard, T. Gallouët, R. Herbin, J.-C. Latché: Convergence of the MAC scheme for the compressible Stokes equations. SIAM Journal on Numerical Analysis, **48**, 2218–2246 (2010).
7. T. Gallouët, L. Gastaldo, R. Herbin, J.-C. Latché: An unconditionally stable pressure correction scheme for compressible barotropic Navier-Stokes equations. Mathematical Modelling and Numerical Analysis, **42**, 303–331 (2008).
8. T. Gallouët, R. Herbin, J.-C. Latché: A convergent Finite Element-Finite Volume scheme for the compressible Stokes problem. Part I: the isothermal case. Mathematics of Computation, **78**, 1333–1352 (2009).
9. T. Gallouët, R. Herbin, J.-C. Latché: Kinetic energy control in explicit Finite-Volume discretizations of the incompressible and compressible Navier-Stokes equations. International Journal of Finite Volumes **2** (2010).

10. T. Gallouët, R. Herbin, J.-C. Latché, T.T. Nguyen: Playing with Burgers equation. Finite Volumes for Complex Applications VI (FVCA VI), these proceedings.
11. L. Gastaldo, R. Herbin, J.-C. Latché: An unconditionally stable Finite Element-Finite Volume pressure correction scheme for the drift-flux model. Mathematical Modelling and Numerical Analysis, **44**, 251–287 (2010).
12. J.-L. Guermond, L. Quartapelle: A Projection FEM for Variable Density Incompressible Flows. Journal of Computational Physics, **165**, 167–188 (2000).
13. F.H. Harlow, J.E. Welsh: Numerical calculation of time-dependent viscous incompressible flow of fluid with free surface. Physics of Fluids, **8**, 2182–2189 (1965).
14. F.H. Harlow, A.A. Amsden: A numerical fluid dynamics calculation method for all flow speeds. Journal of Computational Physics, **8**, 197–213 (1971).
15. R. Herbin, J.-C. Latché: A kinetic energy control in the MAC discretization of compressible Navier-Stokes equations. International Journal of Finite Volumes **2** (2010).
16. R. Herbin, J.-C. Latché, B. Piar: A finite-element finite-volume face centred scheme with non-conforming local refinement. I – Convection-diffusion equation. In preparation (2011).
17. R. Herbin, W. Kheriji, J.-C. Latché: An unconditionally stable pressure correction scheme for the compressible Navier-Stokes equations. In preparation (2011).
18. ISIS: a CFD computer code for the simulation of reactive turbulent flows, https://gforge.irsn.fr/gf/project/isis.
19. W. Kheriji, R. Herbin, J.-C. Latché: Numerical tests of a new pressure correction scheme for the homogeneous model. ECCOMAS CFD 2010, Lisbon, Portugal, June 2010.
20. C. Lapuerta, J.-C. Latché: Discrete artificial boundary conditions for compressible external flows. In preparation (2011).
21. P.-L. Lions: Mathematical Topics in Fluid Mecanics. Volume 2. Compressible Models. Oxford Lecture Series in Mathematics and its Applications, vol. 10 (1998).
22. PELICANS: Collaborative Development Environment. https://gforge.irsn.fr/gf/project/pelicans.
23. R. Rannacher, S. Turek: Simple Nonconforming Quadrilateral Stokes Element. Numerical Methods for Partial Differential Equations, **8**, 97–111 (1992).

The paper is in final form and no similar paper has been or is being submitted elsewhere.

ALE Method for Simulations of Laser-Produced Plasmas

Liska R., Kuchařík M., Limpouch J., Renner O., Váchal P., Bednárik L., and Velechovský J.

Abstract Simulations of laser-produced plasmas are essential for laser-plasma interaction studies and for inertial confinement fusion (ICF) technology. Dynamics of such plasmas typically involves regions of large scale expansion or compression, which requires to use the moving Lagrangian coordinates. For some kind of flows such as shear or vortex the moving Lagrangian mesh however tangles and such flows require the use of arbitrary Lagrangian Eulerian (ALE) method. We have developed code PALE (Prague ALE) for simulations of laser-produced plasmas which includes Lagrangian and ALE hydrodynamics complemented by heat conductivity and laser absorption. Here we briefly review the numerical methods used in PALE code and present its selected applications to modeling of laser interaction with targets.

Keywords Lagrangian coordinates, ALE method, laser-produced plasma
MSC2010: 35L65, 35K05, 65M08

1 Introduction

Understanding of laser-produced plasma behavior and evolution is crucial for studies of intense laser interaction with targets and for inertial confinement fusion (ICF) technology employing high-power laser beams to ignite the fusion reaction of deuterium-tritium fuel. We model the complex problem of laser-produced plasma by a set of hydrodynamical conservation laws for mass, momentum and energy

---

Liska R., Kuchařík M., Limpouch J., Váchal P., Bednárik L, and Velechovský J.
Czech Technical University, Faculty of Nuclear Sciences and Physical Engineering, Břehová 7, 115 19 Prague 1, Czech Republic, e-mail: liska@siduri.fjfi.cvut.cz

Renner O.
Institute of Physics, v.v.i., Academy of Sciences Czech Rep., Na Slovance 2, 182 21 Prague, Czech Republic

of compressible fluid, complemented by laser absorption and heat transfer, which written in Lagrangian coordinates have the form

$$\frac{1}{\rho}\frac{d\rho}{dt} = -\text{div }\mathbf{U}, \tag{1}$$

$$\rho\frac{d\mathbf{U}}{dt} = -\text{grad }p, \tag{2}$$

$$\rho\frac{d\varepsilon}{dt} = -p\text{ div }\mathbf{U} + \text{div}(\kappa\text{ grad }T) - \text{div }\mathbf{I}, \tag{3}$$

where ρ is density, $\mathbf{U}$ velocity, p pressure, ε specific internal energy (energy per unit mass), T temperature, κ heat conductivity, $\mathbf{I}$ laser energy flux density (Poynting vector) and $d/dt = \partial/\partial t + \mathbf{u}\cdot\text{grad}$ is the total Lagrangian time derivative including convective terms. The system is closed by the equation of state coupling density, internal energy, pressure and temperature. Laser-produced plasma is usually modeled in the Lagrangian coordinates moving with the fluid, which are able to deal with moving boundaries and large scale deformation like compression or expansion appearing typically in laser-produced plasmas. Some types of plasma flows such as shear or vortex can result in tangling of the computational mesh. This problem can be avoided by Arbitrary Lagrangian-Eulerian (ALE) method [8], which smooths (rezones) the computational mesh and interpolates (remaps) conservative variables to the smoothed mesh after several Lagrangian time steps. Standard Lagrangian numerical hydrodynamics employs staggered method [3, 4, 6], however one can use composite schemes [18, 26] and recently much attention has been attracted by the cell-centered methods [19, 22, 23].

We have developed a 2D ALE code PALE (Prague ALE) for laser-produced plasma simulations, which uses a 2D quadrilateral, logically rectangular computational mesh, We shortly outline the numerical methods employed in the PALE code for ALE hydrodynamics, heat conductivity and laser absorption. The PALE code capabilities are demonstrated on simulations of laser interaction with targets.

2 Hydrodynamics

The hydrodynamical ALE method consists from Lagrangian, rezone and remap phases. Rezone and remap is applied either regularly after fixed number of Lagrangian time steps or adaptively when quality of the moving mesh becomes bad.

2.1 Staggered Lagrangian Method

We consider a 2D staggered location of physical variables: velocity vector is defined at point (node) p of the computational mesh and is denoted $\mathbf{U}_p = (u_p, v_p)$, specific internal energy ε_c is defined at the center of the cell c and density ρ_{pc}

is defined at the center of the subcell Ω_{pc}. The subcell Ω_{pc} is the quadrilateral whose vertexes are point p, center of cell c and two midpoints of two edges of cell c originating at point p. In the Lagrangian gas dynamics the nodes move with the local fluid velocity, and the mass of a subcell is assumed to be constant in time. Conservative variables are the mass m, momentum $m\mathbf{U}$ and total energy $E = m\left(\varepsilon + \frac{1}{2}\|\mathbf{U}\|^2\right)$. This discretization is based on the philosophy of compatible hydrodynamics algorithms introduced in [4]. The Lagrangian phase is conservative, that is, some discrete form of mass, momentum, and total energy is conserved [4]. The mass of any subcell is given by $m_{pc} = \rho_{pc} V_{pc}$, where V_{pc} is the volume of the subcell. Then masses of the cell and node are defined by summation of subcell masses

$$m_c = \sum_{n \in \mathscr{P}(c)} m_{pc} \quad \text{and} \quad m_p = \sum_{c \in \mathscr{C}(p)} m_{pc}$$

over set of points $\mathscr{P}(c)$ being vertexes of the cell c and set of cells $\mathscr{C}(p)$ sharing the node p. All these masses participate in the Lagrangian phase of an ALE method, subcell mass m_{pc} is assumed to be Lagrangian, so it does not change with time, therefore $\rho_{pc}(t) = m_{pc}/V_{pc}(t)$, which can be considered as a definition of subcell density for given constant subcell mass. Masses of cells and nodes are also Lagrangian because they are sums of subcell masses. As in the subcell density definition, one gets

$$\rho_c(t) = \frac{m_c}{V_c(t)} = \frac{\sum_{p \in \mathscr{P}(c)} m_{pc}}{\sum_{p \in \mathscr{P}(c)} V_{pc}(t)}. \tag{4}$$

The total mass $\mathscr{M}$, which is conserved during the Lagrangian phase, is $\mathscr{M} = \sum_{pc} m_{pc} = \sum_c m_c = \sum_p m_p$. In this part we show how momentum and specific internal energy can be discretized in such a way that mass, momentum and total energy are conserved.

Assume a general force $\mathbf{F}_{pc}$ modeling the action of subcell pc on point p is given, then a general force for point p can be assembled as

$$\mathbf{F}_p = \sum_{c \in \mathscr{C}(p)} \mathbf{F}_{pc}. \tag{5}$$

Then spatial discretization of the momentum equation is defined by

$$m_p \frac{d\mathbf{U}_p}{dt} = \mathbf{F}_p. \tag{6}$$

The discrete total energy over the whole domain is given by the sum of internal energy and kinetic energy

$$\mathscr{E} = \sum_c m_c \varepsilon_c + \sum_p \frac{1}{2} m_p \|\mathbf{U}_p\|^2, \tag{7}$$

and conservation implies $d\mathscr{E}/dt) = 0$. However the differentiation of (7) with respect to time formally gives

$$\frac{d\mathscr{E}}{dt} = \sum_c m_c \frac{d\varepsilon_c}{dt} + \sum_p \underbrace{m_p \frac{dU_p}{dt}}_{=F_p} \cdot U_p,$$

$$= \sum_c \left(m_c \frac{d\varepsilon_c}{dt} + \sum_{p \in \mathscr{P}(c)} \mathbf{F}_{pc} \cdot \mathbf{U}_p \right) = 0. \tag{8}$$

If the sum over cells is set to zero, for each cell this gives an expression for the change in internal energy as

$$m_c \frac{d\varepsilon_c}{dt} = -\sum_{p \in \mathscr{P}(c)} \mathbf{F}_{pc} \cdot \mathbf{U}_p, \tag{9}$$

such that (8) is true and total energy conservation is preserved. This is the semi-discrete form of internal energy equation which was derived from the total energy conservation.

Provided that the subcell force $\mathbf{F}_{pc}$ is known, the numerical scheme is defined by equations for the velocity (6), specific internal energy (9) and density (4) defined from the mesh motion. The mesh motion is modeled by the set of ordinary differential equations $d\mathbf{X}_p/dt = \mathbf{U}_p$ being solved at each mesh point p for its position $\mathbf{X}_p(t)$. Remark that whatever subcell force $\mathbf{F}_{pc}$ one wishes to consider (pressure force, viscosity, elastic-plastic contribution, etc.), the conservation of discrete momentum and total energy as defined by (7) is fulfilled. In other words, the mechanism responsible for the conservativeness is independent of the way the forces are constructed.

The subcell force is an combination of three forces: a pressure force $\mathbf{F}_{pc}^p$ that approximates grad p in the momentum equation (2), a subzonal pressure force $\mathbf{F}_{pc}^{\delta p}$ designed to prevent the Hourglass mesh motion and an artificial viscosity force $\mathbf{F}_{pc}^v$ designed to treat shock waves

$$\mathbf{F}_{pc} = \mathbf{F}_{pc}^p + \mathbf{F}_{pc}^{\delta p} + \mathbf{F}_{pc}^v. \tag{10}$$

The pressure force in subcell Ω_{pc} with boundary $\partial\Omega_{pc}$ is given by

$$\mathbf{F}_{pc}^p = -\int_{\Omega_{pc}} \text{grad } p \, dV = -\int_{\partial\Omega_{pc}} p \mathbf{N} \, dl. \tag{11}$$

The subzonal pressure force $\mathbf{F}_{pc}^{\delta p}$ [6, 20], given by the difference between the subcell pressure and the cell pressure $p_{pc} - p_c$ and the geometry of the cell, acts against the Hourglass mode motion, which might invert the cell (moving cell is being inverted when its node crosses another edge of the cell).

The last part of the subcell force is the artificial viscosity devoted to deal with shock waves. The simplest viscosity in cell c, across which the velocity has a difference $\Delta \mathbf{U}$, is in the compression regime $\Delta \mathbf{U} < 0$ [3]

$$Q_c = c_1 \rho_c a_c |\Delta \mathbf{U}| + c_2 \rho_c (\Delta \mathbf{U})^2. \qquad (12)$$

Constants c_1, c_2 are of the order of unity and a_c is the sound speed. In the expansion regime $\Delta \mathbf{U} \geq 0$ viscosity is set to zero. The artificial viscosity has the dimension of pressure and is generally added to the classical cell pressure producing the viscosity force $\mathbf{F}^v_{pc}$ in the same way as pressure force (11), usually preventing spurious numerical oscillations on shock waves. Many formulations of artificial viscosity in more than one dimension use the form given above with multidimensional modifications as edge artificial viscosity [3, 5] or tensor artificial viscosity [2].

2.2 Rezoning Phase

The rezoning phase of the ALE method consists in moving the nodes of the Lagrangian mesh to improve the geometric quality of the grid while keeping the rezoned grid as close as possible to the Lagrangian grid. This constraint must be taken into account to maintain the accuracy of the computation gained by the Lagrangian phase and to minimize the error of the remap phase. If the Lagrangian phase produces non-valid (inverted) cells then we have to use an untangling procedure [27]. The rezoning phase of the ALE method covers mesh smoothing and untangling.

The mesh resulting from the Lagrangian step can be of low quality and smoothing process changes the mesh in a way to improve it. One of the simplest smoothing methods is Winslow smoothing method [28]. The new positions of the mesh nodes are computed (with possible iteration over l starting at the old Lagrangian mesh) in case of logically rectangular mesh as

$$\begin{aligned}\mathbf{X}^{l+1}_{i,j} = \frac{1}{2(\alpha^l + \gamma^l)} &\left(\alpha^l \left(\mathbf{X}^l_{i,j+1} + \mathbf{X}^l_{i,j-1} \right) + \gamma^l \left(\mathbf{X}^l_{i+1,j} + \mathbf{X}^l_{i-1,j} \right) \right. \\ &\left. - \frac{1}{2} \beta^l \left(\mathbf{X}^l_{i+1,j+1} - \mathbf{X}^l_{i-1,j+1} + \mathbf{X}^l_{i-1,j-1} - \mathbf{X}^l_{i+1,j-1} \right) \right),\end{aligned}$$

where the coefficients $\alpha^l = x_\xi^2 + y_\xi^2$, $\beta^l = x_\xi x_\eta + y_\xi y_\eta$, $\gamma^l = x_\eta^2 + y_\eta^2$, and (ξ, η) are logical coordinates $\xi_i = i/n_x$, $\eta_j = j/n_y$ for $i = 0, \ldots, n_x$ and $j = 0, \ldots, n_y$. The derivatives x_ξ, x_η are approximated by the central differences $(x_\xi)_{i,j} \approx (x_{i+1,j} - x_{i-1,j})/(2\Delta\xi)$, $(x_\eta)_{i,j} \approx (x_{i,j+1} - x_{i,j-1})/(2\Delta\eta)$ and similarly for y.

Further rezoning methods include the condition number smoothing [10] and the Reference Jacobian Method [11].

2.3 Remapping Phase

The remapping stage of the ALE method is in fact a conservative interpolation of the discrete conserved quantities from the old Lagrangian mesh to the new smoother one. The remapping stage consists of three steps: reconstruction, integration and repair. First, the remapped conservative function g (e.g. density ρ) is reconstructed from the discrete values by a piecewise linear function on each old cell, e.g. with the Barth-Jespersen limiter [1]. Then, the reconstructed piecewise linear function is integrated over each new cell $\tilde{c}$ (objects related to the new mesh are accented by a tilde here) to get the total value $G_{\tilde{c}} = \int_{\tilde{c}} g \, dx \, dy$ of the conserved quantity (e.g. mass of the cell) inside the new cell, which defines the remapped density of conserved quantity $g_{\tilde{c}} = G_{\tilde{c}}/V_{\tilde{c}}$, where $V_{\tilde{c}}$ is the volume of the cell $\tilde{c}$.

The natural exact integration of the piecewise linear function over the new cell requires computing intersections of the new cell with all neighboring old cells. For example on logically rectangular mesh see Fig. 1 (a) where the new cell $\tilde{c}_{i,j} = [\tilde{P}_{i,j}, \tilde{P}_{i+1,j}, \tilde{P}_{i+1,j+1}, \tilde{P}_{i,j+1}]$ intersects with nine (3 × 3 patch) old cells $c_{k,l}$, $k = i-1, i, i+1, l = j-1, j, j+1$. The linear reconstruction at the old cell c'

$$g(x, y) = g_{c'} + \left(\frac{\partial g}{\partial x}\right)_{c'} (x - x_{c'}) + \left(\frac{\partial g}{\partial y}\right)_{c'} (y - y_{c'}),$$

(where $(x_{c'}, y_{c'})$ is the centroid of the cell c') inside each such intersection $I_{c'}^{\tilde{c}} = \tilde{c} \cap c'$ results in the contribution

$$G_{I_{c'}^{\tilde{c}}} = g_{c'} \int_{I_{c'}^{\tilde{c}}} dx \, dy + \left(\frac{\partial g}{\partial x}\right)_{c'} \left(\int_{I_{c'}^{\tilde{c}}} x \, dx \, dy - x_{c'} \int_{I_{c'}^{\tilde{c}}} dx \, dy\right) \quad (13)$$

$$+ \left(\frac{\partial g}{\partial y}\right)_{c'} \left(\int_{I_{c'}^{\tilde{c}}} y \, dx \, dy - y_{c'} \int_{I_{c'}^{\tilde{c}}} dx \, dy\right)$$

to the whole integral $G_{\tilde{c}}$. The integrals in this contribution over the polygonal intersection are transformed using Green's theorem into integrals over the edges of the polygonal intersection and computed analytically. The exact integration is computationally rather expensive because it requires finding all cell intersections.

The approximate integration over swept regions [14], which are the regions swept by the cell edges moving from the old mesh to the new position in the new mesh (see Fig. 1 (b)), is much faster. The contribution from each of the four swept regions has similar form as (13) with the intersection $I_{c'}^{\tilde{c}}$ replaced by the swept region. Green's theorem again transforms integrals over polygons into integrals over the edges of the polygon, which can be exactly evaluated. In the swept region method the integrals of the reconstructed function over the swept regions can be interpreted as remap fluxes through the mesh edges and the remapping formula can be written in a conservative flux form.

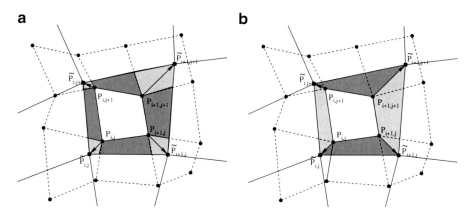

Fig. 1 Old (dashed) and new (solid segments) mesh with intersection regions for the exact integration (a) and swept regions for the approximate integration (b)

The last step of the remapping phase is repair [21] which conservatively redistributes conserved quantities in such a way that the remapping does not introduce any new local extrema. The repair is a post-processing *adhoc* correction. Better treatment, based on flux corrected transport (FCT) and called flux corrected remap [16, 17], guarantees that new local extrema are not introduced.

For the staggered scheme the remapping is however more complicated than what is outlined above as thermodynamical quantities are cell or subcell centered and velocity (thus also momentum) is centered at mesh nodes. This means that the internal energy is defined at cells and the kinetic energy at nodes and one has to be careful to conserve the total energy during the remapping step. The basic idea to treat this issue is to transform all quantities to subcells, remap on the subcell mesh and transform the quantities back to staggered form (density in subcells, internal energy in cells and velocity in nodes). The most accurate method employs the rigorously derived, matrix based, invertible transformation between the nodal and subcell velocities [20].

3 Heat Conductivity

The parabolic part of energy equation (3) is treated separately by splitting from the hyperbolic part of the whole system (1)-(3). It is transformed to the heat equation for temperature $a\partial T/\partial t = \text{div}(\kappa \text{ grad } T)$ where $a = \rho \partial \epsilon / \partial T$. We write the heat equation as the first order system, in so-called flux form

$$a\frac{\partial T}{\partial t} - \text{div } \mathbf{w} = 0, \quad \mathbf{w} = -\kappa \text{ grad } T, \tag{14}$$

introducing the heat flux **w**. The heat equation is solved on domain V with boundary ∂V with Neumann boundary conditions.

We treat the space discretization of the heat equation by the mimetic method [25], which has been generalized to unstructured triangular meshes in [7]. The mimetic method introduces operators of generalized gradient **G** and extended divergence **D**

$$\mathbf{G}T = -\kappa \operatorname{grad} T, \quad \mathbf{D}\mathbf{w} = \begin{cases} \operatorname{div} \mathbf{w} & \text{on} \quad V \\ -(\mathbf{w}, \mathbf{n}) & \text{on} \quad \partial V \end{cases}.$$

The integral properties of these operators are given by divergence Green formula and Gauss theorem. The divergence Green formula

$$\int_V \operatorname{div} \mathbf{w} \, dV - \oint_{\partial V} (\mathbf{w}, \mathbf{n}) \, dS = 0 \tag{15}$$

can be restated as $(\mathbf{D}\mathbf{w}, 1)_H = 0$ where we use the inner product on space H of scalar functions

$$(u, v)_H = \int_V u \, v \, dV + \oint_{\partial V} u \, v \, dS. \tag{16}$$

Gauss theorem

$$\int_V T \operatorname{div} \mathbf{w} \, dV - \oint T(\mathbf{w}, \mathbf{n}) \, dS + \int_V (\mathbf{w}, \kappa^{-1}\kappa \operatorname{grad} T) dV = 0$$

can be restated as $(\mathbf{D}\mathbf{w}, T)_H = (\mathbf{w}, \mathbf{G}T)_{\mathbf{H}}$ where we use also the inner product on space **H** of vector functions

$$(\mathbf{A}, \mathbf{B})_{\mathbf{H}} = \int_V (\kappa^{-1}\mathbf{A}, \mathbf{B}) dV. \tag{17}$$

Gauss theorem states that the generalized gradient is the adjoint operator of the extended divergence $\mathbf{G} = \mathbf{D}^*$. The basic idea of the support operator mimetic method [25] is to mimic these two integral properties also in the discrete case on spaces of discrete functions. We discretize the temperature T inside each computational cell and the vector heat flux **w** at the center of each edge by its projections on the normal of the edge. This discretization of vector heat flux guarantees the continuity of normal flux through each edge. On the spaces of discrete scalar and vector functions the discrete analogs of the inner products (16) and (17) are defined. The discrete divergence is derived in a standard way from the discrete analog of (15) on a computational cell. This gives the discrete operator of extended divergence D inside computational domain, while on the boundary it is given by the discrete heat flux (up to sign). Now the discrete extended gradient G is constructed as the adjoint (in the discrete inner products on the whole domain) of the discrete extended divergence $G = D^*$. The discrete gradient constructed this way has a global stencil.

Now in the heat equation (14) we use this mimetic spatial discretization, i.e. the discrete extended divergence D and the discrete generalized gradient operators G. We employ the implicit scheme written in flux form

$$a\frac{T^{n+1} - T^n}{\Delta t} + D\mathbf{W}^{n+1} = 0, \quad \mathbf{W}^{n+1} - GT^{n+1} = 0. \tag{18}$$

We express $T^{n+1} = T^n - D\mathbf{W}^{n+1}$ and eliminate it from the second equation which gives us

$$(I + \Delta t/a\, GD)\mathbf{W}^{n+1} = GT^n.$$

This system with a global stencil can be transformed into a system with local stencil [25] having symmetric, positive definite matrix. The conjugate gradient method, preconditioned by the altered direction implicit method, is applied as effective iterative solver for this system resulting in the numerical heat fluxes.

For laser plasma these heat fluxes often produce physically unrealistic (too big) heat fluxes which cannot be carried by electrons carrying most of heat energy. Direct decrease of the heat flux magnitude (where needed) leads to temperature oscillations and checker board patterns, thus the heat flux limiting has to be performed differently. In the regions where unlimited heat flux violates physical limits the heat conductivity is decreased by the ratio of the unlimited flux magnitude and the heat flux limit. The heat equation it then solved again with the updated heat conductivity giving the final limited heat fluxes. Finally, having fluxes $\mathbf{W}^{n+1}$ the temperature T^{n+1} is computed from the first equation of (18). The presented numerical method for heat equation works well on bad quality meshes appearing often in Lagrangian simulations and it allows discontinuous diffusion coefficient.

4 Laser Absorption and Cylindrical Geometry

Laser absorption in plasma is modeled by the term div $\mathbf{I}$ in the internal energy equation (3). Important notion for laser absorption is a critical density, which for laser with wavelength λ is proportional to $1/\lambda^2$. The critical density defines the critical surface which is the surface of electron density being equal to critical density. The laser can propagate only in the sub-critical regions of plasma with electron density less than critical density. Typically most of the laser energy is absorbed into plasma around the critical surface. Laser absorption on critical surface assumes that laser propagates without damping and refraction till the critical surface where it is absorbed. The absorption term div $\mathbf{I}$ with absorption coefficient is evaluated in cells at the critical surface and is zero everywhere else.

Ray tracing is a more complicated method for laser absorption modeling. The laser beam is split into many laser rays carrying initially appropriate energy depending on radius and radial laser profile. Propagating of each ray is computed (traced) independently. Inside a cell through which the ray propagates it does not change direction and deposits a part of its energy into plasma internal energy by inverse bremsstrahlung. On the edge the ray refracts according to Snell law with refraction plane being orthogonal to the electron density gradient. A special case is full reflection near the critical surface when the ray on the edge reflects back.

Laser beam has cylindrical symmetry and most simulated problems are cylindrically symmetric (here all problems except oblique incidence on thin foil studied in Sect. 5.1), so one has to include cylindrical $r - z$ geometry. All numerical methods, initially designed in Cartesian geometry, have been generalized into cylindrical geometry with a special boundary condition on the symmetry axis z. In cylindrical geometry for Lagrangian method we employ control volume method [4], rezoning methods need only to change boundary treatment on the symmetry axis z, cylindrical remapping [13] requires additional factor r in the integrals (13, the mimetic method for heat conductivity has been generalized to cylindrical geometry and also laser absorption module supports both cylindrical and Cartesian geometries.

5 Interaction of Laser with Targets

In this section we present selected simulations of laser beam interaction with targets modeled by our ALE code PALE. All simulations correspond to particular experiments performed at PALS laser facility in Prague, which provides a laser beam on the first harmonics with wavelength $\lambda = 1.315$ nm or on the third harmonics with wavelength $\lambda = 438$ nm.

5.1 Oblique Incidence on Thin Foil

We start with oblique incidence of laser beam on a 0.8 μm thin Aluminum foil which is reasonably simple and provides initial insight into laser interactions with matter. This simulation is an initial study to double foil targets which are used for investigation of plasma-wall interactions [24] and which are subject of the next Sect. 5.2. The third harmonics Gaussian laser pulse with energy 36 J, full width half maximum (FWHM) length 250 ps and focal spot radius 40 μm interacts with 30° oblique thin foil (the angle between laser beam axis and normal to the foil is 30°). The simulation starts at time $t = 0$, which is 250 ps before the laser maximum at $t = 250$ ps, and uses Cartesian geometry as the setup is not cylindrically symmetric. Density of the developing laser plasma at three times 150, 200 and 250 ps is presented in Fig. 2 in a logarithmic scale with computational mesh and a magenta curves of the critical surface. Laser is coming from above with the beam axis on the z axis $r = 0$. It propagates through the sub-critical plasma until the critical surface and is absorbed on the critical surface. At time 150 ps in Fig. 2 (a) the laser does not penetrate the foil, while at time 200 ps in Fig. 2 (b) the laser has already burned through the foil and only a small part far from the z axis is still being

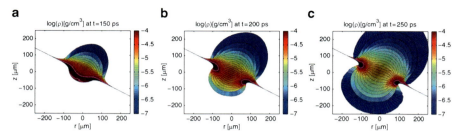

Fig. 2 Density for interaction of oblique laser beam with a thin Aluminum foil at time: (a) 150 ps, (b) 200 ps and (c) 250 ps. Magenta curves denote the position of the critical surface

absorbed at the critical surface. In the beginning of the interaction laser energy is being deposited close to the upper boundary of the foil which starts to expand in the upper right direction creating plasma plume (corona). Before time 150 ps the whole foil in an area around the z axis is heated and secondary plume starts to expand in the lower left direction. This simulation provides an example of a large scale change of computational domain (initial 0.8 μm thin foil expands to plumes of the size around 500 μm at Fig. 2 (c) at time of laser maximum, which is still not the end of simulation), which dictates the use of Lagrangian coordinates moving with the moving plasma. The simulation justifies that even with oblique laser incidence the plasma plumes propagation is orthogonal to the foil. This is going to be used for oblique incidence double foil targets where the laser going through the upper foil does not hit the shorter lower foil, which interacts directly with the plasma plume from the upper foil.

5.2 Double Foil Target

The double foil target shown in Fig. 3 (a) is composed from two parallel foils located at distance $L = 600$ μm. The thickness of the upper Aluminum foil is $d_u = 0.8$ μm and the thickness of the lower Magnesium foil is $d_l = 2$ μm. The double foil target is irradiated by the third harmonics Gaussian laser beam (orthogonal to the foils) with energy 115 J, FWHM length 300 ps, focal spot radius 40 μm and angular beam divergence 15°. The beam is focused on the lower foil. Laser absorption has been modeled in this simulation by ray tracing. Results are presented in Fig. 3(b) and (c) by density and pressure color-maps at time 600 ps. Fig. 3 (b) shows density color-map with selected laser rays in the left part. The thickness of the rays is proportional to the energy they carry, so when the energy goes below a threshold the thickness goes to zero and the ray curve ends. Rays are refracted around the critical surface, some rays are reflected from the z axis. In the beginning the upper foil expands in two plumes similarly as the oblique foil in Fig. 2. The lower foil remains static until the laser burns through the upper foil. After burning through the upper foil the laser

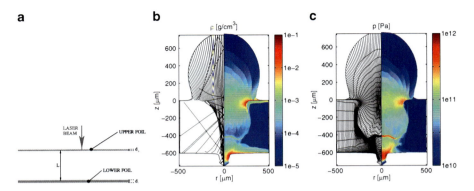

Fig. 3 Experimental setup for double foil target (a), density ρ with laser rays (b) and pressure p with computational mesh (c) for double foil target at $t = 500$ ps

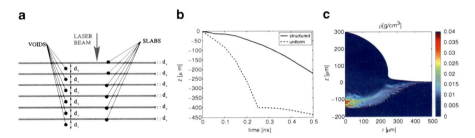

Fig. 4 Structured model of the foam (a), burning of laser through the foam target (b) and density from simulation with the structured target at $t = 400$ ps (c)

reaches also the lower foil from which at first the upper Magnesium plume starts to develop. The lower Aluminum plume collides with the upper Magnesium plume around time 500 ps producing high density and pressure at the colliding area.

5.3 Foam Target

Foam layers are used in the ICF targets for smoothing inhomogeneities in the laser beam by its propagation through the low density foam. Interaction of laser with low density foam is difficult to model because for low density homogeneous material one gets too high speed of laser burning through the foam. This problem can be avoided by introducing structured model of the foam [9] shown in Fig. 4 (a) consisting from a series of parallel high-density slabs separated by low-density voids. When the laser burns through this structured model of foam it is delayed on each slab as it needs some time to burn through the slab. Here, we simulate the interaction of the third harmonic Gaussian laser pulse of 320 ps FWHM duration, energy of 170 J and focal spot radius of 300 μm with 400 μm-thick layer of TAC foam of density 9.1 mg/cm^3

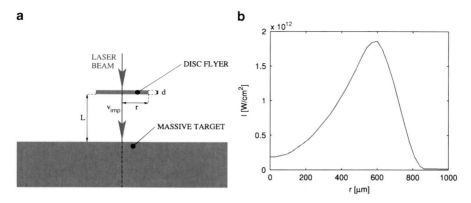

Fig. 5 Experimental setup for high velocity impact (a) and annular radial laser beam intensity profile at time of laser maximum for jet formation (b)

with 2 μm pores. The foam is modeled by uniform density 9.1 mg/cm^3 material and by structured model consisting from a sequence of $d_s = 0.018$ μm thick dense slabs of density $\rho_s = 1$ g/cm^3 separated by $d_v = 1.982$ μm thick voids with density $\rho_v = 1$ mg/cm^3. The time evolution of the depth of the burned region of foam on the z axis is plotted in Fig. 4 (b) for uniform and structured foam model (with density at $t = 400$ ps in Fig. 4 (c)). The speed of burning through for the structured model is reasonably close to the experimental measurement, while this speed for uniform model is more than twice higher.

5.4 High Velocity Impact

The target setup for the high velocity impact problem is shown in Fig. 5(a). A cylindrical Aluminum disc flyer with radius $r = 150$ μm and thickness $d = 11$ μm is placed at distance $L = 200$ μm above an Aluminum massive target, parallel to it. The laser irradiated disc flyer is ablatively accelerated up to very high velocity (40-190 km/s) and impacts the massive target creating a crater in it. Here the third harmonics laser pulse of energy 390 J, FWHM length 400 ps and focal spot radius 125 μm interacts with the target. The simulation is split into two parts: ablative disc flyer acceleration by laser beam and the impact of disc flyer into the massive target. The density for the final time of disc acceleration is presented in Fig. 6 (a) and (b). In the presented case the final time of acceleration is about 1.1 ns, when the high density region, which contains most of the disc mass, momentum and energy and is seen in Fig. 6 (b), approaches the upper boundary of the massive target located at $z = -200$ μm. The average vertical velocity of the impacting disc is 187 km/s. A new mesh is constructed containing this region and the whole massive target. Conservative quantities at the final time of acceleration are remapped to this new

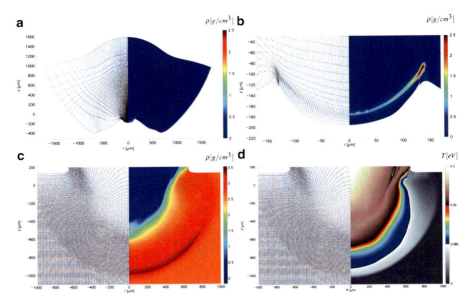

Fig. 6 High velocity impact problem: density (a) and zoomed density (b) of accelerated disc before the impact, density (c) and temperature (d) at the final time 80 ns after the impact (zoomed to crater) – three different color scales in temperature color-map distinguish solid (gray), liquid (blue-red) and gaseous (brown-pink) phase of Aluminum

mesh and serve as initial conditions for the second part of the simulation, disc flyer impact. The impact creates circular shock wave propagating into the massive target and visible in density and temperature color-maps in Figs. 6 (c),(d). A large hot low-density plasma plume is being reflected from the massive target. The impact melts and evaporates part of massive target creating a crater. Solid, liquid and gas phase of Aluminum are distinguished by three color-maps in Fig. 6 (d). Solid-liquid and liquid-gas phase interfaces, given by temperature isolines (corresponding to Aluminum melting and boiling temperature), are visible in Fig. 6 (d) as two white-black (in bottom-up direction on the z axis) interfaces. The crater is defined by gas - liquid phase interface. Simulated craters size and shape correspond reasonably well to experimental data also for other laser energies and other disc flyers [12].

5.5 Jet Formation

In this section we investigate formation of plasma jets by interaction of annular laser beam with a massive Aluminum target. We use Gaussian in time laser pulse on 3-rd harmonics with FWHM length 400 ps and energy 10 J. The radial intensity profile of the annular beam is presented in Fig. 5(b). It has 10% minimum on the z axis at $r = 0$, it is proportional to r^2 for small r and has a smooth maximum

ALE Method for Simulations of Laser-Produced Plasmas 871

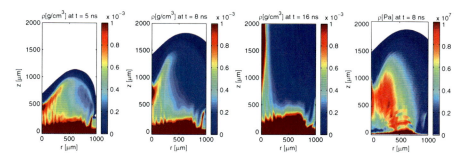

Fig. 7 Plasma jet formation by annular laser beam: (a), (b), (c) density evolution at times 5, 8 and 16 ns; (d) pressure at 8 ns

around $r = 600$ μm. The density evolution at times 5, 8 and 16 ns is presented in Fig. 7 (a),(b),(c). Thanks to the annular radial laser profile the plasma plume develops and expands faster around the radial maximum of intensity at $r = 600\mu$m, than around the z axis at $r = 0$. Such plume development leads to cone profile of higher density region visible in Fig. 7 (a) and (b) at 5 and 8 ns. The cone moves up in z direction and left in r direction towards the z axis and collides on the symmetry axis creating a plasma jet which can be seen in Fig. 7 (c) at 16 ns as high density, high pressure region along the z axis propagating up. Important is the radial pressure gradient on the cone directed inwards towards the z axis which drives the negative radial velocity towards the z axis, which can be seen in Fig. 7 (d). The outlined dynamics of the plasma plume created by annular laser provides pure hydrodynamical mechanism for plasma jets generation [15]. Plasma jets appear not only on the laser plasma micro-scale presented here, but also astrophysics deals with giant jets on macro-scale.

6 Conclusion

Numerical methods for Lagrangian and ALE hydrodynamics, heat conductivity and laser absorption used in our code PALE have been shortly presented. PALE code have been applied to simulate selected problems of laser interaction with targets. For most simulations we have to use the ALE method as pure Lagrangian computation (without any rezoning and remapping) fails due to severe distortion of moving computational mesh. Only the first simulation of oblique incidence on a thin foil presented in section 5.1 has been computed by pure Lagrangian method. PALE code is regularly used for simulations of experiments at PALS laser facility. The simulations provide theoretical backgound for interpretation of experimetal results.

Acknowledgements This research has been supported in part by the Czech Ministry of Education grants MSM6840770022 and LC528, Czech Science Foundation grant GAP205/10/0814 and Czech Technical University grant SGS10/299/OHK4/3T/14.

The authors thank M. Shashkov, B. Wendroff, R. Loubere, P.-H. Maire, V. Kmetik, R. Garimella, M. Berndt for fruitful discussions and constructive comments.

References

1. Barth, T., Jespersen, D.: The design and application of upwind schemes on unstructured meshes. Tech. Rep. AIAA-89-0366, AIAA, NASA Ames Research Center (1989)
2. Campbell, J., Shashkov, M.: A tensor artificial viscosity using a mimetic finite difference algorithm. J. Comput. Phys. **172**(2), 739–765 (2001)
3. Caramana, E., Shashkov, M.J., Whalen, P.: Formulations of artificial viscosity for multi-dimensional shock wave computations. J. Comput. Phys. **144**, 70–97 (1998)
4. Caramana, E.J., Burton, D.E., Shashkov, M.J., Whalen, P.P.: The construction of compatible hydrodynamics algorithms utilizing conservation of total energy. J. Comput. Phys. **146**(1), 227–262 (1998)
5. Caramana, E.J., Loubère, R.: "curl-q": A vorticity damping artificial viscosity for Lagrangian hydrodynamics calculations. J. Comput. Phys. **215**(2), 385–391 (2006)
6. Caramana, E.J., Shashkov, M.J.: Elimination of artificial grid distortion and hourglass-type motions by means of Lagrangian subzonal masses and pressures. J. Comput. Phys. **142**, 521–561 (1998)
7. Ganzha, V., Liska, R., Shashkov, M., Zenger, C.: Mimetic finite difference methods for diffusion equations on unstructured triangular grid. In: M. Feistauer, V. Dolejší, P. Knobloch, K. Najzar (eds.) Numerical Mathematics and Advanced Applications ENUMATH 2003, pp. 368–377. Springer-Verlag, Berlin (2004)
8. Hirt, C., Amsden, A., Cook, J.: An arbitrary Lagrangian-Eulerian computing method for all flow speeds. J. Comput. Phys. **14**, 227–253 (1974). Reprinted in vol. 135(2), 203–216, 1997.
9. Kapin, T., Kuchařík, M., Limpouch, J., Liska, R.: Hydrodynamic simulations of laser interactions with low-density foams. Czechoslovak Journal of Physics **56**, B493–B499 (2006)
10. Knupp, P.: Achieving finite element mesh quality via optimization of the Jacobian matrix norm and associated quantities. Part I – a framework for surface mesh optimization. Int. J. Numer. Meth. Eng. **48**, 401–420 (2000)
11. Knupp, P., Margolin, L., Shashkov, M.: Reference Jacobian optimization-based rezone strategies for arbitrary Lagrangian Eulerian methods. J. Comput. Phys. **176**, 93–128 (2002)
12. Kuchařík, M., Limpouch, J., Liska, R.: Laser plasma simulations by arbitrary Lagrangian Eulerian method. J. de Physique IV **133**, 167–169 (2006)
13. Kuchařík, M., Liska, R., Loubere, R., Shashkov, M.: Arbitrary Lagrangian-Eulerian (ALE) method in cylindrical coordinates for laser plasma simulations. In: S. Benzoni-Gavage, D. Serre (eds.) Hyperbolic Problems: Theory, Numerics, Applications, pp. 687–694. Springer (2008)
14. Kuchařík, M., Shashkov, M., Wendroff, B.: An efficient linearity-and-bound-preserving remapping method. J. Comput. Phys. **188**(2), 462–471 (2003)
15. Limpouch, J., Liska, R., Kuchařík, M., Váchal, P., Kmetík, V.: Laser-driven collimated plasma flows studied via ALE code. In: 37th EPS Conference on Plasma Physics, pp. P4.222, 1–4. European Physical Society, Mulhouse (2010)
16. Liska, R., Shashkov, M., Váchal, P., Wendroff, B.: Optimization-based synchronized flux-corrected conservative interpolation (remapping) of mass and momentum for arbitrary Lagrangian-Eulerian methods. J. Comput. Phys. **229**(5), 1467–1497 (2010)
17. Liska, R., Shashkov, M., Váchal, P., Wendroff, B.: Synchronized flux corrected remapping for ale methods. Computers and Fluids (2011). DOI: 10.1016/j.compfluid.2010.11.013
18. Liska, R., Shashkov, M., Wendroff, B.: Lagrangian composite schemes on triangular unstructured grids. In: M. Kočandrlová, V. Kelar (eds.) Mathematical and Computer Modelling in Science and Engineering, pp. 216–220. Prague (2003)

19. Loubere, R., Ovadia, J., Abgrall, R.: A Lagrangian discontinuous Galerkin type method on unstructured meshes to solve hydrodynamics problems. Int. J. Numer. Meth. Fluids **44**(6), 645–663 (2004)
20. Loubère, R., Shashkov, M.: A subcell remapping method on staggered polygonal grids for arbitrary-Lagrangian-Eulerian methods. J. Comput. Phys. **209**(1), 105–138 (2005)
21. Loubère, R., Staley, M., Wendroff, B.: The repair paradigm: New algorithms and applications to compressible flow. J. Comput. Phys. **211**(2), 385–404 (2006)
22. Maire, P.H.: A high-order cell-centered Lagrangian scheme for two-dimensional compressible fluid flows on unstructured meshes. J. Comput. Phys. **228**(7), 2391–2425 (2009)
23. Maire, P.H., Abgrall, R., Breil, J., Ovadia, J.: A cell-centered Lagrangian scheme for two-dimensional compressible flow problems. SIAM Journal on Scientific Computing **29**(4), 1781–1824 (2007)
24. Renner, O., Liska, R., Rosmej, F.: Laser-produced plasma-wall interaction. Laser and Particle Beams **27**(4), 725–731 (2009)
25. Shashkov, M., Steinberg, S.: Solving diffusion equation with rough coefficients in rough grids. J. Comput. Phys. **129**, 383–405 (1996)
26. Shashkov, M., Wendroff, B.: A composite scheme for gas dynamics in Lagrangian coordinates. J. Comput. Phys. **150**, 502–517 (1999)
27. Váchal, P., Garimella, R., Shashkov, M.: Untangling of 2D meshes in ALE simulations. J. Comput. Phys. **196**(2), 627–644 (2004)
28. Winslow, A.: Equipotential zoning of two-dimensional meshes. Tech. Rep. UCRL-7312, Lawrence Livermore National Laboratory (1963)

The paper is in final form and no similar paper has been or is being submitted elsewhere.

A two-dimensional finite volume solution of dam-break hydraulics over erodible sediment beds

Fayssal Benkhaldoun, Imad Elmahi, Saïda Sari, and Mohammed Seaïd

Abstract Two-dimensional dam-break hydraulics over erodible sediment beds are solved using a well-balanced finite volume method. The governing equations consist of three coupled model components: (i) the shallow water equations for the hydrodynamical model, (ii) a transport equation for the dispersion of suspended sediments, and (iii) an Exner equation for the morphological model. These coupled models form a hyperbolic system of conservation laws with source terms. The proposed finite volume method consists of a predictor stage for the discretization of gradient terms and a corrector stage for the treatment of source terms. The gradient fluxes are discretized using a modified Roe's scheme using the sign of the Jacobian matrix in the coupled system. A well-balanced discretization is used for the treatment of source terms. In this paper, we also describe an adaptive procedure in the finite volume method by monitoring the concentration of suspended sediments in the computational domain during its transport process. The method uses unstructured meshes, incorporates upwinded numerical fluxes and slope limiters to provide sharp resolution of steep sediment concentration and bed-load gradients that may form in the approximate solution.

Keywords Sediment transport, shallow water equations, finite volume method
MSC2010: 35L65, 65L08, 65C20

Fayssal Benkhaldoun and Saïda Sari
LAGA, Université Paris 13, 99 Av J.B. Clement, 93430 Villetaneuse, France,
e-mail: fayssal@math.univ-paris13.fr, sari@math.univ-paris13.fr

Imad Elmahi
ENSAO Complex Universitaire, B.P. 669, 60000 Oujda, Morocco, e-mail: ielmahi@ensa.univ-oujda.ac.ma

Mohammed Seaïd
School of Engineering and Computing Sciences, University of Durham, South Road, Durham DH1 3LE, UK, e-mail: m.seaid@durham.ac.uk

1 Introduction

The main concern of the sediment transport (or morphodynamics) is to determine the evolution of bed levels for hydrodynamics systems such as rivers, estuaries, bays and other nearshore regions where water flows interact with the bed geometry. Example of applications include among others, beach profile changes due to severe wave climates, seabed response to dredging procedures or imposed structures, and harbour siltation. The ability to design numerical methods able to predict the morphodynamics evolution of the coastal seabed has a clear mathematical and engineering relevances. In practice, morphodynamics involve coupling between a hydrodynamics model, which provides a description of the flow field leading to a specification of local sediment transport rates, and an equation for bed level change which expresses the conservative balance of sediment volume and its continual redistribution with time. Here, the hydrodynamic model is described by the shallow water equations, the bed-load is modelled by the Exner equation, and the suspended sediment transport is modelled by an advection equation accounting for erosion and deposition effects. The coupled models form a hyperbolic system of conservation laws with a source term. Nowadays, much effort has been devoted to develop numerical schemes for morphodynamics models able to resolve all hydrodynamics and morphodynamics scales. In the current study, a class of finite volume methods is proposed for numerical simulation of transient flows involving erosion and deposition of sediments. The method consists of a predictor stage where the numerical fluxes are constructed and a corrector stage to recover the conservation equations. The sign matrix of the Jacobian matrix is used in the reconstruction of the numerical fluxes. Most of these techniques have been recently investigated in [1,2] for solving sediment transport models without accounting for erosion and deposition effects. The current study presents an extension of this method to transient flows involving erosion and deposition of sediments. A detailed formulation of the sign matrix and the numerical fluxes is presented. The proposed method also satisfies the property of well-balancing flux-gradient and source-term in the system. Numerical results and comparisons will be shown for several suspended sediment transport problems.

2 The governing equations

In the current study, the sediment transport model consists of three parts: A hydraulic variables describing the motion of water, a concentration variable describing the dispersion of suspended sediments, and a morphology variable which describes the deformation of the bed-load. In the present work we assume that the flow is almost horizontal, the vertical component of the acceleration is vanishingly small, the pressure is taken to be hydrostatic, the free-surface gravity waves are long with respect to the mean flow depth and wave amplitude, and the water-species mixture is vertically homogeneous and non-reactive. The governing equations are obtained by balancing the net inflow of mass, momentum and species through boundaries of a control volume during an infinitesimal time interval while accounting for the accumulation of mass, resultant forces and species within the control volume, compare for example [1, 17] among others. Thus, the equations for mass conservation and momentum flux balance are given by

$$\frac{\partial h}{\partial t} + \frac{\partial (hu)}{\partial x} + \frac{\partial (hv)}{\partial y} = \frac{E-D}{1-p},$$

$$\frac{\partial (hu)}{\partial t} + \frac{\partial}{\partial x}\left(hu^2 + \frac{1}{2}gh^2\right) + \frac{\partial}{\partial y}(huv) = gh\left(-\frac{\partial Z}{\partial x} - S_f^x\right) - \frac{(\rho_s - \rho_w)}{2\rho}gh^2\frac{\partial c}{\partial x}$$
$$- \frac{(\rho_0 - \rho)(E-D)}{\rho(1-p)}u, \quad (1)$$

$$\frac{\partial (hv)}{\partial t} + \frac{\partial}{\partial x}(huv) + \frac{\partial}{\partial y}\left(hv^2 + \frac{1}{2}gh^2\right) = gh\left(-\frac{\partial Z}{\partial y} - S_f^y\right) - \frac{(\rho_s - \rho_w)}{2\rho}gh^2\frac{\partial c}{\partial y}$$
$$- \frac{(\rho_0 - \rho)(E-D)}{\rho(1-p)}v,$$

where t is the time variable, $\mathbf{x} = (x, y)^T$ the space coordinates, $\mathbf{u} = (u, v)^T$ the depth-averaged water velocity, h the water depth, Z the bottom topography, g the gravitational acceleration, p the porosity, ρ_w the water density, ρ_s the sediment density, c is the depth-averaged concentration of the suspended sediment, E and D represent the entrainment and deposition terms in upward and downward directions, respectively. In (1), ρ and ρ_0 are respectively, the density of the water-sediment mixture and the density of the saturated bed defined by

$$\begin{aligned}\rho &= \rho_w(1-c) + \rho_s c, \\ \rho_0 &= \rho_w p + \rho_s(1-p).\end{aligned} \quad (2)$$

The friction slopes S_f^x and S_f^y are defined, using the Manning roughness coefficient n_b, as

$$S_f^x = \frac{n_b^2}{h^{4/3}}u\sqrt{u^2+v^2}, \qquad S_f^y = \frac{n_b^2}{h^{4/3}}v\sqrt{u^2+v^2}. \quad (3)$$

The equation for mass conservation of species is modeled by

$$\frac{\partial (hc)}{\partial t} + \frac{\partial}{\partial x}(huc) + \frac{\partial}{\partial y}(hvc) = E - D. \quad (4)$$

To determine the entrainment and deposition terms in the above equations we assume a non-cohesive sediment and we use empirical relations reported in [8]. Thus,

$$D = w(1 - C_a)^m C_a, \quad (5)$$

where w is the settling velocity of a single particle in tranquil water

$$\omega = \frac{\sqrt{(36v/d)^2 + 7.5\rho_s g d} - 36v/d}{2.8}, \quad (6)$$

with v is the kinematic viscosity of the water, d the averaged diameter of the sediment particle, m an exponent indicating the effects of hindered settling due to high sediment concentrations, C_a the near-bed volumetric sediment concentration, $C_a = \alpha_c c$, where α_c is a coefficient larger than unity. To ensure that the near-bed concentration does not exceed $(1 - p)$, the coefficient α_c is computed by [10]

$$\alpha_c = \min\left(2, \frac{1-p}{c}\right).$$

For the entrainment of a cohesive material the following relation is used

$$E = \begin{cases} \varphi \dfrac{\theta - \theta_c}{h} \bar{u} d^{-0.2}, & \text{if } \theta \geq \theta_c, \\ 0, & \text{otherwise}, \end{cases} \quad (7)$$

where

$$\bar{u} = \sqrt{u^2 + v^2},$$

and φ is a coefficient to control the erosion forces, θ_c is a critical value of Shields parameter for the initiation of sediment motion and θ is the Shields coefficient defined by

$$\theta = \frac{u_*^2}{sgd}, \quad (8)$$

with u_* is the friction velocity defined using the Darcy-Weisbach friction factor f as

$$u_*^2 = \sqrt{\frac{f}{8}} \bar{u}.$$

In (8), s is the submerged specific gravity of sediment given by

$$s = \frac{\rho_s}{\rho_w} - 1.$$

To update the bedload, we consider the Exner equation proposed in [14]

$$\frac{\partial Z}{\partial t} + \frac{A_s}{1-p} \frac{\partial}{\partial x}\left(u(u^2 + v^2)\right) + \frac{A_s}{1-p} \frac{\partial}{\partial y}\left(v(u^2 + v^2)\right) = -\frac{E - D}{1 - p}, \quad (9)$$

where A_s is a coefficient usually obtained from experiments taking into account the grain diameter and the kinematic viscosity of the sediments. For simplicity in the presentation, let us rewrite the equations (1), (4) and (9) in the following vector form

$$\frac{\partial \mathbf{W}}{\partial t} + \frac{\partial \mathbf{F(W)}}{\partial x} + \frac{\partial \mathbf{G(W)}}{\partial y} = \mathbf{S(W)} + \mathbf{Q(W)}, \qquad (10)$$

where $\mathbf{W}$ is the vector of conserved variables, $\mathbf{F}$ and $\mathbf{G}$ are the physical fluxes in x- and y-direction, $\mathbf{S}$ and $\mathbf{Q}$ are the source terms. These variables are defined as

$$\mathbf{W} = \begin{pmatrix} h \\ hu \\ hv \\ hc \\ Z \end{pmatrix}, \quad \mathbf{F(W)} = \begin{pmatrix} hu \\ hu^2 + \frac{1}{2}gh^2 \\ huv \\ huc \\ \frac{A_s}{1-p}u(u^2+v^2) \end{pmatrix}, \quad \mathbf{G(W)} = \begin{pmatrix} hv \\ huv \\ hv^2 + \frac{1}{2}gh^2 \\ hvc \\ \frac{A_s}{1-p}v(u^2+v^2) \end{pmatrix},$$

$$\mathbf{S} = \begin{pmatrix} 0 \\ -gh\frac{\partial Z}{\partial x} - \frac{(\rho_s-\rho_w)}{2\rho}gh^2\frac{\partial c}{\partial x} \\ -gh\frac{\partial Z}{\partial y} - \frac{(\rho_s-\rho_w)}{2\rho}gh^2\frac{\partial c}{\partial y} \\ 0 \\ 0 \end{pmatrix}, \quad \mathbf{Q} = \begin{pmatrix} \frac{E-D}{1-p} \\ -ghS_f^x - \frac{(\rho_0-\rho)(E-D)}{\rho(1-p)}u \\ -ghS_f^y - \frac{(\rho_0-\rho)(E-D)}{\rho(1-p)}v \\ E-D \\ -\frac{E-D}{1-p} \end{pmatrix}.$$

It is worth emphasizing that, using the Exner equation (9) to model the bedload transport, the nonhomegenuous terms in the right-hand side in (10) are not standard source terms but nonconservative products, since they include derivatives of two of the variables. The presence of these terms in sediment transport system can cause sever difficulties in their numerical approximations. In principle, the nonhomegenuous term in these equations can be viewed as a source term and/or a nonconservative term. In the approach presented in this study these terms are considered and discretized as source terms.

3 The finite volume method

The governing sediment transport equations (10) are formulated in Cartesian coordinates and will be discretized into the unstructured grids by the finite volume method. The unstructured grids are polygons and the number of edges of the grids is not limited in theory, but only triangular grids are considered in the current study. Hence, we divide the time interval into sub-intervals $[t_n, t_{n+1}]$ with stepsize Δt and discretize the spatial domain in conforming triangular elements $\mathcal{T}_i$. Each triangle represents a control volume and the variables are located at the geometric centres of the cells. Hence, using the control volume depicted in Fig. 1, a finite volume discretization of (10) yields

$$\mathbf{W}_i^{n+1} = \mathbf{W}_i^n - \frac{\Delta t}{|\mathcal{T}_i|} \sum_{j \in N(i)} \int_{\Gamma_{ij}} \mathcal{F}(\mathbf{W}^n; \mathbf{n}) \, d\sigma + \frac{\Delta t}{|\mathcal{T}_i|} \int_{\mathcal{T}_i} \mathbf{S}(\mathbf{W}^n) \, dV$$

$$+ \frac{\Delta t}{|\mathcal{T}_i|} \int_{\mathcal{T}_i} \mathbf{Q}(\mathbf{W}^n) \, dV, \tag{11}$$

where $N(i)$ is the set of neighboring triangles of the cell $\mathcal{T}_i$, $\mathbf{W}_i^n$ is an averaged value of the solution $\mathbf{W}$ in the cell $\mathcal{T}_i$ at time t_n,

$$\mathbf{W}_i = \frac{1}{|\mathcal{T}_i|} \int_{\mathcal{T}_i} \mathbf{W} \, dV,$$

where $|\mathcal{T}_i|$ denotes the area of $\mathcal{T}_i$ and $\mathcal{S}_i$ is the surface surrounding the control volume $\mathcal{T}_i$. Here, Γ_{ij} is the interface between the two control volumes $\mathcal{T}_i$ and $\mathcal{T}_j$, $\mathbf{n} = (n_x, n_y)^T$ denotes the unit outward normal to the surface $\mathcal{S}_i$, and

$$\mathcal{F}(\mathbf{W}; \mathbf{n}) = \mathbf{F}(\mathbf{W}) n_x + \mathbf{G}(\mathbf{W}) n_y.$$

To deal with the source terms $\mathbf{Q}$, a standard splitting procedure (see for instance [2]) is employed for the discrete system (11) as

$$\mathbf{W}_i^* = \mathbf{W}_i^n - \frac{\Delta t}{|\mathcal{T}_i|} \sum_{j \in N(i)} \int_{\Gamma_{ij}} \mathcal{F}(\mathbf{W}^n; \mathbf{n}) \, d\sigma + \frac{\Delta t}{|\mathcal{T}_i|} \int_{\mathcal{T}_i} \mathbf{S}(\mathbf{W}^n) \, dV,$$

$$\mathbf{W}_i^{n+1} = \mathbf{W}_i^* + \frac{\Delta t}{|\mathcal{T}_i|} \int_{\mathcal{T}_i} \mathbf{Q}(\mathbf{W}^*) \, dV. \tag{12}$$

Note that the time splitting (12) is only first-order accurate. A second-order splitting for the system (11) can be derived analogously using the Strang method [16]. The finite volume discretization (11) is complete once the gradient fluxes $\mathcal{F}(\mathbf{W}; \mathbf{n})$ and a discretization of source terms $\mathbf{Q}(\mathbf{W}^n)$ and $\mathbf{S}(\mathbf{W}^n)$ are well defined.

For the discretization of the gradient fluxes we consider a modified Roe's method studied in [3–6] among others. The method consists of the predictor-corrector

A two-dimensional finite volume solution

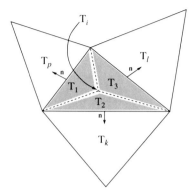

Fig. 1 A generic control volume $\mathcal{T}_i$ and notations

procedure

$$\mathbf{U}_{ij}^n = \frac{1}{2}\left(\mathbf{U}_i^n + \mathbf{U}_j^n\right) - \frac{1}{2}\operatorname{sgn}\left[\mathbf{A}_\eta\left(\overline{\mathbf{U}}\right)\right]\left(\mathbf{U}_j^n - \mathbf{U}_i^n\right),$$

$$\mathbf{W}_i^{n+1} = \mathbf{W}_i^n - \frac{\Delta t}{|\mathcal{T}_i|}\sum_{j \in N(i)} \mathscr{F}\left(\mathbf{W}_{ij}^n; \eta_{ij}\right)|\Gamma_{ij}| + \Delta t \mathbf{S}_i^n,$$

(13)

where

$$\mathbf{U} = \begin{pmatrix} h \\ u_\eta \\ u_\tau \\ c \\ Z \end{pmatrix}, \quad \mathbf{A}_\eta(\mathbf{U}) = \begin{pmatrix} u_\eta & h & 0 & 0 & 0 \\ g & u_\eta & 0 & \dfrac{(\rho_s - \rho_w)}{2\rho}gh & g \\ 0 & 0 & u_\eta & 0 & 0 \\ 0 & 0 & 0 & u_\eta & 0 \\ 0 & \dfrac{A_s}{1-p}(3u_\eta^2 + u_\tau^2) & 2\dfrac{A_s}{1-p}u_\eta u_\tau & 0 & 0 \end{pmatrix}.$$

the normal velocity $u_\eta = un_x + vn_y$ and tangential velocity $u_\tau = un_y - vn_x$. The sign matrix of the Jacobian is defined as

$$\operatorname{sgn}\left[\nabla \mathbf{F}_\eta\left(\overline{\mathbf{U}}\right)\right] = \mathscr{R}(\overline{\mathbf{U}})\operatorname{sgn}\left[\Lambda(\overline{\mathbf{U}})\right]\mathscr{R}^{-1}(\overline{\mathbf{U}}),$$

with $\Lambda(\overline{\mathbf{U}})$ is the diagonal matrix of eigenvalues, and $\mathscr{R}(\overline{\mathbf{U}})$ is the right eigenvector matrix. These matrices can be explicitly expressed using the associated eingenvalues of $\mathbf{A}_\eta(\mathbf{U})$. The sign matrix can be formulated in the same manner as in [3–6] and details are omitted here. In (13), $\overline{\mathbf{U}}$ is the Roe's average state given by

$$\overline{\mathbf{U}} = \begin{pmatrix} \dfrac{h_i + h_j}{2} \\ \dfrac{u_i \sqrt{h_i} + u_j \sqrt{h_j}}{\sqrt{h_i} + \sqrt{h_j}} \eta_x + \dfrac{v_i \sqrt{h_i} + v_j \sqrt{h_j}}{\sqrt{h_i} + \sqrt{h_j}} \eta_y \\ -\dfrac{u_i \sqrt{h_i} + u_j \sqrt{h_j}}{\sqrt{h_i} + \sqrt{h_j}} \eta_y + \dfrac{v_i \sqrt{h_i} + v_j \sqrt{h_j}}{\sqrt{h_i} + \sqrt{h_j}} \eta_x \\ \dfrac{c_i \sqrt{h_i} + c_j \sqrt{h_j}}{\sqrt{h_i} + \sqrt{h_j}} \\ \dfrac{Z_i + Z_j}{2} \end{pmatrix}. \quad (14)$$

Next we discuss the treatment of source terms $\mathbf{S}_i^n$ in the proposed finite volume scheme and also the extension of the scheme to a second-order accuracy. An adaptive procedure is also described in this section.

3.1 Treatment of the source term

The treatment of the source terms in the shallow water equations presents a challenge in many numerical methods. In our scheme, the source term approximation $\mathbf{S}_i^n$ in the corrector stage is reconstructed such that the still-water equilibrium (C-property) is satisfied. Here, a numerical scheme is said to satisfy the C-property for the equations (10) if the condition

$$E - D = 0, \quad u = 0, \quad Z = \bar{Z}(x), \quad h + Z = H, \quad \rho = C, \quad (15)$$

holds for stationary flows at rest. In (15), H and C are nonnegative constants. Therefore, the treatment of source terms in (13) is reconstructed such that the condition (15) is preserved at the discretized level. Remark that the last condition in (15) means that at the equilibrium the sediment medium is assumed to be saturated. Furthermore, from the density equation (2), a constant density is equivalent to a constant concentration c. Hence, $\mathbf{S}_i^n$ should be consistent discretization of the source term in (13) defined as

$$\mathbf{S}_i^n = \begin{pmatrix} 0 \\ -g\bar{h}_{xi}^n \sum_{j \in N(i)} Z_{ij} n_{xij} |\Gamma_{ij}| - \dfrac{(\rho_s - \rho_w)}{2\rho} g \left(\bar{h}_{xi}^n\right)^2 \sum_{j \in N(i)} c_{ij} n_{xij} |\Gamma_{ij}| \\ -g\bar{h}_{yi}^n \sum_{j \in N(i)} Z_{ij} n_{yij} |\Gamma_{ij}| - \dfrac{(\rho_s - \rho_w)}{2\rho} g \left(\bar{h}_{yi}^n\right)^2 \sum_{j \in N(i)} c_{ij} n_{yij} |\Gamma_{ij}| \\ 0 \\ 0 \end{pmatrix}.$$

$$(16)$$

A two-dimensional finite volume solution

The approximations $\bar{h}^n_{xi}$ and $\bar{h}^n_{yi}$ are reconstructed using a technique recently developed in [3] for the proposed finite volume method to satisfy the well-known C-property. In this section we briefly describe the formulation of this procedure and more details can be found in [3]. Hence, at the stationary state, the numerical flux in the corrector stage yields

$$\sum_{j \in N(i)} \mathscr{F}\left(\mathbf{W}^n_{ij}; \mathbf{n}_{ij}\right) = \begin{pmatrix} 0 \\ -g \int_{T_i} h \frac{\partial Z}{\partial x} dV - g \frac{(\rho_s - \rho_w)}{2\rho} \int_{T_i} h^2 \frac{\partial c}{\partial x} dV \\ -g \int_{T_i} h \frac{\partial Z}{\partial y} dV - g \frac{(\rho_s - \rho_w)}{2\rho} \int_{T_i} h^2 \frac{\partial c}{\partial y} dV \\ 0 \\ 0 \end{pmatrix},$$

which is equivalent to

$$\begin{pmatrix} 0 \\ \sum_{j \in N(i)} \frac{1}{2} g \left(h^n_{ij}\right)^2 N_{xij} \\ \sum_{j \in N(i)} \frac{1}{2} g \left(h^n_{ij}\right)^2 N_{yij} \\ 0 \\ 0 \end{pmatrix} = \begin{pmatrix} 0 \\ -g \int_{T_i} h \frac{\partial Z}{\partial x} dV - g \frac{(\rho_s - \rho_w)}{2\rho} \int_{T_i} h^2 \frac{\partial c}{\partial x} dV \\ -g \int_{T_i} h \frac{\partial Z}{\partial y} dV - g \frac{(\rho_s - \rho_w)}{2\rho} \int_{T_i} h^2 \frac{\partial c}{\partial y} dV \\ 0 \\ 0 \end{pmatrix}. \tag{17}$$

where $N_{xij} = n_{xij} |\Gamma_{ij}|$ and $N_{yij} = n_{yij} |\Gamma_{ij}|$. Next, to approximate the source terms we proceed as follows. First we decompose the triangle T_i into three sub-triangles as depicted in Fig. 1. Then, the source term is approximated as

$$\int_{T_i} h \frac{\partial Z}{\partial x} dV = \int_{T_1} h \frac{\partial Z}{\partial x} dV + \int_{T_2} h \frac{\partial Z}{\partial x} dV + \int_{T_3} h \frac{\partial Z}{\partial x} dV, \tag{18}$$

where

$$\int_{T_1} h \frac{\partial Z}{\partial x} dV = h_1 \int_{T_1} \frac{\partial Z}{\partial x} dV,$$

with h_1 is an average value of h on the sub-triangle T_1. Hence,

$$\int_{T_1} h \frac{\partial Z}{\partial x} dV = h_1 \sum_{j \in N(1)} \int_{\Gamma_{1j}} Z n_x \, d\sigma,$$

$$= h_1 \sum_{j \in N(1)} Z_{1j} \, N_{x1j},$$

$$= h_1 \sum_{j \in N(1)} \frac{Z_1 + Z_j}{2} N_{x1j}. \quad (19)$$

Again, using the stationary flow condition $h_1 + Z_1 = h_j + Z_j = H = constant$, one gets

$$h_1 + Z_1 + h_j + Z_j = 2H \quad \text{and} \quad \frac{Z_1 + Z_j}{2} = H - \frac{h_1 + h_j}{2}.$$

Thus, (19) gives

$$\int_{T_1} h \frac{\partial Z}{\partial x} dV = h_1 \sum_{j \in N(1)} \left(H - \frac{h_1 + h_j}{2} \right) N_{x1j}.$$

Using the fact that $\sum_{j \in N(1)} N_{x1j} = 0$,

$$\int_{T_1} h \frac{\partial Z}{\partial x} dV = -\frac{h_1}{2} \sum_{j \in N(1)} h_j \, N_{x1j},$$

$$= -\frac{h_1}{2} \left(h_p N_{x1p} + h_2 N_{x12} + h_3 N_{x13} \right).$$

A similar procedure leads to the following approximations of the other terms in (18)

$$\int_{T_2} h \frac{\partial Z}{\partial x} dV = -\frac{h_2}{2} \left(h_k N_{x2k} + h_1 N_{x21} + h_3 N_{x23} \right),$$

$$\int_{T_3} h \frac{\partial Z}{\partial x} dV = -\frac{h_3}{2} \left(h_l N_{x3l} + h_1 N_{x31} + h_2 N_{x32} \right).$$

Notice that h_p, h_k and h_l are the average values of h respectively, on the triangle T_p, T_k and T_l, see Fig. 1. Summing up, the discretization (18) gives

$$\int_{T_i} h \frac{\partial Z}{\partial x} dV = -\frac{h_1}{2} h_p N_{x1p} - \frac{h_2}{2} h_k N_{x2k} - \frac{h_3}{2} h_l N_{x3l}.$$

For this reconstruction, the source terms in (16) result in

$$\sum_{j \in N(i)} \left(h_{ij}^n\right)^2 N_{xij} = h_1 \left(h_p N_{x1p}\right) + h_2 \left(h_k N_{x2k}\right) + h_3 \left(h_l N_{x3l}\right),$$

$$\sum_{j \in N(i)} \left(h_{ij}^n\right)^2 N_{yij} = h_1 \left(h_p N_{y1p}\right) + h_2 \left(h_k N_{y2k}\right) + h_3 \left(h_l N_{y3l}\right). \tag{20}$$

Here, (20) forms a linear system of two equations for the three unknowns h_1, h_2 and h_3. To complete the system we add the natural conservation equation

$$h_1 + h_2 + h_3 = 3h_i.$$

Analogously, the bottom values Z_j, $j = 1, 2, 3$ are reconstructed in each sub-triangle of T_i as

$$Z_j + h_j^n = Z_i + h_i^n, \quad j = 1, 2, 3.$$

Finally, the source terms in (18) are approximated as

$$h_1 \int_{T_1} \frac{\partial Z}{\partial x} dV = h_1 \left(\frac{Z_1 + Z_p}{2} N_{x1p} + \frac{Z_1 + Z_2}{2} N_{x12} + \frac{Z_1 + Z_3}{2} N_{x13} \right),$$

$$h_2 \int_{T_1} \frac{\partial Z}{\partial x} dV = h_2 \left(\frac{Z_2 + Z_k}{2} N_{x2k} + \frac{Z_2 + Z_1}{2} N_{x21} + \frac{Z_2 + Z_3}{2} N_{x23} \right), \tag{21}$$

$$h_3 \int_{T_1} \frac{\partial Z}{\partial x} dV = h_3 \left(\frac{Z_3 + Z_l}{2} N_{x3l} + \frac{Z_3 + Z_1}{2} N_{x31} + \frac{Z_3 + Z_2}{2} N_{x32} \right),$$

with a similar equation for the other source terms in the y-direction.

4 Numerical results

We present numerical results for a test problem of partial dam-break over erodible bed. In all the computations reported herein, the Courant number Cr is set to 0.8 and the time stepsize Δt is adjusted at each step according to the stability condition

$$\Delta t = Cr \min_{\Gamma_{ij}} \left(\frac{|T_i| + |T_j|}{2 |\Gamma_{ij}| \max_p |(\lambda_p)_{ij}|} \right),$$

where Γ_{ij} is the edge between two triangles T_i and T_j. The water density $\rho_w = 1000 \, kg/m^3$ and the gravitational acceleration is fixed to $g = 9.81 \, m/s^2$.

We consider a 200 m long and 200 m wide flat reservoir with two different constant levels of water separated by a dam. At $t = 0$ part of the dam breaks instantaneously. The dam is 10 m thick and the breach is assumed to be 75 m wide, as shown in Fig. 2. Initially, $u(x, y, 0) = v(x, y, 0) = 0 \, m/s$

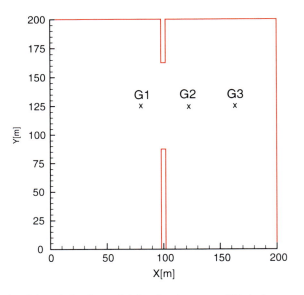

Fig. 2 Computational domain for the partial dam-break over erodible bed

$$h(x,y,0) = \begin{cases} 10\,m, & \text{if } x < 100\,m, \\ 1\,m, & \text{otherwise}, \end{cases} \qquad c(x,y,0) = \begin{cases} 0.01, & \text{if } x < 100\,m, \\ 0, & \text{otherwise}. \end{cases}$$

The selected values for the evaluation of the present finite volume model are summarized in Table 1. At $t = 0$ the dam collapses and the flow problem consists of a shock wave traveling downstream and a rarefaction wave traveling upstream.

Table 1 Reference parameters used for the dam-break problem

Quantity	Reference value	Quantity	Reference value
ρ_s	$1500\,kg/m^3$	ν	$1.2 \times 10^{-6}\,m^2/s$
p	0.28	n_b	$0.015\,s/m^{1/3}$
φ	$0.015\,m^{1.2}$	θ_c	0.045
m	2	d	$1\,mm$

In Fig. 3 we present the water free-surface and bed-load, the adapted meshes and snapshots of the water depth obtained for the partial dam-break over fixed bed at times $t = 2, 4, 6$ and $8\,s$. The results obtained for the partial dam-break over erodible bed are presented in Fig. 4. By using adaptive meshes, high resolution is automatically obtained in those regions where the gradients of the water depth are steep such as the moving fronts. Apparently, the overall flow pattern for this

A two-dimensional finite volume solution 887

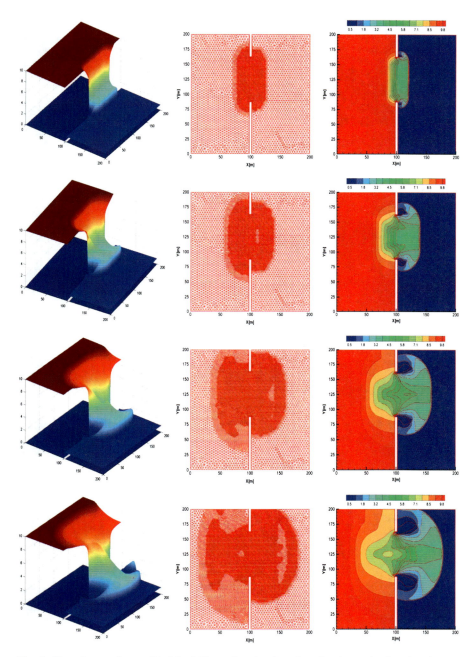

Fig. 3 Water free-surface and bed-load (first column), adapted meshes (second column) and water free-surface contours (third column) for the partial dam-break over fixed bed at different simulation times. From top to bottom $t = 2, 4, 6$ and 8 s

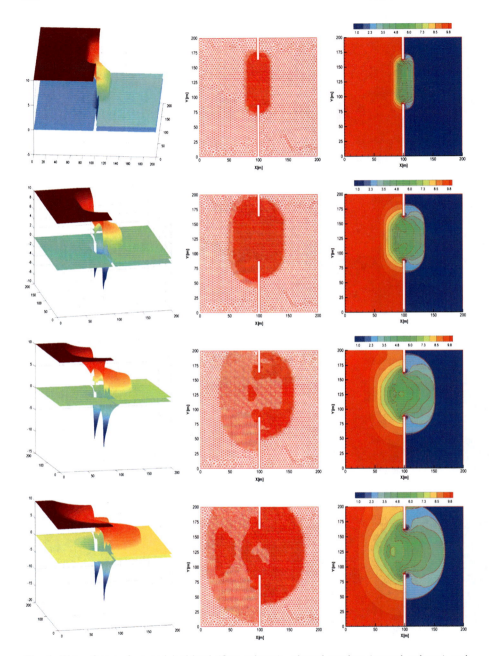

Fig. 4 Water free-surface and bed-load (first column), adapted meshes (second column) and water free-surface contours (third column) for the partial dambreak over erodible bed at different simulation times. From top to bottom $t = 2, 4, 6$ and 8 s

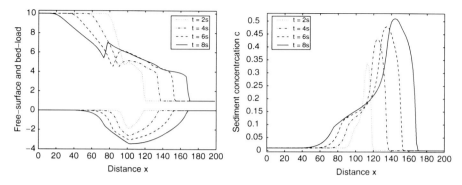

Fig. 5 Cross sections at $y = 125\ m$ of the water free-surface and bed-load (left plot) and sediment concentration (right plot) at four instants

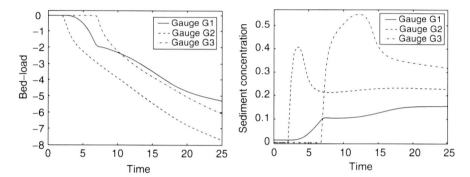

Fig. 6 Time evolution of the water free-surface and bed-load (left plot) and sediment concentration (right plot) at the three gauges G1, G2 and G3 presented in Fig. 2

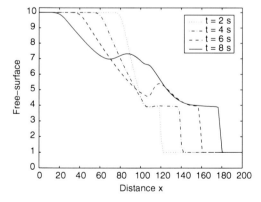

Fig. 7 Cross sections at $y = 125\ m$ of the water free-surface for the partial dam-break over fixed bed

example is preserved with no excessive numerical diffusion in the results by finite volume method using adaptive mesh. The adaptive finite volume method performs well for this test problem since it does not diffuse the moving fronts and no spurious oscillations have been observed when the water flows over the movable bed.

In order to quantify the results for this test example we display in Fig. 5 cross sections at $y = 125\ m$ of the water free-surface and bed-load and sediment concentration at four instants shown in Fig. 4. The results for the partial dam-break over fixed bed are depicted in Fig. 7. Figure 6 exhibites the time evolution of the water free-surface, bed-load and sediment concentration at the three gauges G1, G2 and G3 presented in Fig. 2. As can be observed from these results, the erosion effects on the bed are clearly visible for the considered sediment conditions. The inclusion of Exner equation in the model creates a very active sediment exchange between the water flow and the bed load, and also produces a sharp spatial gradient of sediment concentration, which justifies its incorporation in the momentum equations (10). Apparently, the overall flow and sediment features for this example are preserved with no spurious oscillations appearing in the results obtained using the adaptive finite volume method. Obviously, the computed results verify the stability and the shock capturing properties of the proposed finite volume method.

References

1. M.B. Abbott, Computational hydraulics: Elements of the theory of free surface flows, Fearon-Pitman Publishers, London, 1979.
2. S.J. Billett, E.F. Toro, On WAF-Type Schemes for Multidimensional Hyperbolic Conservation Laws, J. Comp. Physics. **130**, 1–24 (1997).
3. F. Benkhaldoun, I. Elmahi, M. Seaïd, "A new finite volume method for flux-gradient and source-term balancing in shallow water equations", Computer Methods in Applied Mechanics and Engineering. **199** pp:49-52 (2010).
4. F. Benkhaldoun, S. Sahmim, M. Seaïd, A two-dimensional finite volume morphodynamic model on unstructured triangular grids. Int. J. Num. Meth. Fluids. 63 (2010) 1296–1327.
5. F. Benkhaldoun, S. Sahmim, M. Seaïd, Solution of the sediment transport equations using a finite volume method based on sign matrix. SIAM J. Sci. Comp. 31 (2009) 2866–2889.
6. F. Benkhaldoun, I. Elmahi, M. Seaïd. "Well-balanced finite volume schemes for pollutant transport by shallow water equations on unstructured meshes", J. Comp. Physics. **226** pp:180-203 (2007).
7. K. Bloundi and J. Duplay, Heavy metals distribution in sediments of nador lagoon (Morocco), *Geophysical Research Abstracts*, 5, pp. 11744, 2003.
8. Z. Cao and P. Carling. Mathematical modelling of alluvial rivers: reality and myth. Part I: General overview. Water Maritime Engineering. **154**, 207-220 (2002)
9. Z. Cao and G. Pender. Numerical modelling of alluvial rivers subject to interactive sediment mining and feeding. Advances in Water Resources. **27**, 533-546 (2004)
10. Z. Cao, G. Pender, S. Wallis and P. Carling. Computational dam-break hydraulics over erodible sediment bed. J. Hydraulic Engineering. **67**, 689-703 (2004)
11. Z. Cao, G. Pender and P. Carling. Shallow water hydrodynamic models for hyperconcentrated sediment-laden floods over erodible bed. Advances in Water Resources. **29**, 546-557 (2006)
12. N.S. Cheng, Simplified settling velocity formula for sediment paticle. J. Hydraulic Engineering ASCE. 123 (1997) 149–152.

13. J. Fredsøe, R. Deigaard, Mechanics of Coastal Sediment Transport. *Advanced Series on Ocean Engineering - Vol. 3*, 1992.
14. A.J. Grass, Sediment Transport by Waves and Currents. (SERC London Cent. Mar. Technol. Report No: FL29, 1981).
15. G. Simpson and S. Castelltort. Coupled model of surface water flow, sediment transport and morphological evolution. Computers & Geosciences. **32**, 1600-1614 (2006)
16. G. Strang. On the Construction and the Comparison of Difference Schemes. SIAM J. Numer. Anal. **5**, 506517 (1968)
17. C.Y. Yang,: Sediment Transport: Theory and Practice. McGraw-Hill, New York (1996)
18. R.L. Soulsby: Dynamics of marine sands, a manual for practical applications. HR Wallingford, Report SR 466 (1997)

The paper is in final form and no similar paper has been or is being submitted elsewhere.

Part III
Benchmark Papers

3D Benchmark on Discretization Schemes for Anisotropic Diffusion Problems on General Grids

Robert Eymard, Gérard Henry, Raphaèle Herbin, Florence Hubert, Robert Klöfkorn, and Gianmarco Manzini

Abstract We present a number of test cases and meshes that were designed as a benchmark for numerical schemes dedicated to the approximation of three-dimensional anisotropic and heterogeneous diffusion problems. These numerical schemes may be applied to general, possibly non conforming, meshes composed of tetrahedra, hexahedra and quite distorted general polyhedra. A number of methods were tested among which conforming finite element methods, discontinuous Galerkin finite element methods, cell-centered finite volume methods, discrete duality finite volume methods, mimetic finite difference methods, mixed finite element methods, and gradient schemes. We summarize the results presented by the participants to the benchmark, which range from the number of unknowns, the approximation errors of the solution and its gradient, to the minimum and maximum values and energy. We also compare the performance of several iterative or direct linear solvers for the resolution of the linear systems issued from the presented schemes.

Keywords Anisotropic and heterogeneous medium, diffusion problem, numerical schemes for general polyhedral meshes, non-conforming meshes, 3D benchmark.
MSC2010: 65N08, 65N30, 65Y20, 76S05

Robert Eymard
Université Paris-Est, France, e-mail: Robert.Eymard@univ-mlv.fr

Gérard Henry, Raphaèle Herbin, Florence Hubert
Université Aix-Marseille, France, e-mail: Gerard.Henry@latp.univ-mrs.fr, Raphaele.Herbin@latp.univ-mrs.fr, Florence.Hubert@latp.univ-mrs.fr

Robert Klöfkorn
Universität Freiburg, Germany, e-mail: robertk@mathematik.uni-freiburg.de

Gianmarco Manzini
IMATI-CNR and CeSNA-IUSS Pavia, Italy, e-mail: Marco.Manzini@imati.cnr.it

1 Introduction

The two-dimensional (2D) anisotropy benchmark organized in 2007-2008 [21] provided a better understanding of the relative properties of a huge number of numerical schemes in terms of robustness, accuracy, problem size (number of degrees of freedom and matrix size), quality of the numerical approximation (maximum/minimum principles), etc. Nonetheless, a direct extrapolation of these results to three-dimensional (3D) problems is not possible because of the much higher complexity of the meshes involved in a 3D calculation and the larger size of the resulting linear systems. Hence, a new benchmark was organized between the end of 2010 and the beginning of 2011 with the additional goal of comparing CPU times versus accuracy.

A number of anisotropic and heterogeneous diffusion problems, associated with general, possibly non-conforming, 3D grids, were proposed in order for the participants to test a variety of numerical schemes. The participants were expected to provide information about the results obtained in these test cases and to use a set of solvers made available by the benchmark organizers for the linear systems arising from the discretization. In order to ensure a fair comparison of CPU times, all linear systems were solved by the same program implemented sequentially on the same computer, located at Université Aix-Marseille, France.

In most test cases the domain Ω is the unit cube; the boundary of Ω is denoted by Γ. We consider the steady diffusion problem with either homogeneous or non-homogeneous Dirichlet conditions on the boundary that is formulated in strong form as:

$$-\nabla \cdot (\mathbf{K}\nabla u) = f \quad \text{on } \Omega, \tag{1}$$

$$u = \bar{u} \quad \text{on } \Gamma, \tag{2}$$

where $\mathbf{K} : \Omega \to \mathbb{R}^{3\times 3}$ is the diffusion tensor, f is the source term and $\bar{u}$ is the Dirichlet boundary condition. The tensor fields $\mathbf{K}$ that we consider in the benchmark test cases are, as usual, strongly elliptic in Ω, i.e., each $\mathbf{K}$ is given by a field of symmetric matrices whose eigenvalues are uniformly bounded from above and from below by two strictly positive values. The data f and $\bar{u}$ of the problem are determined in accordance with the given exact solution and the diffusion field of each test case.

The paper is organized as follows. In Sect. 2, we present the five test cases, each one being specified by the shape of the computational domain, the exact solution, the diffusion field, and the set of meshes to be used. In Sect. 3, we briefly describe the linear solvers that were proposed for the resolution of the linear systems issued from the different numerical schemes. In Sect. 4, we list the participants to the benchmark and the numerical method that they used. In Sect. 5, we present the nature of the results obtained from the participants.

Final conclusions are drawn in Sect. 6. The tables and figures of results are given in Sect 7.

3D Benchmark on Discretization Schemes

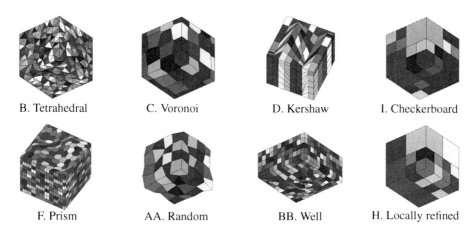

Fig. 1 The different meshes.

2 The test cases and the meshes

The test cases are summarized in Table 1, where we specify, for each test case, the shape of the computational domain Ω, the label of the permeability tensor, the label of the exact solution and the name of the mesh family. For more details about the meshes and other data, see at the URL:

```
http://www.latp.univ-mrs.fr/latp_numerique/?q=node/4,
```

where mesh data files can be downloaded.

Table 1 The test cases

Test Case	Domain	Permeability $\mathbf{K}(x, y, z)$	Solution $u(x, y, z)$	Meshes
Test 1 Mild anisotropy	Unit cube	$\mathbf{K}_1(x, y, z)$	$u_1(x, y, z)$	Tetrahedral (B) Voronoi (C) Kershaw (D) Checkerboard (I)
Test 2 Heterogeneity and anisotropy	Unit cube	$\mathbf{K}_2(x, y, z)$	$u_2(x, y, z)$	Prism (F)
Test 3 Random meshes	Determined by the mesh	$\mathbf{K}_3(x, y, z)$	$u_3(x, y, z)$	Random (AA)
Test 4 The well	Ω_4	$\mathbf{K}_4(x, y, z)$	$u_4(x, y, z)$	Well (BB)
Test 5 Locally refined	Unit cube	$\mathbf{K}_5(x, y, z)$	$u_5(x, y, z)$	Locally refined (H)

The meshes are presented in Fig. 1.

The data labeled in Table 1 (permeability tensor and exact solution for all test cases and computational domain for Test Cases 4 and 5) are as follows.

1. **Test Case 1.** We consider a constant, anisotropic permeability tensor and a regular solution that implies a non-homogeneous Dirichlet condition on the domain boundary Γ:

$$\mathbf{K}_1(x, y, z) = \begin{pmatrix} 1 & 0.5 & 0 \\ 0.5 & 1 & 0.5 \\ 0 & 0.5 & 1 \end{pmatrix}$$

$$u_1(x, y, z) = 1 + \sin(\pi x) \sin\left(\pi\left(y + \frac{1}{2}\right)\right) \sin\left(\pi\left(z + \frac{1}{3}\right)\right)$$

2. **Test Case 2.** We consider a smoothly variable permeability tensor and a regular solution that implies a non-homogeneous Dirichlet condition on the domain boundary Γ:

$$\mathbf{K}_2(x, y, z) = \begin{pmatrix} y^2 + z^2 + 1 & -xy & -xz \\ -xy & x^2 + z^2 + 1 & -yz \\ -xz & -yz & x^2 + y^2 + 1 \end{pmatrix}$$

$$u_2(x, y, z) = x^3 y^2 z + x \sin(2\pi xz) \sin(2\pi xy) \sin(2\pi z)$$

3. **Test Case 3.** We consider a constant, anisotropic permeability tensor and a regular solution on the domain, whose definition results from each of the considered meshes:

$$\mathbf{K}_3(x, y, z) = \begin{pmatrix} 1 & 0 & 0 \\ 0 & 1 & 0 \\ 0 & 0 & 10^3 \end{pmatrix}$$

$$u_3(x, y, z) = \sin(2\pi x) \sin(2\pi y) \sin(2\pi z)$$

Since the meshes which are used for this test case (random meshes) have boundary vertices which are not located exactly on the boundary of the unit cube, the boundary conditions are non-homogeneous Dirichlet boundary conditions.

4. **Test Case 4.** The computational domain is given by $\Omega_4 = P \setminus W$, where P is the parallelepiped $]-15, 15[\times]-15, 15[\times]-7.5, 7.5[$ and W is a slanted circular cylinder with radius $r_w = 0.1$. The axis of this well is a straight line located in the $x0z$ plane, passing by the origin, with an angle (in degrees) $\theta = -70°$ with the x axis, as shown in Fig. 2.

We consider the constant permeability tensor, which is slightly anisotropic in the third coordinate direction, given by

$$\mathbf{K}_4 = \begin{pmatrix} 1 & 0 & 0 \\ 0 & 1 & 0 \\ 0 & 0 & 0.2 \end{pmatrix}.$$

3D Benchmark on Discretization Schemes

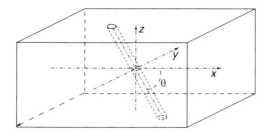

Fig. 2 The circular slanted well

The exact solution $u_4(x, y, z)$ is detailed in [1]: once a stretching of the axes has been performed so as to obtain an isotropic problem, we seek an exact solution that is constant on the well boundary. The solution simulates the pressure field that would be obtained for the same infinite slanted circular well in an infinite domain for a given constant flow rate q across any section of the well.

5. **Test Case 5**. The domain $\Omega = [0, 1]^3$ is split into four subdomains $\Omega = \cup_{i=1}^{4} \Omega_i$, which are given by

$$\Omega_1 = \{(x, y, z) \in [0, 1]^3 \text{ such that } y \leq 0.5, z \leq 0.5\}$$
$$\Omega_2 = \{(x, y, z) \in [0, 1]^3 \text{ such that } y > 0.5, z \leq 0.5\}$$
$$\Omega_3 = \{(x, y, z) \in [0, 1]^3 \text{ such that } y > 0.5, z > 0.5\}$$
$$\Omega_4 = \{(x, y, z) \in [0, 1]^3 \text{ such that } y \leq 0.5, z > 0.5\}$$

The permeability tensor and the exact solution are given by:

$$\mathbf{K}_5(x, y, z) = \begin{pmatrix} a_x^i & 0 & 0 \\ 0 & a_y^i & 0 \\ 0 & 0 & a_z^i \end{pmatrix} \text{ for } (x, y, z) \in \Omega_i \text{ with}$$

i	1	2	3	4
a_x^i	1	1	1	1
a_y^i	10	0.1	0.01	100
a_z^i	0.01	100	10	0.1
α_i	0.1	10	100	0.01

$$u_5(x, y, z) = \alpha_i \sin(2\pi x) \sin(2\pi y) \sin(2\pi z)$$

The permeability tensor $\mathbf{K}_5$ is discontinuous across the internal planes separating the unit cube in four subdomains and the exact solution u_5 is designed to be continuous and to ensure the conservation of the normal flux across such planes. Note that the homogeneous Dirichlet boundary condition is imposed in this test case.

3 Linear solvers used for the linear system benchmark

In order to access the different linear solver packages: UMFPACK [17, 18], DUNE-ISTL [7, 12], and PETSc [4, 5], all participants were asked to store their resulting linear systems for each test/mesh using a Compressed Row Storage (CRS) format, using an open source software package, which is available on line. All packages were installed on the 1 node Sun Fire X2270, equipped with 2 Quad-core processors (Intel, X5570, 2.93 GHz) and 24 GB memory (1333 MHz DDR3) and run sequentially.

Let us now briefly describe the available linear solvers and preconditioning methods.

1. **The direct solver library UMFPACK.** Written in ANSI/ISO C, UMFPACK is a set of routines for solving unsymmetric sparse linear systems $Ax = b$, using the Unsymmetric MultiFrontal method (see [17, 18] for details). For the benchmark, version 5.4.0 was used.

2. **The Iterative Solver Template Library – DUNE-ISTL** is a DUNE module [7, 12], which provides C++ programmed iterative solvers of linear systems stemming from finite element discretizations. The efficiency of the solvers is enhanced by taking into account the specific block recursive structure of matrices and vectors. For the benchmark version 2.0 has been used. The following solvers and preconditioning methods are used:

    - *Iterative solvers*: Conjugate Gradient, BiCG-stab, GMRES;
    - *Preconditioning*: Jacobi, ILU-0, ILU-n, $n = 1, ..., 4$, Algebraic Multi Grid.

3. **The Portable, Extensible Toolkit for Scientific Computation – PETSc** [4, 5] is a suite of data structures and routines for the scalable (parallel) solution of scientific applications modeled by partial differential equations. The program code is written in ANSI C. For the benchmark, version 3.1-p5 was used. The following iterative solvers and preconditioning methods are used:

    - *Iterative solvers*: Conjugate Gradient, BiCG-stab, GMRES;
    - *Preconditioning*: Jacobi, ILU-n, $n = 0, ..., 4$.

4. **Condition number calculation.** For the approximate calculation of the condition number of a given matrix, the Krylov-Schur method from the Scalable Library for Eigenvalue Problem Computations (SLEPc) package version 3.1-p4 [22] was used. SLEPc is written in ANSI C and built on top of PETSc.

5. **CPU time measurement.** The measurement of the CPU time spent for the solution process is based on the getrusage routines. The setup of the matrices (for the different solvers) is not included in the CPU time measurement in any case. The CPU time needed for the solution of the system with the iterative solvers (DUNE-ISTL and PETSc) is calculated by adding the time spent for building the preconditioner and the time spent in the linear solver. The CPU time with UMFPACK is not provided because the size of the matrices was too large for a direct solver in several cases.

4 The participating schemes and teams

Even though the benchmark is associated with the FVCA6 conference, the call for submission was by no means restricted to finite volume schemes, and, indeed, many types of schemes were submitted.

Cell-centered schemes

- MPFA-O: a Multi-Point Flux Approximation O-scheme programmed by the benchmark organizers for completeness purposes.
- LS-FVM: *The cell-centered finite volume method using least squares vertex reconstruction (diamond scheme)*, by Y. Coudière and G. Manzini [14].

Discontinuous Galerkin schemes

- CDG2: *The Compact Discontinuous Galerkin 2 Scheme*, R. Klöfkorn, [23].
- SWPG: *Symmetric Weighted Interior Penalty Discontinuous Galerkin Scheme*, by P. Bastian [6].

Discrete duality finite volume schemes

- CEVEDDFV-A: *A version of the DDFV scheme with cell/vertex unknowns on general meshes*, by B. Andreianov, F. Hubert and S. Krell [3].
- CEVEDDFV-B: *CeVe-DDFV, a discrete duality scheme with cell/vertex unknowns*, by Y. Coudière and C. Pierre [15].
- CEVEFE-DDFV: *CeVeFE-DDFV, a discrete duality scheme with cell/vertex/-face+edge unknowns*, by Y. Coudière, F. Hubert and G. Manzini [13].

Finite element schemes

- FEM: *Finite elements of order one* (FEM1) *and two* (FEM2) *provided by* P. Bastian with the DUNE environment [8, 9].
- MELODIE, *A linear finite element solver*, by H. Amor, M. Bourgeois, and G. Mathieu [2].

Mixed or hybrid methods

- MFD-GEN: *Mimetic finite difference method for generalized polyhedral meshes*, by K. Lipnikov and G. Manzini [24].
- MFD-PLAIN: *A mimetic finite difference method*, by P. Bastian, O. Ippisch, and S. Marnach, [10].
- MFMFE: *A multipoint flux mixed finite element method on general hexahedra*, by M. F. Wheeler, G. Xue and I. Yotov [25].
- CHMFE: *A composite hexahedral mixed finite element*, by I. Ben Gharbia, J. Jaffré, N. Suresh Kumar and J. E. Roberts [11].

Gradient schemes

- SUSHI: *The SUSHI scheme*, by R. Eymard, T. Gallouët and R. Herbin, [19].
- VAG and VAGR: *The VAG scheme*, by R. Eymard, C. Guichard and R. Herbin, [20].

Nonlinear schemes The schemes are nonlinear in order to ensure the positivity of the scheme (that is, if the right hand side is positive then the solution is positive) or the discrete maximum principle (that is, if the linear system stems from the discretization of an elliptic equation satisfying the maximum principle, then its solution is also bounded by the bounds of the continuous system).

- FVMON: *A monotone nonlinear finite volume method for diffusion equations on polyhedral meshes*, by A. Danilov and Y. Vassilevski, [16].

The choice of categories that we considered above is neither exhaustive nor unique. In fact, most of these categories intersect: schemes are not so easy to classify, and some schemes are known to be identical in special cases and when using some special meshes. We refer to the above-cited papers for the details of the schemes and their implementation. Our purpose is to give here a synthesis of the results presented by the participants.

5 Results obtained by the participants

5.1 Results provided by the participants

The results obtained by the participants are presented in the contributed papers in several tables.

First table: it reports the data related to the size of the discrete problem produced by a numerical scheme and some information about the quality of the numerical approximation. In particular, the minimum and maximum values of the discrete solution at cell-centers are compared with the same kind of values for the exact solution, and an estimate of ngrad $\sim \int_\Omega \|\nabla u\|$ allows us to evaluate possible oscillations of the approximation.

i	number of mesh
nu	number of unknowns of the linear system
nmat	number of non zero terms in the matrix
umin	minimum value of the approximate solution at the cell centers
uemin	minimum value of the exact solution at the cell centers
umax	maximum value of the approximate solution at the cell centers
uemax	maximum value of the exact solution at the cell centers
normg	L^1 norm of the euclidean norm of the approximate gradient

Second table: it provides information about the accuracy of the schemes, which is measured for all the test cases versus nu, the number of unknowns, by the following quantities:

3D Benchmark on Discretization Schemes

i	number of mesh
nu	number of unknowns of the linear system
erl2	relative L^2 norm of the error with respect to the L^2 norm of the exact solution.
ratiol2	order of convergence of the L^2 norm of the error on the solution between mesh i and i-1.
ergrad	relative H^1 semi-norm of the error with respect to the H^1 semi-norm of the exact solution.
ratiograd	order of convergence of the H^1 norm of the error on the solution between mesh i and i-1.
ener	relative energy norm of the error with respect to the energy norm of the exact solution.
ratioener	order of convergence of the energy norm of the error on the solution between mesh i and i-1.

where, denoting *err* the numerical error,

- the relative L^2 norm of the error is given by:

$$\text{erl2} \approx \left(\int_\Omega |err|^2 / \int_\Omega |u|^2 \right)^{\frac{1}{2}};$$

- the relative L^2 norm of the gradient of the error is given by:

$$\text{ergrad} \approx \left(\int_\Omega |\nabla err|^2 / \int_\Omega |\nabla u|^2 \right)^{\frac{1}{2}};$$

- the relative energy norm of the error is given by:

$$\text{ener} \approx \left(\int_\Omega \mathbf{K} \nabla err \cdot \nabla err / \int_\Omega \mathbf{K} \nabla u \cdot \nabla u \right)^{\frac{1}{2}}.$$

and the convergence rates are defined, for $i \geq 2$, by:

$$\text{ratiol2}(i) = -3 \frac{\log(\text{erl2}(i)/\text{erl2}(i-1))}{\log(\text{nu}(i)/\text{nu}(i-1))};$$

$$\text{ratiograd}(i) = -3 \frac{\log(\text{ergrad}(i)/\text{ergrad}(i-1))}{\log(\text{nu}(i)/\text{nu}(i-1))};$$

$$\text{ratioener}(i) = -3 \frac{\log(\text{ener}(i)/\text{ener}(i-1))}{\log(\text{nu}(i)/\text{nu}(i-1))}.$$

Matrices and right-hand sides were uploaded by the participants on the computer dedicated to the bench, in order to compare CPU time and memory.

5.2 Comparisons

- **Maximum principle**. For all test cases, we collect the values of u_{min}, u_{max} for the coarsest and finest grids handled by the participants, in Tables 2, 3 and 4 (Test Case 1), 5 (Test Case 2), 6 (Test Case 3), 7 (Test Case 4) and 8 (Test Case 5). We colored in red (resp. purple) the values that are below (resp. above) the minimum value of the exact solution.
- **Accuracy**. In Figs. 3-10, we report the log-log curves of the approximation errors measured by the benchmark participants for their numerical schemes. Each figure refers to a specific combination "test case + mesh family"; the upper left-most plot reports `erl2`, the upper right-most plot reports `ergrad`, the lower left-most plot reports `normg`, and the lower right-most plot reports `ener`. The convergence rates in these log-log plots are reflected by the slopes of the convergence curves.
- **Condition number**. We report the condition number (see Sect. 3) of the matrices involved in the numerical discretizations of first two test cases in Tables 9, 10, 11 and 12 (Test Case 1) and in Table 13 (Test Cases 2). The condition numbers in each table are calculated for the first mesh and the two next mesh refinements. The eigensolver tolerance was set to 10^{-8} for all matrices.
- **Cost of the resolution**. The cost of the resolution of the linear systems is shown in Figs. 11-18, where the L^2 error is plotted with respect to the CPU time and the used memory. The CPU time was measured for the linear system with the right hand side $b = A\mathbf{1}$, where $\mathbf{1}$ is the vector with all components equal to 1. The stopping criterion for all the iterative methods is: residual $\leq 10^{-10}$. For the sake of simplicity, all methods, including conjugate gradient methods, have been applied to symmetric and non-symmetric matrices.

6 Conclusion

This paper proposes a comparison of sixteen numerical schemes (and variants) which were tested on a family of three-dimensional anisotropic diffusion problems. The tests presented here involve both a wide class of diffusion tensors (anisotropic and at time heterogeneous and/or discontinuous) and a wide class of conforming and non-conforming meshes with very general polyhedral cells.

The number of results which were obtained on this benchmark is impressive with respect to the difficulty of the exercise and the time constraint. In fact, additional results are available on the bench web site:

```
http://www.latp.univ-mrs.fr/latp_numerique/?q=node/4.
```

and will be updated. The benchmark was found to be most useful to the participants to compare their schemes to reference solutions. The participation to the 3D benchmark was an opportunity for several participants to learn more about the efficient implementation of their schemes. Indeed, several variants of the schemes were thus developed. Last but not least, a user-friendly comparison platform was developed for this benchmark, which allows anyone to link to the solver and preconditioner of his choice; this possibility has already been used by other users than the 3D benchmark. The platform which was developed for the 3D benchmark should proof useful for further investigations on numerical schemes for various models.

7 Tables and figures of results

Table 2 Maximum principle for Test 1: mild anisotropy on tetrahedral meshes

Scheme	umin coarse	umax coarse	umin fine	umax fine
CDG2K1	−1.54E-02	2.017	−6.63E-04	2.002
CDG2K2	0.00	1.999	0.00E+00	1.999
CEVEDDFV-A	0.706E-02	1.992	0.140E-02	1.999
CEVEDDFV-B	1.34E-02	1.99	1.30E-03	2.00
CEVEFE-DDFV	6.09E-03	1.988	1.93E-03	1.999
FEM1	8.34E-02	1.932	6.35E-03	1.990
FEM2	2.13E-02	1.989	1.84E-03	1.997
FVMON	0.028	1.997	0.003	1.998
LS-FVM	2.03E-02	1.989	1.83E-03	1.997
MELODIE	7.69E-02	1.935	6.19E-03	1.991
MPFA-O	−1.13E-02	2.01	−1.46E-03	2.00
MFD-PLAIN	2.33E-03	1.994	1.66E-03	1.998
MFD-GEN	2.26E-02	1.986	1.75E-03	1.997
SWPG-1	5.32E-02	1.965	3.69E-03	1.994
SWPG-2	2.11E-02	1.989	1.84E-03	1.997
SWPG-3	2.04E-02	1.989	1.83E-03	1.997
SWPG-4	2.03E-02	1.989	1.83E-03	1.997
SUSHI	3.21E-02	1.98	1.74E-03	2.00
VAG	6.77E-02	1.94	4.62E-03	1.99
VAGR	5.77E-02	1.95	3.63E-03	1.99

Table 3 Maximum principle for Test 1: mild anisotropy on Kershaw meshes

Scheme	umin (coarse)	umax (coarse)	umin(fine)	umax(fine)
CDG2Legk1	−2.95E-02	2.016	−5.37E-04	2.000
CDG2Legk2	0.00	1.997	0.00	1.999
CDG2Tetk1	−2.81E-02	2.012	−4.65E-04	2.000
CDG2Tetk2	0.00	1.995	0.00	1.999
CeVeDDFV-A	2.28E-02	1.989	3.82E-04	2.000
CeVeDDFV-B	7.16E-02	1.94	4.61E-04	2.00
CeVeFE-DDFV	5.67E-02	1.940	6.52E-04	2.000
CHMFE	−0.032	1.94685	−0.008	2.00061
FEM1	1.77E-01	1.786	2.94E-03	1.996
FEM2	3.29E-02	1.941	7.11E-04	1.999
FVMON	0.112	1.942	0.003	1.997
LS-FVM	3.03E-02	1.958	7.14E-04	1.999
MELODIE	1.34E-01	1.833	2.04E-03	1.997
MFD-gen	−2.52E-02	1.973	2.71E-04	1.999
MFD-plain	−6.03E-01	2.100	1.65E-04	2.000
MFMFE-ns	−1.26E-03	2.01	5.00E-05	2.00
MFMFE-s	4.66E-03	1.97	7.49E-05	2.00
MPFA-O	−3.76E-02	2.05	−1.06E-03	2.00
SWPG-1	9.58E-02	1.850	1.71E-03	1.997
SWPG-2	3.12E-02	1.944	7.11E-04	1.999
SWPG-3	2.91E-02	1.955	1.75E-03	1.997
SWPG-4	3.02E-02	1.958	1.75E-03	1.997
SUSHI	−2.14E-03	1.91	8.51E-04	2.00
VAG	1.43E-01	1.93	1.07E-03	2.00
VAGR	7.80E-02	1.96	−2.64E-04	2.00

Table 4 Maximum principle for Test 1: mild anisotropy on Checkerboard meshes

Scheme	umin (coarse)	umax(coarse)	umin(fine)	umax(fine)
CDG2K1	0.00	1.901	−5.50E-04	2.000
CDG2K2	−3.34E-02	2.050	0.000	1.999
CDG2LEGK1	−7.94E-02	2.081	−3.06E-04	2.000
CDG2LEGK2	0.00	1.998	0.00	1.999
CDG2TETK2	0.00	2.003	0.00	1.999
CEVEDDFV-A	0.341E-01	1.966	0.134E-03	2.000
CEVEDDFV-B	1.46E-01	1.86	5.01E-04	2.00
CEVEFE-DDFV	8.58E-02	1.903	2.88E-04	2.000
FEM1	3.26E-01	1.671	1.54E-03	1.998
FVMON	0.122	1.905	0.001	2.000
LS-FVM	1.54E-01	1.846	6.36E-04	1.999
MFD-GEN	2.91E-01	1.880	2.15E-03	1.999
MFD-PLAIN	1.27E-01	1.883	−3.52E-03	2.004
SWPG-1	2.35E-01	1.784	6.36E-04	1.999
SWPG-2	1.82E-01	1.812	6.37E-04	1.999
SWPG-3	1.61E-01	1.839	6.36E-04	1.999
SWPG-4	1.55E-01	1.845	6.36E-04	1.999
SUSHI	1.05E-01	1.87	3.83E-04	2.00
VAG	−1.95	2.50	−3.06E-02	2.03
VAGR	−9.81E-02	2.08E+00	−4.33E-03	2.00

Table 5 Maximum principle for Test 2: heterogeneous anisotropy on Prismatic meshes

Scheme	umin (coarse)	umax(coarse)	umin(fine)	umax(fine)
CEVEDDFV-A	−.856	1.044	−.862	1.049
CEVEDDFV-B	−8.53E-01	9.85E-01	−8.58E-01	1.03
CEVEFE-DDFV	−8.55E-01	1.014	−8.60E-01	1.040
FVMON	−0.854	1.002	−0.858	1.034
LS-FVM	−8.42E-01	0.978	−8.57E-01	1.033
MFD-GEN	−0.873	0.832	−0.890	0.963
MPFA-O	−9.23E-01	1.07	−8.63E-01	1.05
SUSHI	−8.22E-01	9.82E-01	−8.55E-01	1.03
VAG	−9.49E-01	1.23	−8.53E-01	1.05
VAGR	−8.73E-01	1.10E+00	−8.53E-01	1.04

Table 6 Maximum principle for Test 3: flow on random meshes

Scheme	umin(coarse)	umax(coarse)	umin(fine)	umax(fine)
CDG2LEGK1	−1.143	1.244	−1.009	1.000
CDG2LEGK2	−1.015	1.034	−1.00E+00	1.00
CDG2TETK1	−1.261	1.167	−1.008	1.002
CDG2TETK2	−1.238	1.295	−1.000	1.000
CEVEDDFV-A	−.202E+01	1.969	−.101E+01	1.014
CEVEDDFV-B	−1.58	1.54	−1.01	1.01
CEVEFE-DDFV	−4.25E+01	49.169	−2.67	2.725
FEM1	−3.73E-01	0.313	−9.90E-01	0.989
FEM2	−7.48E-01	0.679	−9.96E-01	0.996
FVMON	−0.905	0.759	−0.989	1.001
LS-FVM	−7.56E-01	0.711	−9.96E-01	0.996
MELODIE	−0.665	0.685	−0.988	0.991
MFD-GEN	−1.268	1.430	−1.027	1.021
MFD-PLAIN	−1.02E+00	1.045	−1.00	1.000
MFMFE-S	−6.20	5.75	−1.06	1.04
MPFA-O	−9.79	1.22E+01	−2.61E+01	2.44E+01
SUSHI	−7.51E-01	7.58E-01	−9.90E-01	9.89E-01
SWPG-1	−4.34E-01	0.355	−9.90E-01	0.989
SWPG-2	−7.50E-01	0.676	−9.96E-01	0.996
SWPG-3	−7.53E-01	0.684	−9.96E-01	0.996
SWPG-4	−7.59E-01	0.691	−9.85E-01	0.982
VAG	−1.31	1.50	−1.00	1.00
VAGR	−1.51	1.68E+00	−1.01	1.01

3D Benchmark on Discretization Schemes

Table 7 Maximum principle for Test 4: the flow around the well

Scheme	umin(coarse)	umax(coarse)	umin(fine)	umax(fine)
CDG2L$_{EG}$K1	0.00	5.406	0.00	5.410
CDG2L$_{EG}$K2	0.00	5.408	0.00	5.411
CDG2T$_{ET}$K1	0.00	5.406	0.00	5.410
CDG2T$_{ET}$K2	−5.92E-03	5.414	0.00	5.414
C$_E$V$_E$DDFV-A	−.438E-01	5.415	−.198E-02	5.415
C$_E$V$_E$DDFV-B	4.85E-01	5.32	5.80E-02	5.36
C$_E$V$_E$FE-DDFV	3.83E-01	5.317	5.66E-02	5.361
FEM1	3.73E-01	5.317	5.66E-02	5.361
FEM2	4.12E-01	5.317	5.65E-02	5.361
FVMON	0.518	5.318	0.059	5.361
LS-FVM	4.57E-01	5.317	5.75E-02	5.361
MELODIE	0.189	5.360	0.029	5.39
MFD-$_{GEN}$	5.37E-01	5.317	5.91E-02	5.361
MFD-$_{PLAIN}$	5.74E-01	5.317	5.91E-02	5.361
MPFA-O	4.36E-01	5.39	−1.49E-03	5.40
SUSHI	4.26E-01	5.32	5.78E-02	5.36
SWPG-1	3.52E-01	5.316	5.55E-02	5.361
SWPG-2	4.13E-01	5.317	5.65E-02	5.361
SWPG-3	4.15E-01	5.317	5.65E-02	5.361
SWPG-4	4.14E-01	5.317	8.99E-02	5.339
VAG	3.89E-01	5.32	5.69E-02	5.36
VAGR	3.89E-01	5.32	5.69E-02	5.36

Table 8 Maximum principle for Test 5: discontinuous anisotropy

Scheme	umin(coarse)	umax(coarse)	umin(fine)	umax(fine)
CDG2L$_{EG}$K1	−12.747	12.747	−100.241	100.241
CDG2L$_{EG}$K2	−94.815	94.815	−99.987	99.987
C$_E$V$_E$FE-DDFV	−6.34E+01	64.462	−1.02E+02	102.394
FEM1	−1.87E-02	0.019	−9.78E+01	97.772
FVMON	−246.736	246.736	−99.719	99.719
LS-FVM	−1.00E+02	1.00E+02	−9.86E+01	98.562
MFD-$_{GEN}$	−1.66E+02	1.66E+02	−9.95E+01	9.95E+01
MFD-$_{PLAIN}$	−2.51E+02	250.808	−9.89E+01	98.887
SWPG-1	−5.46E+01	54.594	−9.78E+01	97.780
SWPG-2	−1.18E+02	118.325	−9.86E+01	98.563
SWPG-3	−1.05E+02	104.586	−9.86E+01	98.562
SUSHI	−2.49E+02	2.49E+02	−9.89E+01	9.89E+01
VAG	−7.65E+02	7.65E+02	−9.93E+01	9.93E+01
VAGR	−7.39E+02	7.39E+02	−1.00E+02	1.00E+02

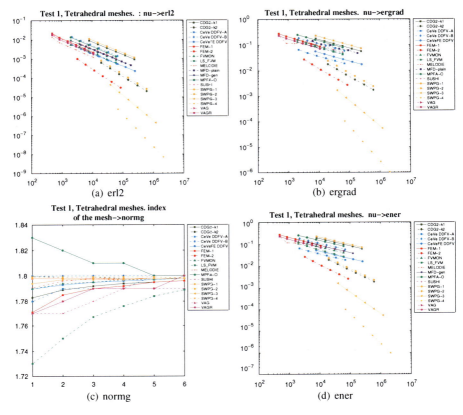

Fig. 3 Accuracy of the schemes for Test Case 1 on tetrahedral meshes. Plot (*a*) shows the relative L^2-norm of the error, plot (*b*) shows the relative H^1-seminorm of the error, plot (*c*) the L^1-norm of the numerical gradient, and (*d*) the energy norm of the error

3D Benchmark on Discretization Schemes

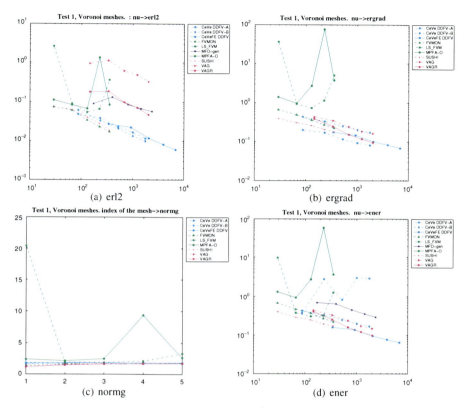

Fig. 4 Accuracy of the schemes for Test Case 1 on Voronoi meshes. Plot (*a*) shows the relative L^2-norm of the error, plot (*b*) shows the relative H^1-seminorm of the error, plot (*c*) the L^1-norm of the numerical gradient, and (*d*) the energy norm of the error

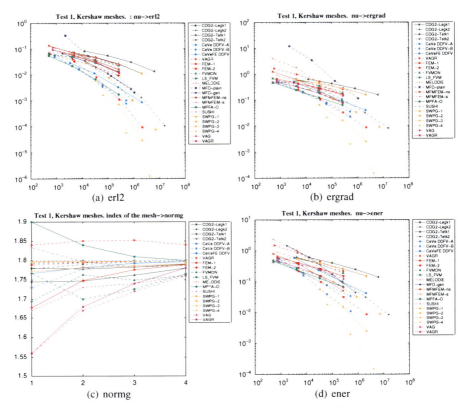

Fig. 5 Accuracy of the schemes for Test Case 1 on Kershaw meshes. Plot (*a*) shows the relative L^2-norm of the error, plot (*b*) shows the relative H^1-seminorm of the error, plot (*c*) the L^1-norm of the numerical gradient, and (*d*) the energy norm of the error

3D Benchmark on Discretization Schemes

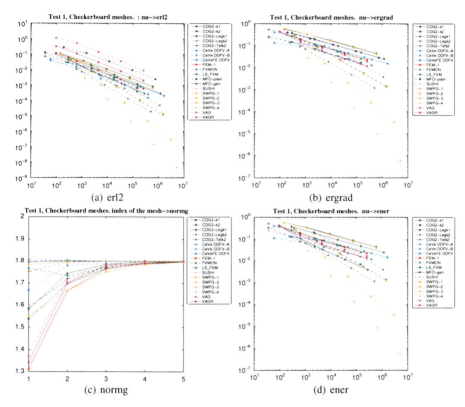

Fig. 6 Accuracy of the schemes for Test Case 1 on Checkerboard meshes. Plot (*a*) shows the relative L^2-norm of the error, plot (*b*) shows the relative H^1-seminorm of the error, plot (*c*) the L^1-norm of the numerical gradient, and (*d*) the energy norm of the error

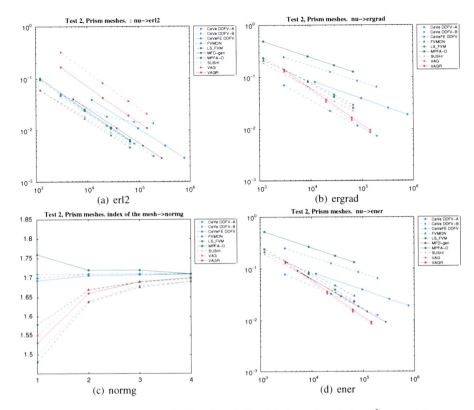

Fig. 7 Accuracy of the schemes for Test Case 2. Plot (*a*) shows the relative L^2-norm of the error, plot (*b*) shows the relative H^1-seminorm of the error, plot (*c*) the L^1-norm of the numerical gradient, and (*d*) the energy norm of the error

3D Benchmark on Discretization Schemes

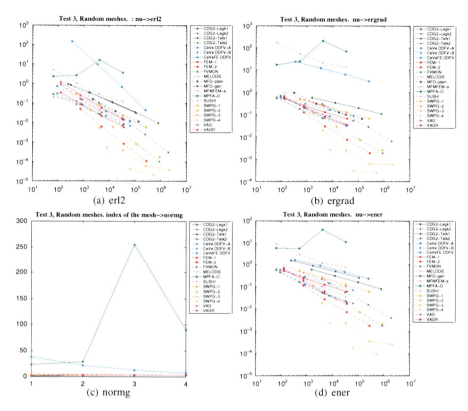

Fig. 8 Accuracy of the schemes for Test Case 3. Plot (*a*) shows the relative L^2-norm of the error, plot (*b*) shows the relative H^1-seminorm of the error, plot (*c*) the L^1-norm of the numerical gradient, and (*d*) the energy norm of the error

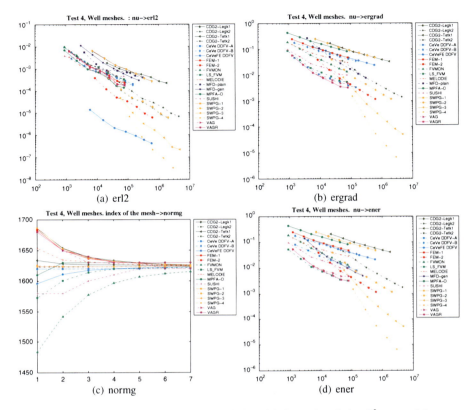

Fig. 9 Accuracy of the schemes for Test Case 4. Plot (*a*) shows the relative L^2-norm of the error, plot (*b*) shows the relative H^1-seminorm of the error, plot (*c*) the L^1-norm of the numerical gradient, and (*d*) the energy norm of the error

3D Benchmark on Discretization Schemes

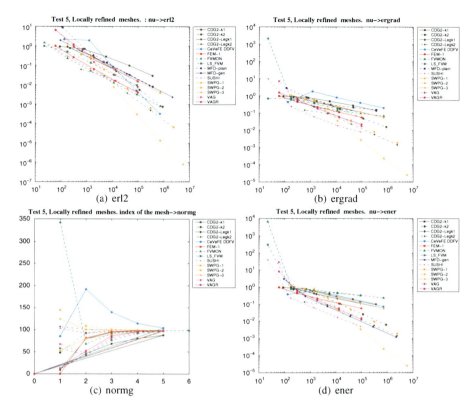

Fig. 10 Accuracy of the schemes for Test Case 5. Plot (*a*) shows the relative L^2-norm of the error, plot (*b*) shows the relative H^1-seminorm of the error, plot (*c*) the L^1-norm of the numerical gradient, and (*d*) the energy norm of the error

Table 9 Matrix condition numbers for the first three meshes in the solution of Test Case 1 using Tetrahedral meshes.

Scheme	Condition number		
	i=1	i=2	i=3
CDG2-k1	6.96E+03	8.36E+03	1.81E+04
CDG2-k2	2.37E+04	2.80E+04	5.98E+04
CeVe DDFV-A	2.67E+02	3.72E+02	9.15E+02
CeVe DDFV-B	2.75E+02	3.91E+02	9.34E+02
CeVeFE DDFV	9.36E+02	1.31E+03	3.04E+03
FEM-1	2.68E+01	4.79E+01	7.39E+01
FEM-2	2.24E+02	3.66E+02	5.96E+02
FVMON	8.90E+02	9.56E+02	2.09E+03
LS-FVM	2.80E+02	3.91E+02	9.62E+02
MELODIE	1.18E+02	6.04E+01	1.65E+02
MFD-gen	1.34E+03	1.46E+03	3.32E+03
MFD-plain	1.46E+03	2.47E+03	3.87E+03
MPFA-O	3.28E+01	5.01E+01	8.64E+01
SUSHI	8.34E+02	1.32E+03	2.67E+03
SWPG-1	1.29E+04	1.53E+04	3.33E+04
SWPG-2	6.37E+04	7.42E+04	1.60E+05
SWPG-3	1.82E+05	2.13E+05	4.64E+05
SWPG-4	4.15E+05	4.86E+05	1.06E+06
VAG	2.68E+01	4.79E+01	7.39E+01
VAGR	2.68E+01	4.79E+01	7.39E+01

Table 10 Matrix condition numbers for the first three meshes in the solution of Test Case 1 using Voronoi meshes.

Scheme	Condition number		
	i=1	i=2	i=3
CeVe DDFV-A	9.51E+01	1.24E+02	3.33E+02
CeVe DDFV-B	5.07E+01	9.40E+01	2.05E+02
CeVeFE DDFV	1.05E+03	2.00E+05	1.98E+05
FVMON	1.03E+01	9.97E+00	1.58E+02
MPFA-O	5.78E+01	8.32E+01	–
SUSHI	1.45E+01	1.12E+01	3.07E+01
VAG	6.51E+01	7.95E+02	4.19E+02
VAGR	1.82E+01	3.68E+01	8.36E+01

3D Benchmark on Discretization Schemes

Table 11 Matrix condition numbers for the first three meshes in the solution of Test Case 1 using Kershaw meshes.

Scheme	Condition number		
	i=1	i=2	i=3
CDG2-Legk1	3.06E+04	1.84E+05	1.01E+06
CDG2-Legk2	1.99E+05	1.04E+06	–
CDG2-Tetk1	1.41E+05	6.14E+05	2.62E+06
CDG2-Tetk2	5.22E+05	2.17E+06	–
CeVe DDFV-A	6.67E+02	3.25E+03	1.54E+04
CeVe DDFV-B	7.08E+02	3.85E+03	1.84E+04
CeVeFE DDFV	3.80E+03	1.99E+04	9.77E+04
FEM-1	1.54E+02	1.12E+03	7.50E+03
FEM-2	2.55E+03	1.55E+04	9.58E+04
FVMON	3.31E+02	2.07E+03	8.65E+03
LS-FVM	2.86E+02	1.37E+03	9.76E+03
MELODIE	5.27E+02	2.27E+03	1.28E+04
MFD-gen	2.10E+03	7.53E+03	4.17E+04
MFD-plain	2.65E+03	1.29E+04	7.47E+04
MFMFEM-ns	1.12E+02	9.19E+02	6.88E+03
MFMFEM-s	2.02E+02	1.25E+03	7.77E+03
MPFA-O	8.19E+01	8.12E+02	5.31E+02
SUSHI	1.08E+03	2.51E+03	1.47E+04
VAG	1.80E+02	1.08E+03	7.28E+03
VAGR	1.76E+02	1.19E+03	7.62E+03

Table 12 Matrix condition numbers for the first three meshes in the solution of Test Case 1 using Checkerboard meshes.

Scheme	Condition number		
	i=1	i=2	i=3
CeVe DDFV-A	1.52E+01	5.20E+01	2.00E+02
CeVe DDFV-B	9.82E+00	3.39E+01	1.29E+02
CeVeFE DDFV	5.72E+01	2.31E+02	9.29E+02
FVMON	8.00E+00	2.62E+01	9.44E+01
MFD-plain	3.06E+01	1.71E+02	8.09E+02
SUSHI	6.96E+00	2.47E+01	9.83E+01
SWPG-1	–	1.50E+02	6.55E+02
VAG	3.41E+00	2.01E+01	1.46E+02
VAGR	2.62E+00	1.83E+01	1.42E+02

Table 13 Matrix condition numbers for the first three meshes in the solution of Test Case 2 using Prismatic meshes.

Scheme	Condition number		
	i=1	i=2	i=3
CeVe DDFV-A	2.08E+02	1.03E+03	2.51E+03
CeVe DDFV-B	1.31E+02	7.16E+02	1.79E+03
CeVeFE DDFV	1.17E+03	5.65E+03	1.35E+04
FVMON	7.23E+01	3.49E+02	8.41E+02
LS-FVM	9.77E+01	5.13E+02	1.29E+03
MPFA-O	8.65E+01	4.90E+02	1.27E+03
SUSHI	1.02E+02	5.26E+02	1.30E+03
VAG	7.44E+01	4.48E+02	1.42E+03
VAGR	9.57E+01	5.41E+02	1.42E+03

3D Benchmark on Discretization Schemes

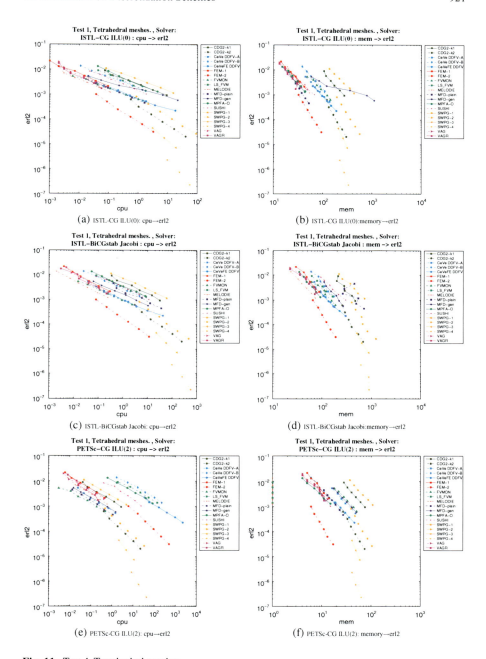

Fig. 11 Test 1-Tetrahedral meshes

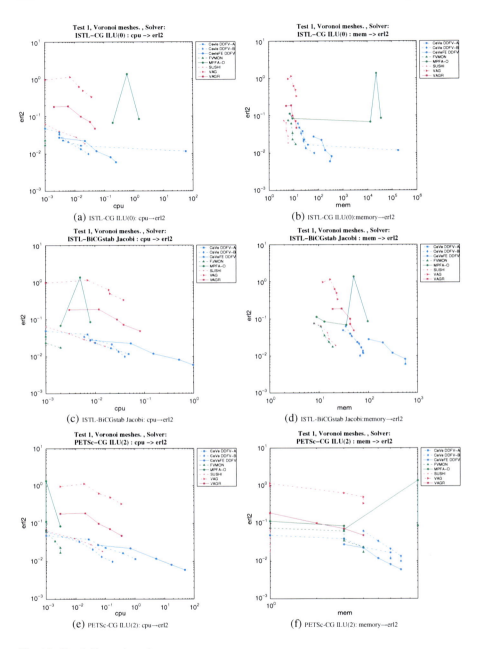

Fig. 12 Test 1-Voronoi meshes

3D Benchmark on Discretization Schemes

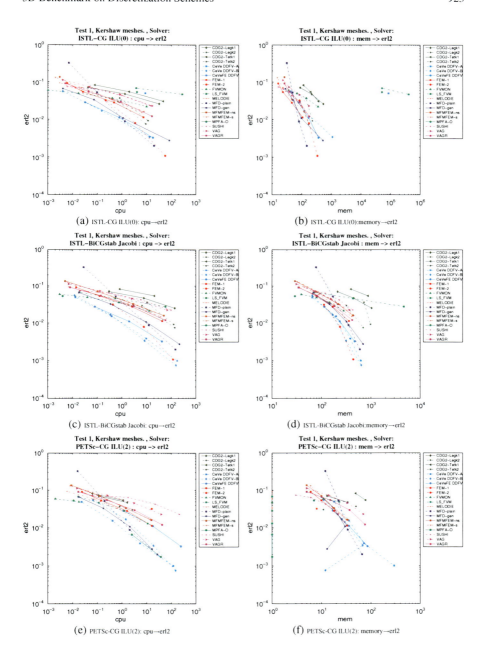

Fig. 13 Test 1-Kershaw meshes

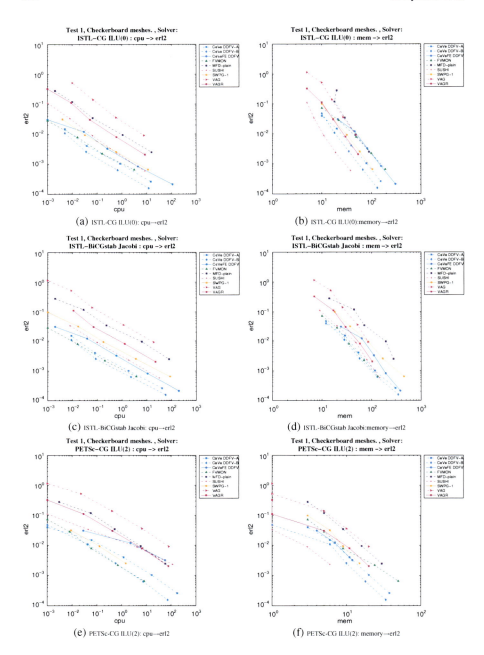

Fig. 14 Test 1-Checkerboard meshes

3D Benchmark on Discretization Schemes

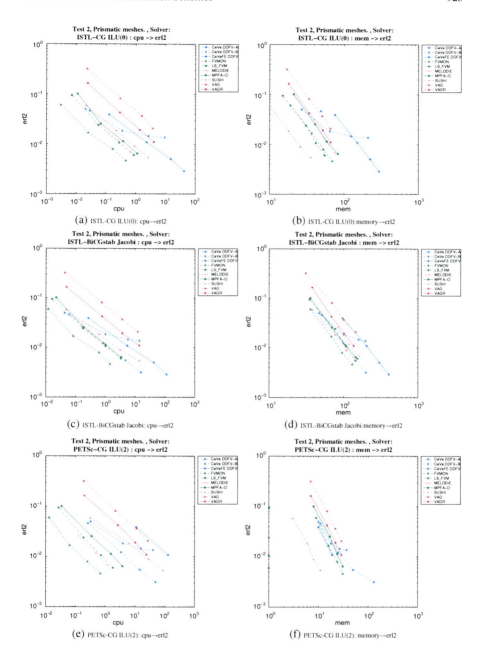

Fig. 15 Test 2-Prismatic meshes

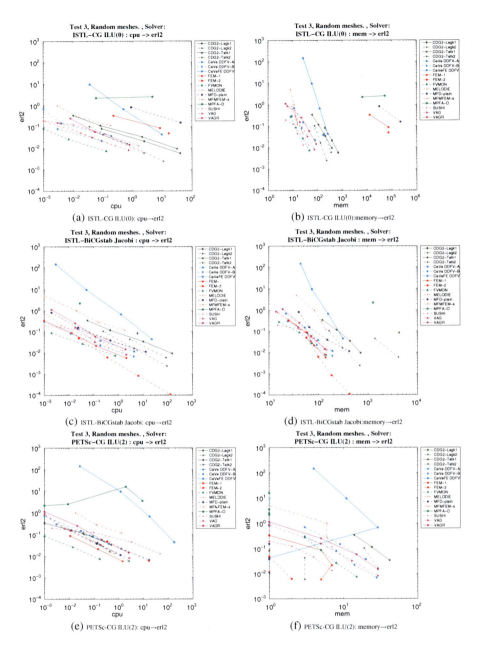

Fig. 16 Test 3-Random meshes

3D Benchmark on Discretization Schemes

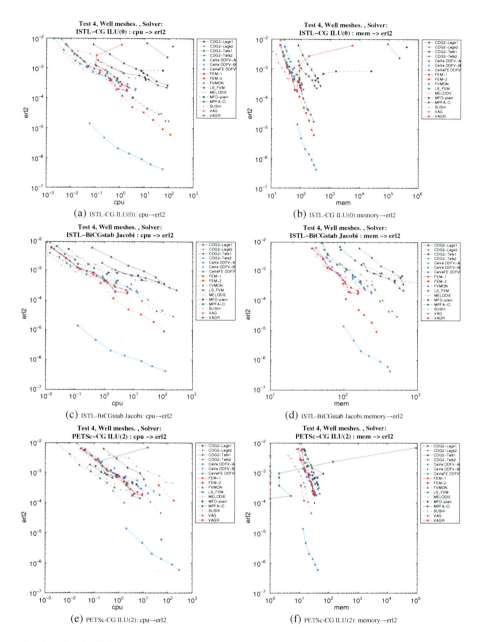

Fig. 17 Test 4- Well meshes

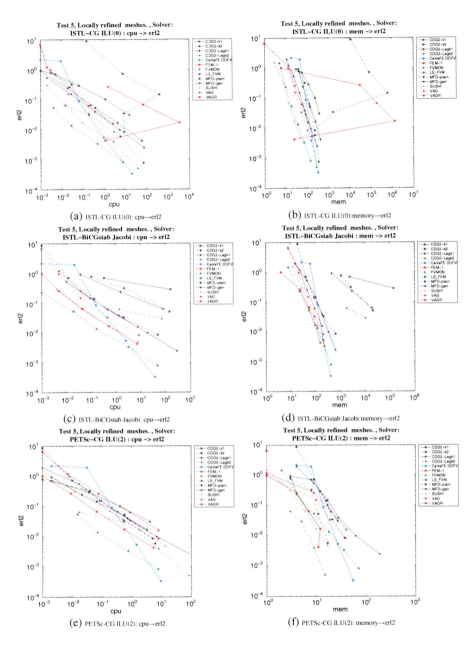

Fig. 18 Test 5-Locally refined grid

Acknowledgements The present paper is a synthesis of the work of several people:

- we first warmly thank all participants for their active collaboration in this benchmark, and for letting us use their results in this synthesis,
- we are also very grateful to P. Bastian, C. Guichard, K. Lipnikov, R. Masson and S. Minjeaud for their help in the benchmark design.

This work was supported by GDR MOMAS, CNRS/PACEN and the ANR VFSitCom project. The work of the last author was also partially supported by the Italian MIUR through the program PRIN2008.

References

1. I. Aavatsmark and R. Klausen. Well index in reservoir simulation for slanted and slightly curved wells in 3D grids. *SPE Journal*, 8:41–48, 2003.
2. H. Amor, M. Bourgeois, and G. Mathieu. Benchmark 3D: a linear finite element solver. In *these proceedings*, 2011.
3. B. Andreianov, F. Hubert, and S. Krell. Benchmark 3D: a version of the DDFV scheme with cell/vertex unknowns on general meshes. In *these proceedings*, 2011.
4. S. Balay, J. Brown, K. Buschelman, W. D. Gropp, D. Kaushik, M. G. Knepley, L. C. McInnes, B. F. Smith, and H. Zhang. PETSc Web page, 2011. http://www.mcs.anl.gov/petsc.
5. S. Balay, W. D. Gropp, L. C. McInnes, and B. F. Smith. Efficient management of parallelism in object oriented numerical software libraries. In E. Arge, A. M. Bruaset, and H. P. Langtangen, editors, *Modern Software Tools in Scientific Computing*, pages 163–202. Birkhäuser Press, 1997.
6. P. Bastian. Benchmark 3D: Symmetric weighted interior penalty discontinuous Galerkin scheme. In *these proceedings*, 2011.
7. P. Bastian, M. Blatt, A. Dedner, C. Engwer, J. Fahlke, C. Gräser, R. Klöfkorn, M. Nolte, M. Ohlberger, and O. Sander. DUNE Web page, 2011. http://www.dune-project.org.
8. P. Bastian, M. Blatt, A. Dedner, M. Engwer, R. Klöfkorn, M. Ohlberger, and O. Sander. A generic grid interface for parallel and adaptive scientific computing. Part I: abstract framework. *Computing*, 82(2-3):103–119, 2008.
9. P. Bastian, M. Blatt, A. Dedner, M. Engwer, R. Klöfkorn, M. Ohlberger, and O. Sander. A generic grid interface for parallel and adaptive scientific computing. Part II: implementation and tests in DUNE. *Computing*, 82(2-3):121–138, 2008.
10. P. Bastian, O. Ippisch, and S. Marnach. Benchmark 3D: A mimetic finite difference method. In *these proceedings*, 2011.
11. I. Ben Gharbia, J. Jaffré, S. N. Kumar, and J. E. Roberts. Benchmark 3D: a composite hexahedral mixed finite element. In *these proceedings*, 2011.
12. M. Blatt and P. Bastian. The iterative solver template library. In B. Kågström, E. Elmroth, J. Dongarra, and J. Wasniewski, editors, *Applied Parallel Computing. State of the Art in Scientific Computing*, volume 4699 of *Lecture Notes in Computer Science*, pages 666–675. Springer Berlin / Heidelberg, 2007.
13. Y. Coudière, F. Hubert, and G. Manzini. Benchmark 3D: CeVeFE-DDFV, a discrete duality scheme with cell/vertex/face+edge unknowns. In *these proceedings*, 2011.
14. Y. Coudière and G. Manzini. The cell-centered finite volume method using least squares vertex reconstruction (diamond scheme). In *these proceedings*, 2011.
15. Y. Coudière and C. Pierre. Benchmark 3D: CeVe-DDFV, a discrete duality scheme with cell/vertex unknowns. In *these proceedings*, 2011.
16. A. Danilov and Y. Vassilevski. Benchmark 3D: A monotone nonlinear finite volume method for diffusion equations on polyhedral meshes. In *these proceedings*, 2011.

17. T. A. Davis. Algorithm 832: UMFPACK V4.3 – an unsymmetric-pattern multifrontal method. *ACM Trans. Math. Softw.*, 30(2):196–199, 2004.
18. T. A. Davis. UMFPACK Web page, 2011. http://www.cise.ufl.edu/research/sparse/umfpack/.
19. R. Eymard, T. Gallouët, and R. Herbin. Benchmark 3D: the SUSHI scheme. In *these proceedings*, 2011.
20. R. Eymard, C. Guichard, and R. Herbin. Benchmark 3D: the VAG scheme. In *these proceedings*, 2011.
21. R. Herbin and F. Hubert. Benchmark on discretization schemes for anisotropic diffusion problems on general grids. In *Finite volumes for complex applications V*, pages 659–692. ISTE, London, 2008.
22. V. Hernandez, J. E. Roman, and V. Vidal. SLEPc: A scalable and flexible toolkit for the solution of eigenvalue problems. *ACM Transactions on Mathematical Software*, 31(3):351–362, Sept. 2005.
23. R. Klöfkorn. Benchmark 3D: The compact discontinuous Galerkin 2 scheme. In *these proceedings*, 2011.
24. K. Lipnikov. and G. Manzini. Benchmark 3D: Mimetic finite difference method for generalized polyhedral meshes. In *these proceedings*, 2011.
25. M. F. Wheeler, G. Xue, and I. Yotov. Benchmark 3D: A multipoint flux mixed finite element method on general hexahedra. In *these proceedings*, 2011.

The paper is in final form and no similar paper has been or is being submitted elsewhere.

Benchmark 3D: a linear finite element solver

Hanen Amor, Marc Bourgeois, and Gregory Mathieu

1 Introduction

In the present paper[1], we address some of the benchmark problems defined for the Finite Volume for Complex Applications conference (FVCA6 [1]). The tests, which are described in [2], consist in solving the following anisotropic diffusion problem :

$$\begin{cases} -\nabla.(K\nabla u) = f & \text{on} \quad \Omega \\ u = \bar{u} & \text{on} \quad \Gamma_d \end{cases} \quad (1)$$

where u is the unknown, Ω is in most cases a unit cube, $K : \Omega \to \mathbb{R}^{3\times 3}$ is the diffusion tensor, f the source term and $\bar{u}$ the Dirichlet boundary conditions.
For this benchmark, the computations were performed with MELODIE (Modèle d'Evaluation à LOng terme des Déchets Irradiants Enterrés) software, which is devoted to simulate the migration of a plume of radionuclides in a 3-dimensional geological media.

2 Presentation of MELODIE

The MELODIE [3] software, is developped by IRSN, and constantly upgraded, to assess the long-term containment capabilities of radioactive waste repositories. This software is designed to model a disposal site taking into account all the main

---

[1]The model of this paper is provided by the benchmark organization. The results of this benchmark will be detailed and discussed by Florence HUBERT and Raphaële HERBIN in a paper gathering all the contributions.

Hanen Amor, Marc Bourgeois, and Gregory Mathieu
Institut de Radioprotection et de Sûreté Nucléaire - DSU/SSIAD/BERIS - BP 17 - 92262
Fontenay-aux-Roses Cedex,
email: {hanen.amor,marc.bourgeois,gregory.mathieu}@irsn.fr

physical and chemical characteristics of the disposal components. The model is adapted to large scales of time and space required for simulation.

The MELODIE software models water flow and the phenomena involved in the transport of radionuclides in saturated porous media in 2 dimensions and in 3 dimensions; physical and chemical interactions are represented by a retardation factor integrated in the computational equations. These equations are discretised using a so-called FVFE method -Finite Volume Finite Element-, which is based on a Galerkin method to discretise time and variables, together with a finite volume method using the Godunov scheme for the convection term. The FVFE method is used to convert partial differential equations into a finite number of algebraic equations that match the number of nodes in the mesh used to model the considered site. It also serves to stabilise the numerical scheme. The present benchmark adresses only diffusive problems, which are therefore solved by using a standard P1 finite element method.

3 Numerical results

Numerical results presented in that contribution concern the tests 1, 3 and 4. The error generated by the P1 method has been evaluated by defining the quantity: $e = u - u_h$, where u is the analytical solution and u_h is the numerical solution. Then the error can be computed as follows.

3.1 Discrete L^2 and H^1 norms

The continuous L^2 and H^1 norms of a function u are given by

$$\|u\|_{L^2(\Omega)} = \left(\int_\Omega u^2\right)^{1/2}, \quad \|u\|_{H^1(\Omega)} = \|u\|_{L^2(\Omega)} + \|\nabla u\|_{L^2(\Omega)}$$

where Ω is an open bounded in $\mathbf{R}^3$. In most of the test cases, the domain Ω is a unit cube. To compute those norms, we perform the L^2 and H^1 semi-norm of the function u on a tetrahedron T:

$$\|u\|_{L^2(T)} = \left(\int_T u^2\right)^{1/2} \text{ and } \|\nabla u\|_{H^1(T)} = \left(\int_T \nabla u^2\right)^{1/2}$$

The numerical quadrature used to approximate this integral, are given by the following formula:

- in the case where the values of the function u are known on the vertices

$$\int_T u^2 dx \simeq \frac{1}{4} V_T \sum_{i=1}^{4} u(\overline{s_i})^2$$

Benchmark 3D: a linear finite element solver

- in the case where values of the gradient of the function u are known on the centre of gravity

$$\int_T \nabla u . \nabla u dx \simeq V_T \nabla u(G_T) . \nabla u(G_T)$$

The previous formula are adapted to calculate the relative L^2 norm of the error : erl2, the relative L^2 norm of a gradient of the error : ergrad and the relative L^2 norm of the energy norm : ener.

3.2 Expected results

- **Test 1 Mild anisotropy,** $u(x, y, z) = 1 + \sin(\pi x) \sin\left(\pi \left(y + \frac{1}{2}\right)\right) \sin\left(\pi \left(z + \frac{1}{3}\right)\right)$
min = 0, max = 2, **Tetrahedral meshes**

i	nu	nmat	umin	uemin	umax	uemax	normg
1	488	6072	7.69E-02	8.29E-02	1.935	1.935	1.791
2	857	11269	2.76E-02	2.83E-02	1.955	1.955	1.796
3	1601	21675	3.07E-02	3.07E-02	1.970	1.969	1.798
4	2997	41839	1.81E-02	1.77E-02	1.984	1.983	1.797
5	5692	81688	1.32E-02	1.37E-02	1.990	1.990	1.798
6	10994	160852	6.19E-03	6.49e-03	1.991	1.991	1.798

i	nu	erl2	ratiol2	ergrad	ratiograd	ener	ratioener
1	488	1.35E-02	-	2.32E-01	-	2.29E-01	
2	857	7.01E-03	3.531	1.17E-01	1.370	1.17E-01	1.362
3	1601	4.56E-03	2.052	1.14E-01	1.052	1.14E-01	1.082
4	2997	3.01E-03	1.998	1.13E-01	1.155	1.11E-01	1.170
5	5692	1.87E-03	2.219	9.03E-02	1.067	8.90E-02	1.035
6	10994	1.22E-03	1.941	7.05E-02	1.128	6.92E-02	1.148

- **Test 1 Mild anisotropy,** $u(x, y, z) = 1 + \sin(\pi x) \sin\left(\pi \left(y + \frac{1}{2}\right)\right) \sin\left(\pi \left(z + \frac{1}{3}\right)\right)$
min = 0, max = 2, **Kershaw meshes**

i	nu	nmat	umin	uemin	umax	uemax	normg
1	729	9097	1.34E-01	8.76E-02	1.833	1.883	1.834
2	4913	66961	3.12E-02	1.92E-02	1.955	1.970	1.797
3	35937	513313	8.55E-03	6.64E-03	1.988	1.992	1.783
4	274625	4018753	2.04E-03	1.92E-03	1.997	1.998	1.787

i	nu	erl2	ratiol2	ergrad	ratiograd	ener	ratioener
1	729	9.41E-02	-	8.88E-01	-	9.05E-01	-
2	4913	5.88E-02	0.737	5.70E-01	0.696	5.710E-01	0.723
3	35937	3.35E-02	0.849	3.49E-01	0.741	3.47E-01	0.750
4	274625	1.52E-02	1.158	1.99E-01	0.823	2.02E-01	0.793

- **Test 3 Flow on random meshes,** $u(x, y, z) = \sin(2\pi x) \sin(2\pi y) \sin(2\pi z)$, min $= -1$, max $= 1$, **Random meshes**

i	nu	nmat	umin	uemin	umax	uemax	normg
1	125	1333	-0.665	-0.338	0.685	0.363	6.004
2	729	9097	-0.885	-0.784	0.812	0.751	3.867
3	4913	66961	-0.970	-0.943	0.949	0.925	3.666
4	35937	513313	-0.988	-0.982	0.991	0.984	3.613

i	nu	erl2	ratiol2	ergrad	ratiograd	ener	ratioener
1	125	8.34E-01	-	9.77E-01	-	9.44E-01	-
2	729	1.97E-01	2.456	4.84E-01	1.193	4.17E-01	1.390
3	4913	5.16E-02	2.107	2.48E-01	1.051	2.10E-01	1.078
4	35937	1.33E-02	2.040	1.22E-01	1.067	1.03E-01	1.066

- **Test 4 Flow around a well, Well meshes,** min $= 0$, max $= 5.415$

i	nu	nmat	umin	uemin	umax	uemax	normg
1	1248	15886	0.189	0.189	5.360	5.360	1653.52
2	2800	37836	0.120	0.119	5.368	5.368	1634.57
3	5889	81531	0.078	0.076	5.345	5.345	1631.27
4	12582	178018	0.060	0.058	5.349	5.349	1628.68
5	25300	363768	0.046	0.045	5.377	5.377	1626.49
6	45668	662730	0.037	0.036	5.380	5.380	1625.64
7	79084	1154172	0.029	0.028	5.39	5.39	1624.98

i	nu	erl2	ratiol2	ergrad	ratiograd	ener	ratioener
1	1248	3.30E-03	-	1.52E-01	-	1.48E-01	-
2	2800	1.49E-03	2.941	9.83E-02	1.621	9.40E-02	1.703
3	5889	8.99E-04	2.058	6.93E-02	1.409	6.52E-02	1.472
4	12582	6.11E-04	1.522	5.36E-02	1.014	4.93E-02	1.104
5	25300	4.09E-04	1.722	4.26E-02	0.983	3.94E-02	0.960
6	45668	2.69E-04	2.130	3.50E-02	0.997	3.26E-02	0.954
7	79084	2.58E-04	0.224	3.05E-02	0.753	2.84E-02	0.751

4 Comments

The computations for the post-processing purpose of the relative L^2 error norm and the relative H^1 error semi-norm have been done using the continuous solution and a numerical quadrature rule presented in the section 3.1. As can be seen, for the test 1 using the Tetrahedral meshes, for the test 3 using the Random meshes and for the test 4 using the Well meshes, the theoretical results are recovered, since a convergence of order 2 for the L^2-norm and a convergence of order 1 for the H^1-norm are obtained. For the test 1 using the Kershaw meshes, the theoretical results are not recovered. In fact, a convergence of order 1 for the L^2-norm and a convergence of order 1/2 for the H^1-norm are obtained. Those decreases in the rates of convergence orders are due to the characteristics of the Kershaw meshes that present strong anisotropy.

In addition, FVFE method implemented in MELODIE is only available for tetrahedrons and conform meshes. In the benchmark, hexahedral meshes are divided in tetrahedrons without changing the number of vertices. For the tests 2 and 5, that kind of adaptation is not possible due to the specific shape of the meshes. It is the reason why those cases were not considered.

In this benchmark, the system obtained after assembling of the discretized equations on each element is linear. Within MELODIE, this linear system is solved by using a bi-conjugate gradient method with an incomplete Gauss-type preconditioning. That method is specifically suitable for resolution of non-symmetrical system. Thereby, among the solvers proposed in the benchmark, our choice was the Petsc bi-conjugate gradient (using various preconditioning) complying with the implemented method in MELODIE.

References

1. website : http://fvca6.fs.cvut.cz/
2. website : http://www.latp.univ-mrs.fr/latp_numerique/
3. website : http://www.irsn.fr/FR/Larecherche/outils-scientifiques/Codes-de-calcul/Pages/Le-logiciel-MELODIE-3133.aspxl

The paper is in final form and no similar paper has been or is being submitted elsewhere.

Benchmark 3D: a version of the DDFV scheme with cell/vertex unknowns on general meshes

Boris Andreianov, Florence Hubert, and Stella Krell

1 DDFV methods in 2D and in 3D. A 3D CeVe-DDFV scheme

This paper gives numerical results for a 3D extension of the 2D DDFV scheme. Our scheme is of the same inspiration as the one called CeVe-DDFV ([9]), with a more straightforward dual mesh construction. We sketch the construction in which, starting from a given 3D mesh (which can be non conformal and have arbitrary polygonal faces), one defines a dual mesh and a diamond mesh, reconstructs a discrete gradient, and proves the discrete duality property. Details can be found in [1].

DDFV ("Discrete Duality Finite Volume") scheme was introduced in 2D by Hermeline in [15] and by Domelevo and Omnès in [13]. To handle anisotropic problems or nonlinear problems, or in order to work on general distorted meshes, full gradient reconstruction from point values is a popular strategy. It is well known that reconstruction of a discrete gradient is facilitated by adding unknowns that are new with respect to those of standard cell-centered finite volume schemes. The 2D DDFV method consists in adding new unknowns at the vertices of the initial mesh (this initial mesh is often called the primal one), and in use of new control volumes (called dual cells, or co-volumes) around these points. A family of diamond cells is naturally associated to this construction, each diamond being built on two neighbor cell centers x_K, x_L and the two vertices of the edge $\kappa|L$ that separates them. On a diamond, one can construct a discrete gradient direction per direction (cell-cell and vertex-vertex), following the idea of [8]. It turns out that this discrete gradient

---

Boris Andreianov
CNRS UMR 6623, Besançon, France, e-mail: boris.andreianov@univ-fcomte.fr

Florence Hubert
LATP, Université de Provence, Marseille, France, e-mail: fhubert@cmi.univ-mrs.fr

Stella Krell
INRIA, Lille, France, e-mail: stella.krell@inria.fr

is related by a discrete analogue of integraton-by-parts formula, called "discrete duality", to the classical discrete finite volume divergence associated with these two families of meshes. This duality property greatly simplifies the theoretical analysis of finite volume schemes based on the DDFV construction, see e.g. [2,5]. This 2D strategy reveals to be particularly efficient in terms of gradient approximation (see [7, 14]) and has been extended to a wide class of PDE problems (see [1, 5, 6, 18, 19] and references therein).

The 3D CeVe-DDFV scheme we present here also keeps unknowns only at the cell centers and the vertices of the primal mesh, and it uses the primal mesh, a dual mesh and a diamond mesh; as in the 2D case, a diamond is constructed from two neighbor cell centers x_K, x_L and from l vertices of the edge $K|L$ that separates them ($l \geq 3$). The price to pay is that the gradient reconstruction becomes more intricate. As in 2D, one direction per diamond is reconstructed using the two cell center unknowns at the nodes x_K, x_L; two complementary directions of the gradient in $K|L$ are reconstructed simultaneously, by a suitable interpolation of the vertex values in each face $K|L$ of the primal mesh. While the case $l = 3$ (meshes with triangular faces) offers no choice, in general we have to fix a formula for interpolation that is consistent with affine functions and which leads to discrete duality (with respect to appropriately defined dual cells). This was achieved independently in [17] and in [1, 3, 4], with two different approaches (the above description stems from the point of view developed in [1, 3, 4]).

Several 3D DDFV constructions exist. The CeVe-DDFV scheme by Pierre et al. (see [12]) was the pioneering work in 3D; a particular feature of this method was in the double covering of the domain by the dual mesh. This approach led to a method that is only slightly different from ours; we refer to the benchmark paper [9] in the same collection. Next, Hermeline in [16] introduced the important idea to associate additional unknowns with the face centers of the primal mesh. In the subsequent work [17] of Hermeline, elimination of these unknowns eventually led to the same method that the one we describe. Many numerical tests are given in [16, 17]. Finally, Coudière and Hubert in [10] introduced edge unknowns, instead of eliminating face unknowns. This idea assessed a new strategy of 3D DDFV approximation; we call it CeVeFE-DDFV because with respect to the primal mesh, cell, vertex and face+edge unknowns are used. Let us point out the differences with respect to CeVe-DDFV strategies. In [10], each diamond is constructed on two cell centers x_K, x_L, on two vertices x_{K^*}, x_{L^*} in the face $K|L$, and one face center $x_{K|L} \in K|L$ and one edge center $x_{K^*|L^*} \in [x_{K^*}, x_{L^*}]$. Then the gradient is reconstructed per direction (cell-cell, vertex-vertex and face-edge), as in 2D. The edge and face centers are the centers for a new, third mesh. The CeVeFE-DDFV method is the object of the benchmark paper [11] in the same collection.

Let us present the construction of our 3D CeVe-DDFV scheme. The primal mesh needs not be conformal; there is no restriction on number of faces or face edges. For simplicity, let us assume that the primal mesh volumes are convex; that their centers belong to the volumes; and the face centers belong to the faces. These restrictions can be relaxed, see [1]; but let us stress that the edge points must be the middlepoints.

Benchmark 3D: a version of the DDFV scheme

Notation. We use a triple $\mathfrak{T} = \left(\overline{\mathfrak{M}^o}, \overline{\mathfrak{M}^*}, \mathfrak{D}\right)$ of partitions of Ω into polyhedra.

- $\mathfrak{M}^o$ denotes the initial mesh[1], called *primal mesh*. We call $\partial\mathfrak{M}^o$ the set of all faces of this mesh that are included in $\partial\Omega$. These faces are considered as flat *boundary (primal) control volumes*. We denote by $\overline{\mathfrak{M}^o}$ the union $\mathfrak{M}^o \cup \partial\mathfrak{M}^o$.
  - **Center:** To any (primal) control volume $K \in \overline{\mathfrak{M}^o}$, we associate a point $x_K \in K$.
  - **Vertex:** A generic vertex of $\overline{\mathfrak{M}^o}$ is denoted by x_{K^*}.
  - **Neighbors:** given $K \in \mathfrak{M}^o$, all control volumes $L \in \overline{\mathfrak{M}^o}$ such that K and L have a common face (or part of a face) form the set $\mathcal{N}(K)$ of neighbors of K.
  - **Face:** for all $L \in \mathcal{N}(K)$, by $K|L$ we denote $\partial K \cap \partial L$ which is a face (or a part of a face) of the mesh $\mathfrak{M}^o$; it is supplied with a *face center* $x_{K|L} \in K|L$.
  - **Edge:** An egde $[x_{K^*}, x_{L^*}]$ of $\overline{\mathfrak{M}^o}$ is defined by two neighbor vertices x_{K^*}, x_{L^*}; it is marked with the center $x_{K^*|L^*}$ that must be its middlepoint $(x_{K^*} + x_{L^*})/2$.
  - **Element:** An element $T = T_{K^*;L^*}^{K;L}$ is the tetrahedron $(x_K, x_{K|L}, x_{K^*|L^*}, x_{K^*})$: here K is a primal volume ; $K|L$ is a face of K ; and $[x_{K^*}, x_{L^*}]$ is an edge of $K|L$ (see Fig. 1). The set of all elements is denoted by $\mathscr{T}$. If x_K is a vertice of $T \in \mathscr{T}$, then we say that T is associated[2] with the volume K, and we write $T \sim K$.

- $\mathfrak{M}^*$ denotes the *dual mesh* constructed as follows. A generic vertex x_{K^*} of $\overline{\mathfrak{M}^o}$ is associated with the polyhedron $K^* \in \overline{\mathfrak{M}^*}$ made of all elements $T \in \mathscr{T}$ that share the vertex x_{K^*} (we write $T \sim K^*$). If $x_{K^*} \in \Omega$, we say that K^* is a *dual control volume* and write $K^* \in \mathfrak{M}^*$; and if $x_{K^*} \in \partial\Omega$, we say that K^* is a *boundary dual control volume* and write $K^* \in \partial\mathfrak{M}^*$. Thus $\overline{\mathfrak{M}^*} = \mathfrak{M}^* \cup \partial\mathfrak{M}^*$.

- $\mathfrak{D}$ is the *diamond mesh*. For $K \in \mathfrak{M}^o, L \in \mathcal{N}(K)$, the union of the convex hull of x_K and $K|L$ with the convex hull of x_L and $K|L$ is called *diamond*, denoted by $D^{K|L}$.

For expression of the discrete operators one needs a convention on diamond orientation, subdiamonds and other objects and notation of [1]; we give them via Fig. 1.

Discrete space and discrete operators; the discrete duality feature.

- A *discrete function on* Ω is a set $w^{\mathfrak{T}} = \left(w^{\mathfrak{M}^o}, w^{\mathfrak{M}^*}\right)$ consisting of two sets of real values $w^{\mathfrak{M}^o} = (w_K)_{K \in \mathfrak{M}^o}$ and $w^{\mathfrak{M}^*} = (w_{K^*})_{K^* \in \mathfrak{M}^*}$.

- A *discrete function on* $\overline{\Omega}$ is a set $w^{\overline{\mathfrak{T}}} = \left(w^{\mathfrak{M}^o}, w^{\mathfrak{M}^*}; w^{\partial\mathfrak{M}^o}, w^{\partial\mathfrak{M}^*}\right) \equiv \left(w^{\mathfrak{T}}; w^{\partial\mathfrak{T}}\right)$, $w^{\mathfrak{M}^o} = (w_K)_{K \in \mathfrak{M}^o}$, $w^{\mathfrak{M}^*} = (w_{K^*})_{K^* \in \mathfrak{M}^*}$, $w^{\partial\mathfrak{M}^o} = (w_K)_{K \in \partial\mathfrak{M}^o}$, $w^{\partial\mathfrak{M}^*} = (w_{K^*})_{K^* \in \partial\mathfrak{M}^*}$.

- A *discrete field on* Ω is a set $\overrightarrow{\mathscr{F}^{\mathfrak{T}}} = \left(\overrightarrow{\mathscr{F}_D}\right)_{D \in \mathfrak{D}}$ of vectors of $\mathbb{R}^3$.

- We write $\mathbb{R}^{\mathfrak{T}}, \mathbb{R}^{\overline{\mathfrak{T}}}, (\mathbb{R}^3)^{\mathfrak{D}}$, respectively, for the sets of discrete functions/fields.

---

[1]This means, $\mathfrak{M}^o$ is one of the meshes provided by the benchmark organizers.

[2]Because we have made the assumption that $x_{K|L} \in K|L$, the relation $T \sim K$ simply means that T is included in K. The same observation applies to the notation $T \sim K^*$. See [1] for generalizations.

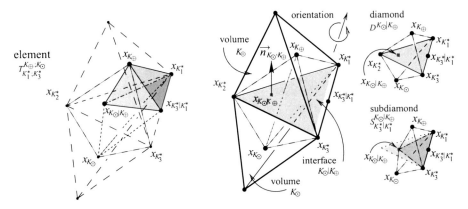

Fig. 1 Element (left). Oriented diamond, subdiamond and related notation, cf. [1] (right)

- *Discrete divergence* is the operator acting from $(\mathbb{R}^3)^{\mathfrak{D}}$ to $\mathbb{R}^{\mathfrak{T}}$, given by

$$\mathrm{div}^{\mathfrak{T}}: \overrightarrow{\mathscr{F}^{\mathfrak{T}}} \mapsto \left(\left(\mathrm{div}_K \overrightarrow{\mathscr{F}^{\mathfrak{T}}}\right)_{K \in \mathfrak{M}^0}, \left(\mathrm{div}_{K^*} \overrightarrow{\mathscr{F}^{\mathfrak{T}}}\right)_{K^* \in \mathfrak{M}^*} \right) =: \mathrm{div}^{\mathfrak{T}} \overrightarrow{\mathscr{F}^{\mathfrak{T}}}, \quad (1)$$

where the entries $\mathrm{div}_K \overrightarrow{\mathscr{F}^{\mathfrak{T}}}$, $\mathrm{div}_{K^*} \overrightarrow{\mathscr{F}^{\mathfrak{T}}}$ of the discrete function $\mathrm{div}^{\mathfrak{T}} \overrightarrow{\mathscr{F}^{\mathfrak{T}}}$ on Ω are

$$\mathrm{div}_K \overrightarrow{\mathscr{F}^{\mathfrak{T}}} = \frac{1}{\mathrm{Vol}(\kappa)} \sum_{T \sim K} m_T \overrightarrow{\mathscr{F}}_T \cdot \overrightarrow{n}_T, \quad \mathrm{div}_{K^*} \overrightarrow{\mathscr{F}^{\mathfrak{T}}} = \frac{1}{\mathrm{Vol}(\kappa^*)} \sum_{T \sim K^*} m_T^* \overrightarrow{\mathscr{F}}_T \cdot \overrightarrow{n}_T^*, \quad (2)$$

$\overrightarrow{n}_T, \overrightarrow{n}_T^*$ being the exterior normal vectors to $\partial \kappa, \partial \kappa^*$. Formulae (2) stem from the standard procedure of finite volume discretization, applied on $\mathfrak{M}^0$ and on $\mathfrak{M}^*$.

- *Discrete gradient* is the operator acting from $\mathbb{R}^{\mathfrak{T}}$ to $(\mathbb{R}^3)^{\mathfrak{D}}$, given by

$$\overrightarrow{\nabla}^{\mathfrak{T}}: w^{\mathfrak{T}} \mapsto \left(\overrightarrow{\nabla}_D w^{\mathfrak{T}} \right)_{D \in \mathfrak{D}} =: \overrightarrow{\nabla}^{\mathfrak{T}} w^{\mathfrak{T}} \quad (3)$$

where the entry $\overrightarrow{\nabla}_D w^{\mathfrak{T}}$ of the discrete field $\overrightarrow{\nabla}^{\mathfrak{T}} w^{\mathfrak{T}}$ corresponding to $D = D^{K_\ominus | K_\oplus}$ (see Fig. 1) is reconstructed from the values $w_{K_\ominus}, w_{K_\oplus}$ at the neighbor centers $x_{K_\ominus}, x_{K_\oplus}$ (they give the projection on $\overline{x_{K_\ominus} x_{K_\oplus}}$) and the values $(w_{K^*_i})_{i=1}^l$ at the l vertices of the interface $\kappa_\ominus | \kappa_\oplus$ (they give the projection on the plane $\kappa_\ominus | \kappa_\oplus$)[3] with

---

[3] When $l = 3$, one simply uses the three-point interpolation in the plane $\kappa_\ominus | \kappa_\oplus$ to reconstruct this projection. Clearly, the interpolation is exact for affine functions. In general, the reconstruction (3), which is exact for affine functions, is based upon the 2D identity given in [3] and [1, Appendix].

Benchmark 3D: a version of the DDFV scheme 941

$$\overrightarrow{\nabla}_D w^{\mathfrak{T}} = \frac{1}{6\operatorname{Vol}(D)} \sum_{i=1}^{l} \left\{ \frac{\langle \overrightarrow{x_{K_\odot} x_{K_\oplus}}, \overrightarrow{x_{K_\odot | K_\oplus} x_{K_i^* | K_{i+1}^*}}, \overrightarrow{x_{K_i^*} x_{K_{i+1}^*}} \rangle}{\overrightarrow{x_{K_\odot} x_{K_\oplus}} \cdot \overrightarrow{n}_{K_\odot, K_\oplus}} (w_{K_\oplus} - w_{K_\odot}) \overrightarrow{n}_{K_\odot, K_\oplus} \right.$$

$$\left. + 2(w_{K_{i+1}^*} - w_{K_i^*}) \left[\overrightarrow{x_{K_\odot} x_{K_\oplus}} \times \overrightarrow{x_{K_\odot | K_\oplus} x_{K_i^* | K_{i+1}^*}} \right] \right\}. \quad (4)$$

- Pick $[\![w^{\mathfrak{T}}, v^{\mathfrak{T}}]\!] := \frac{1}{3} \sum_{K \in \mathfrak{M}^o} \operatorname{Vol}(K) \, w_K v_K + \frac{2}{3} \sum_{K^* \in \mathfrak{M}^*} \operatorname{Vol}(K^*) \, w_{K^*} v_{K^*}$ for scalar product on $\mathbb{R}^{\mathfrak{T}}$ and $\{\!\{ \overrightarrow{\mathcal{F}^{\mathfrak{T}}}, \overrightarrow{\mathcal{G}^{\mathfrak{T}}} \}\!\} := \sum_{D \in \mathfrak{D}} \operatorname{Vol}(D) \overrightarrow{\mathcal{F}_D} \cdot \overrightarrow{\mathcal{G}_D}$ for scalar product on $(\mathbb{R}^3)^{\mathfrak{D}}$.

And now, one can mimic the identity $-\int_\Omega (\operatorname{div} \overrightarrow{\mathcal{F}}) w = \int_\Omega \overrightarrow{\mathcal{F}} \cdot \overrightarrow{\nabla} w$ for $w|_{\partial \Omega} = 0$:

Proposition 1 (the *discrete duality* property; see [1,3], see also [17]). *For all $\overrightarrow{\mathcal{F}^{\mathfrak{T}}} \in (\mathbb{R}^3)^{\mathfrak{D}}$ and all $w^{\mathfrak{T}} \in \mathbb{R}^{\mathfrak{T}}$ with $w^{\partial \mathfrak{T}} = 0$, $[\![-\operatorname{div}^{\mathfrak{T}} \overrightarrow{\mathcal{F}^{\mathfrak{T}}}, w^{\mathfrak{T}}]\!] = \{\!\{ \overrightarrow{\mathcal{F}^{\mathfrak{T}}}, \overrightarrow{\nabla}^{\mathfrak{T}} w^{\mathfrak{T}} \}\!\}$.*

The scheme. In this benchmark, one approximates the linear diffusion problem $-\operatorname{div}[\mathbf{A}(\cdot) \overrightarrow{\nabla} u] = f(\cdot)$ with Dirichlet boundary condition $u|_{\partial \Omega} = \bar{u}(\cdot)$, $\mathbf{A}(\cdot)$ being a heterogeneous anisotropic diffusion tensor and $f(\cdot)$ being a source term. Let $\mathbb{P}^{\mathfrak{T}}$ denote the projection on the DDFV mesh $\mathfrak{T}$ (i.e. the components of $\mathbb{P}^{\mathfrak{T}} f$ are the mean values of $f \in L^1(\Omega)$ per primal and per dual volumes); $\mathbb{P}^{\partial \mathfrak{T}}$ is the projection on the boundary part of the mesh. Let $\overrightarrow{\mathbb{P}^{\mathfrak{T}}}$ denote the projection on the diamond mesh $\mathfrak{D}$. For general data, the heterogeneity of the matrix $\mathbf{A}(\cdot)$ is taken into account by using the diamond-wise projection $\mathbf{A}^{\mathfrak{T}} := \overrightarrow{\mathbb{P}^{\mathfrak{T}}} \mathbf{A}(\cdot)$; similarly, we use $f^{\mathfrak{T}} = \mathbb{P}^{\mathfrak{T}} f(\cdot)$ as the discrete source term. The boundary condition is given by the projection $\mathbb{P}^{\partial \mathfrak{T}} \bar{u}(\cdot)$.

For a fully practical discretization of $\mathbf{A}(\cdot)$ and $f(\cdot)$ (which are continuous in all the tests we perform), for every element (recall that diamonds, primal volumes and dual volumes of a DDFV mesh are unions of elements, see Fig. 1) we take the mean value of the four vertices of the element. The point values of the exact solution u_e in the centers of the boundary volumes are used as discrete boundary conditions.

Given a DDFV mesh $\mathfrak{T}$ of Ω the method writes as:

$$\text{Find } u^{\mathfrak{T}} \text{ s.t. } -\operatorname{div}^{\mathfrak{T}} \left[\mathbf{A}^{\mathfrak{T}} \overrightarrow{\nabla}^{\mathfrak{T}} u^{\mathfrak{T}} \right] = f^{\mathfrak{T}} \text{ with } u^{\mathfrak{T}} = (u^{\mathfrak{T}}; \mathbb{P}^{\partial \mathfrak{T}} \bar{u}).$$

Convergence. From the discrete duality (Prop. 1) which is a cornerstone of DDFV schemes, and from consistency properties of the projection, gradient and divergence operators (see [2]; cf. [5] for analogous properties in 2D) one easily derives that the scheme is well posed for $l \leq 4$.[4] Given a family $(\mathfrak{T}_h)_h$ of CeVe-DDFV meshes, the associated discrete solutions $u^{\overline{\mathfrak{T}}_h}$ enjoy a uniform discrete H^1 estimate, and they converge to the exact solution u as the size h of the mesh tends to zero. Convergence analysis requires mild proportionality assumptions on the meshes $\mathfrak{T}_h$ in use, see [2].

2 Numerical results

In this section, we describe the results obtained on Tests 1–4 of the benchmark. Notice that, while the method converges for merely L^∞ uniformly elliptic tensor $\mathbf{A}(\cdot)$, it is not designed for a smart handling of a *piecewise* continuous $\mathbf{A}(\cdot)$.[5] Therefore, we skip Test 5 that involves piecewise constant $\mathbf{A}(\cdot)$. We refer to Coudière, Pierre, Rousseau and Turpault [12] and to Hermeline [17] for 3D CeVe-DDFV constructions efficiently taking into account discontinuities of $\mathbf{A}(\cdot)$.

Choice of the cell and face points. We pick for x_K, the isobarycenter of the cell K, and for $x_{K|L}$, the isobarycenter of the face $K|L$.

Measure of errors and convergence orders. To put the discrete and the exact solutions "at the same level", we use the projection $\mathbb{P}^\mathfrak{T} u_e$ of the exact solution and the associated discrete gradient reconstruction $\overrightarrow{\nabla}^\mathfrak{T} \mathbb{P}^{\overline{\mathfrak{T}}} u_e$, where $\mathbb{P}^{\overline{\mathfrak{T}}} \cdot = (\mathbb{P}^\mathfrak{T} \cdot ; \mathbb{P}^{\partial \mathfrak{T}} \cdot)$. The L^2 norms of the errors $e^\mathfrak{T} := u^\mathfrak{T} - \mathbb{P}^\mathfrak{T} u_e$ and $\overrightarrow{\nabla}^\mathfrak{T} e^{\overline{\mathfrak{T}}} := \overrightarrow{\nabla}^\mathfrak{T} u^{\overline{\mathfrak{T}}} - \overrightarrow{\nabla}^\mathfrak{T} \mathbb{P}^{\overline{\mathfrak{T}}} u_e$ are measured in terms of the scalar products $[\![\cdot,\cdot]\!]$ on $\mathbb{R}^\mathfrak{T}$, $\{\!\{\mathbf{A}^\mathfrak{T}\cdot,\cdot\}\!\}$ and $\{\!\{\cdot,\cdot\}\!\}$ on $(\mathbb{R}^3)^{\mathfrak{D}}$: the relative error indicators $erl2$ and $ener$, $ergrad$ we use are defined, respectively, as

$$\left(\frac{[\![e^\mathfrak{T}, e^\mathfrak{T}]\!]}{[\![\mathbb{P}^\mathfrak{T} u_e, \mathbb{P}^\mathfrak{T} u_e]\!]}\right)^{1/2} \text{ and as } \left(\frac{\{\!\{\mathbf{A}^\mathfrak{T} \overrightarrow{\nabla}^\mathfrak{T} e^{\overline{\mathfrak{T}}}, \overrightarrow{\nabla}^\mathfrak{T} e^{\overline{\mathfrak{T}}}\}\!\}}{\{\!\{\mathbf{A}^\mathfrak{T} \overrightarrow{\nabla}^\mathfrak{T} \mathbb{P}^{\overline{\mathfrak{T}}} u_e, \overrightarrow{\nabla}^\mathfrak{T} \mathbb{P}^{\overline{\mathfrak{T}}} u_e\}\!\}}\right)^{1/2}, \left(\frac{\{\!\{\overrightarrow{\nabla}^\mathfrak{T} e^{\overline{\mathfrak{T}}}, \overrightarrow{\nabla}^\mathfrak{T} e^{\overline{\mathfrak{T}}}\}\!\}}{\{\!\{\overrightarrow{\nabla}^\mathfrak{T} \mathbb{P}^{\overline{\mathfrak{T}}} u_e, \overrightarrow{\nabla}^\mathfrak{T} \mathbb{P}^{\overline{\mathfrak{T}}} u_e\}\!\}}\right)^{1/2}.$$

---

[4] The restriction on the number l of face vertices is only needed for justifying a discrete Poincaré inequality; yet this property is immaterial, e.g., for the associated evolution problem. In practice, in the below tests values $l = 3, 4, 6$ were used, and no particular problem for $l = 6$ is reported.

[5] In 2D, a scheme called m-DDFV, specifically designed to handle *discontinuous* diffusion tensors, was designed by Boyer and Hubert in [6]. There is a clear difference in convergence orders for the basic DDFV version [5] and the m-DDFV version [6] (see the 2D benchmark paper [7]).

- **Test 1 Mild anisotropy,** $u_e(x, y, z) = 1 + \sin(\pi x) \sin\left(\pi \left(y + \frac{1}{2}\right)\right) \sin\left(\pi \left(z + \frac{1}{3}\right)\right)$
min = 0, max = 2, **Tetrahedral meshes**

i	nu	nmat	umin	uemin	umax	uemax	normg
1	2187	21287	0.706E-02	0.706E-02	1.992	1.992	0.178E+01
2	4301	44813	0.706E-02	0.706E-02	1.997	1.996	0.179E+01
3	8584	94088	0.278E-02	0.278E-02	1.993	1.993	0.179E+01
4	17102	195074	0.792E-03	0.792E-03	1.997	1.997	0.179E+01
5	34343	405077	0.140E-02	0.140E-02	1.999	1.999	0.180E+01
6	69160	838856	0.140E-02	0.140E-02	1.999	1.999	0.180E+01

i	nu	erl2	ratiol2	ergrad	ratiograd	ener	ratioener
1	2187	0.539E-02	-	0.654E-01	-	0.649E-01	-
2	4301	0.331E-02	2.165	0.488E-01	1.297	0.491E-01	1.239
3	8584	0.206E-02	2.069	0.381E-01	1.077	0.383E-01	1.079
4	17102	0.135E-02	1.841	0.301E-01	1.018	0.302E-01	1.026
5	34343	0.846E-03	1.998	0.240E-01	0.973	0.242E-01	0.955
6	69160	0.539E-03	1.934	0.190E-01	1.012	0.191E-01	1.008

- **Test 1 Mild anisotropy,** $u_e(x, y, z) = 1 + \sin(\pi x) \sin\left(\pi \left(y + \frac{1}{2}\right)\right) \sin\left(\pi \left(z + \frac{1}{3}\right)\right)$
min = 0, max = 2, **Voronoi meshes**

i	nu	nmat	umin	uemin	umax	uemax	normg
1	87	1433	0.667E-01	0.667E-01	1.904	1.904	0.159E+01
2	235	4393	0.432E-02	0.432E-02	1.997	1.997	0.172E+01
3	527	10777	0.280E-01	0.280E-01	1.990	1.990	0.176E+01
4	1013	21793	0.108E-02	0.108E-02	2.003	1.995	0.177E+01
5	1776	40998	0.113E-01	0.113E-01	2.000	1.996	0.178E+01

i	nu	erl2	ratiol2	ergrad	ratiograd	ener	ratioener
1	87	0.484E-01	-	0.204E+00	-	0.374E+00	-
2	235	0.388E-01	0.666	0.173E+00	0.496	0.277E+01	-6.049
3	527	0.231E-01	1.925	0.118E+00	1.402	0.838E+00	4.445
4	1013	0.167E-01	1.484	0.940E-01	1.060	0.299E+01	-5.843
5	1776	0.117E-01	1.937	0.818E-01	0.742	0.291E+01	0.147

- **Test 1 Mild anisotropy**, $u_e(x, y, z) = 1+\sin(\pi x) \sin\left(\pi\left(y + \frac{1}{2}\right)\right) \sin\left(\pi\left(z + \frac{1}{3}\right)\right)$
min = 0, max = 2, **Kershaw meshes**

i	nu	nmat	umin	uemin	umax	uemax	normg
1	855	13819	2.28E-02	2.28E-02	1.989	1.989	1.730
2	7471	138691	2.52E-03	2.52E-03	1.994	1.994	1.778
3	62559	1237459	1.99E-03	1.99E-03	1.999	1.999	1.794
4	512191	10443763	3.82E-04	3.82E-04	2.000	2.000	1.797

i	nu	erl2	ratiol2	ergrad	ratiograd	ener	ratioener
1	855	0.501E-01	-	0.484E+00	-	0.558E+00	-
2	7471	0.156E-01	1.611	0.209E+00	1.160	0.159E+00	1.735
3	62559	0.392E-02	1.954	0.677E-01	1.594	0.395E-01	1.970
4	512191	0.101E-02	1.936	0.223E-01	1.585	0.109E-01	1.835

- **Test 1 Mild anisotropy**, $u_e(x, y, z) = 1+\sin(\pi x) \sin\left(\pi\left(y + \frac{1}{2}\right)\right) \sin\left(\pi\left(z + \frac{1}{3}\right)\right)$
min = 0, max = 2, **Checkerboard meshes**

i	nu	nmat	umin	uemin	umax	uemax	normg
1	59	703	0.341E-01	0.341E-01	1.966	1.966	0.167E+01
2	599	9835	0.856E-02	0.856E-02	1.991	1.991	0.178E+01
3	5423	101539	0.214E-02	0.214E-02	1.998	1.998	0.179E+01
4	46175	917395	0.535E-03	0.535E-03	1.999	1.999	0.180E+01
5	381119	7788403	0.134E-03	0.134E-03	2.000	2.000	0.180E+01

i	nu	erl2	ratiol2	ergrad	ratiograd	ener	ratioener
1	59	0.396E-01	-	0.136E+00	-	0.116E+00	-
2	599	0.149E-01	1.266	0.928E-01	0.499	0.818E-01	0.449
3	5423	0.400E-02	1.792	0.497E-01	0.849	0.448E-01	0.820
4	46175	0.103E-02	1.905	0.256E-01	0.931	0.232E-01	0.920
5	381119	0.259E-03	1.954	0.130E-01	0.965	0.118E-01	0.961

- **Test 2 Heterogeneous anisotropy**, min $= -0.862$, max $= 1.0487$
$u_e(x, y, z) = x^3 y^2 z + x \sin(2\pi xz) \sin(2\pi xy) \sin(2\pi z)$, **Prism meshes**

i	nu	nmat	umin	uemin	umax	uemax	normg
1	3010	64158	-.856E+00	-.856E+00	1.044	1.044	0.170E+01
2	24020	555528	-.859E+00	-.859E+00	1.047	1.047	0.171E+01
3	81030	1924098	-.861E+00	-.861E+00	1.049	1.049	0.171E+01
4	192040	4619868	-.862E+00	-.862E+00	1.049	1.049	0.171E+01

i	nu	erl2	ratiol2	ergrad	ratiograd	ener	ratioener
1	3010	0.467E-01	-	0.711E-01	-	0.785E-01	-
2	24020	0.123E-01	1.931	0.224E-01	1.667	0.328E-01	1.262
3	81030	0.554E-02	1.960	0.116E-01	1.634	0.190E-01	1.348
4	192040	0.314E-02	1.973	0.728E-02	1.607	0.127E-01	1.389

- **Test 3 Flow with strong anisotropy on random meshes,** min = 0, max = 1, $u_e(x, y, z) = \sin(\pi x) \sin(\pi y) \sin(\pi z)$, **Random meshes**

i	nu	nmat	umin	uemin	umax	uemax	normg
1	91	1063	-.202E+01	-.978E+00	1.969	0.931	0.392E+01
2	855	13819	-.116E+01	-.994E+00	1.206	0.982	0.363E+01
3	7471	138691	-.105E+01	-.995E+00	1.029	0.991	0.362E+01
4	62559	1237459	-.101E+01	-.998E+00	1.014	0.998	0.360E+01

i	nu	erl2	ratiol2	ergrad	ratiograd	ener	ratioener
1	91	0.713E+00	-	0.716E+00	-	0.439E+00	-
2	855	0.152E+00	2.068	0.199E+00	1.712	0.130E+00	1.633
3	7471	0.384E-01	1.906	0.854E-01	1.174	0.417E-01	1.568
4	62559	0.119E-01	1.656	0.542E-01	0.640	0.183E-01	1.165

- **Test 4 Flow around a well,** min = 0, max = 5.415, **Well meshes**

i	nu	nmat	umin	uemin	umax	uemax	normg
1	1482	23942	-.438E-01	-.438E-01	5.415	5.415	0.162E+04
2	3960	70872	-.239E-01	-.239E-01	5.415	5.415	0.162E+04
3	9229	173951	-.132E-01	-.132E-01	5.415	5.415	0.162E+04
4	21156	412240	-.661E-02	-.661E-02	5.415	5.415	0.162E+04
5	44420	882520	-.411E-02	-.411E-02	5.415	5.415	0.162E+04
6	82335	1654893	-.281E-02	-.281E-02	5.415	5.415	0.162E+04
7	145079	2937937	-.198E-02	-.198E-02	5.415	5.415	0.162E+04

i	nu	erl2	ratiol2	ergrad	ratiograd	ener	ratioener
1	1482	0.564E-02	-	0.473E-01	-	0.817E-01	-
2	3960	0.218E-02	2.897	0.205E-01	2.556	0.487E-01	1.578
3	9229	0.964E-03	2.898	0.108E-01	2.255	0.296E-01	1.770
4	21156	0.645E-03	1.454	0.748E-02	1.344	0.205E-01	1.320
5	44420	0.427E-03	1.664	0.546E-02	1.274	0.144E-01	1.443
6	82335	0.291E-03	1.864	0.396E-02	1.560	0.108E-01	1.391
7	145079	0.205E-03	1.848	0.337E-02	0.858	0.794E-02	1.624

3 Comments

Let us summarize the observations; footnotes provide comments of theoretical order.

Choice of the solvers. The following results have been performed either with the direct solvers given by the UMFPACK library, or with the BiCGStab algorithm with ILU(0) preconditionning delivered in the HSL library. A comparison between ISTL-CG with ILU(0) preconditionning and PETSC-CG with ILU(2) preconditionning shows that, whenever ISTL-CG/ILU(0) algorithm converges, much less CPU time and much less memory is used than for the PETSC-CG/ILU(2) algorithm.

Convergence orders observed[6]. Even if the orders present serious oscillations for some cases (e.g., in Test 3 and in Test 1 on Voronoï meshes), orders slightly below h^2 (superconvergence) for the solution in the L^2 norm are observed quite systematically. One exception is Test 4, where an order intermediate between $h^{3/2}$ and h^2 seems to appear; this may be related to the presence of a singularity in the well center.

Regarding the gradient norm, convergence orders close to h are seen in Test 1 on tetrahedral, Voronoï, checkerboard meshes. On Kershaw meshes in Test 1 and prism meshes of Test 2, more structured though distorted, an $h^{3/2}$ convergence order can be observed. For random meshes of Test 3, orders degrade quickly but the numerical evidence (four meshes only) seems insufficient. The well meshes of Test 4 appear as rather structured but having a singularity; the effect of singularity grows as the mesh becomes finer, and the convergence order falls from h^2 to $h^{3/2}$ and then to h. Yet from Tests 3 and 4 with stronger anisotropy of $\mathbf{A}(\cdot)$, it becomes clear that more adequate norm for measuring gradient convergence is the energy norm. In Test 4 we observe an accurate $h^{3/2}$ convergence and in Test 3, an order $h^{3/2}$ can be conjectured.

Violation (and fulfillment) of the maximum principle[7]. We observe that violation of discrete maximum principle does not occur systematically (or if it occurs, it is of imperceptible magnitude, even on coarse meshes). No overshoot/undershoot is reported on Kershaw, checkerboard and prism meshes for Test 1, nor on the well meshes of Test 4; a very slight overshoot can be seen in Test 1 on tetrahedral meshes. On the contrary, random meshes of Test 3, and also the finest ones among the Voronoï meshes of Test 1, exhibit a perceptible violation of the maximum principle which is nonetheless reduced as the mesh size diminishes[8]. Difficulties on these two

---

[6]For regular enough $\mathbf{A}(\cdot)$ and u_e, order h can be proved for both solution and its gradient in L^2.

[7]In principle, DDFV methods are not designed in order to respect the discrete maximum principle; and the convergence analysis exploits rather the variational structure, well preserved by the method (this is one of the benefits from the discrete duality of Prop. 1). Let us point out that for isotropic problems on primal meshes satisfying the orthogonality condition (e.g., Delaunay tetrahedral meshes with the choice of circumcenters for the cell centers x_K - note that x_K may fall out of K), the discrete maximum principle is easily shown for the CeVe-DDFV scheme under study ([4]).

[8]In theory, one can prove convergence in L^q for $q < 6$; nothing guarantees convergence in L^∞.

kinds of meshes can be explained by their poor shape regularity (e.g., fine Voronoï meshes in Test 1 present a dramatic contrast of size between neighbor cells).

Influence of the mesh type and quality on convergence orders[9]. Among the different mesh properties that could influence the numerical behavior, restrictions on l appear as immaterial (the best convergence orders are achieved for prism meshes of Test 2 having up to $l = 6$ face vertices). While conformity is not needed for the method, non-conformal meshes bring more distorted cells and diamonds. We have seen that bad shape conditioning may induce violation of the maximum principle. In Test 1, presence of neighbor cells with considerable contrast in size (for Voronoï meshes and for non-conformal checkerboard meshes) degrades convergence orders for the gradient, in contrast to rather gradually distorted Kershaw and prism meshes.

References

1. B. Andreianov, M. Bendahmane, F. Hubert and S. Krell. On 3D DDFV discretization of gradient and divergence operators. I. Meshing, operators and discrete duality. Preprint HAL (2011), http://hal.archives-ouvertes.fr/hal-00355212.
2. B. Andreianov, M. Bendahmane and F. Hubert. On 3D DDFV discretization of gradient and divergence operators. II. Discrete functional analysis tools and applications to degenerate parabolic problems. Preprint HAL (2011), http://hal.archives-ouvertes.fr/hal-00567342.
3. B. Andreianov, M. Bendahmane, and K. Karlsen. A gradient reconstruction formula for finite-volume schemes and discrete duality. In R. Eymard and J.-M. Hérard, editors, *Finite Volume For Complex Applications, Problems And Perspectives. 5th International Conference*, (2008),161–168. London (UK) Wiley.
4. B. Andreianov, M. Bendahmane, and K.H. Karlsen. Discrete duality finite volume schemes for doubly nonlinear degenerate hyperbolic-parabolic equations. *J. Hyperbolic Diff. Equ.* (2010), **7**(1):1–67.
5. B. Andreianov, F. Boyer, and F. Hubert. Discrete duality finite volume schemes for Leray-Lions type elliptic problems on general 2D-meshes. *Numer. Methods PDE*, (2007), **23**(1):145–195.
6. F. Boyer and F. Hubert. Finite volume method for 2D linear and nonlinear elliptic problems with discontinuities. *SIAM J. Num. Anal.*, (2008), **46**(6):3032–3070.
7. F. Boyer and F. Hubert. Benchmark on anisotropic problems, the ddfv discrete duality finite volumes and m-ddfv schemes. In R. Eymard and J.-M. Hérard, editors, *Finite Volume For Complex Applications, Problems And Perspectives. 5th International Conference*, (2008), 735–750. London (UK) Wiley.
8. Y. Coudière, J.-P. Vila, and P. Villedieu. Convergence rate of a finite volume scheme for a two dimensional convection-diffusion problem. *M2AN Math. Modelling Num. Anal.*, (1999), **33**(3):493–516.
9. Y. Coudière and C. Pierre. Benchmark 3D: CeVe-DDFV, a discrete duality scheme with cell/vertex unknowns. *This volume*, (2011).
10. Y. Coudière and F. Hubert. A 3D discrete duality finite volume method for nonlinear elliptic equation. preprint (2009), http://hal.archives-ouvertes.fr/docs/00/45/68/37/PDF/ddfv3d.pdf

---

[9]Recall that conformity of meshes is not required by the method; and the construction allows for unrestricted number l of face vertices. Yet it is a well-known difficulty for the analysis of the scheme that the discrete Poincaré inequality cannot be proved for $l > 4$, see [2, 17].

11. Y. Coudière, F. Hubert and G. Manzini. Benchmark 3D: CeVeFE-DDFV, a discrete duality scheme with cell/vertex/face+edge unknowns. *This volume*, (2011).
12. Y. Coudière, C. Pierre, O. Rousseau, and R. Turpault. A 2d/3d discrete duality finite volume scheme. Application to ecg simulation. *Int. Journal on Finite Volumes*, (2009), **6**(1).
13. K. Domelevo and P. Omnès. A finite volume method for the Laplace equation on almost arbitrary two-dimensional grids. *M2AN Math. Model. Numer. Anal.*, (2005), **39**(6):1203–1249.
14. R. Herbin and F. Hubert. Benchmark on discretization schemes for anisotropic diffusion problems on general grids. In R. Eymard and J.-M. Hérard, editors, *Finite Volume For Complex Applications, Problems And Perspectives. 5th International Conference*, (2008), 659–692. London (UK) Wiley.
15. F. Hermeline. A finite volume method for the approximation of diffusion operators on distorted meshes. *J. Comput. Phys.*, (2000), **160**(2):481–499.
16. F. Hermeline. Approximation of 2-d and 3-d diffusion operators with variable full tensor coefficients on arbitrary meshes. *Comput. Methods Appl. Mech. Engrg.*, (2007), **196**(21-24): 2497–2526.
17. F. Hermeline. A finite volume method for approximating 3d diffusion operators on general meshes. *Journal of computational Physics*, (2009), **228**(16):5763–5786.
18. S. Krell. *Schémas Volumes Finis en mécanique des fluides complexes*. Ph.D. Thesis, Univ. de Provence, Marseilles, (2010), http://tel.archives-ouvertes.fr/tel-00524509.
19. S. Krell. Stabilized DDFV schemes for Stokes problem with variable viscosity on general 2D meshes. *Num. Meth. PDEs*, (2010), available on-line: http://dx.doi.org/10.1002/num.20603.

The paper is in final form and no similar paper has been or is being submitted elsewhere.

Benchmark 3D: Symmetric Weighted Interior Penalty Discontinuous Galerkin Scheme

Peter Bastian

1 Weighted Interior Penalty Discontinuous Galerkin Schemes

Consider the stationary diffusion equation with Dirichlet boundary conditions

$$-\nabla \cdot (K \nabla u) = f \qquad \text{in } \Omega, \qquad (1a)$$

$$u = g \qquad \text{on } \partial\Omega. \qquad (1b)$$

with Ω a domain in $\mathbb{R}^n$, $n = 1, 2, 3$, and K a uniformly symmetric positive definite permeability tensor. The weak formulation of (1) consist of finding $u \in u_g + V$, $V = H_0^1(\Omega)$, u_g an extension of the Dirichlet data such that

$$(K \nabla u, \nabla v)_{0,\Omega} = (f, v)_{0,\Omega}$$

where $(.,.)_{0,\omega}$ denotes the L^2-scalar product on a domain ω.

Discontinuous Galerkin (DG) methods are a class of numerical schemes that has been studied extensively in the last two decades, see [8]. We use the weighted interior penalty discontinuous Galerkin (WIPG) schemes introduced in [6].

Let $\{\mathcal{T}_h\}_{h>0}$ denote a family of triangulations of the domain Ω. An element of the triangulation is denoted by T, h_T is its diameter, $|T|$ its volume and n_T its unit outer normal vector. F is called an "interior face" independent of the dimension if there are two elements $T^-(F), T^+(F) \in \mathcal{T}_h$ with $T^-(F) \cap T^+(F) = F$ and F has nonzero measure. All interior faces are collected in the set $\mathcal{F}_h^i$. The intersection of $T \in \mathcal{T}_h$ with the boundary $\partial\Omega$ of non-zero measure is called a boundary face. All boundary faces are collected in the set $\mathcal{F}_h^{\partial\Omega}$ and we set $\mathcal{F} = \mathcal{F}_h^i \cup \mathcal{F}_h^{\partial\Omega}$. The diameter of a face is denoted by h_F and its volume by $|F|$. With each $F \in \mathcal{F}$ we

---

Peter Bastian
Interdisciplinary Center for Scientific Computing, University of Heidelberg, Im Neuenheimer Feld 368, D-69120 Heidelberg, e-mail: peter.bastian@iwr.uni-heidelberg.de

associate a unit normal vector n_F (depending on position if F is curved) oriented from $T^-(F)$ to $T^+(F)$ for an interior face and coinciding with the exterior unit normal for a boundary face.

The DG approximation space is

$$V_h = \{v \in L^2(\Omega) : \forall T \in \mathcal{T}_h, v|_T \in \mathcal{P}\}$$

where $\mathcal{P}$ is either $\mathbb{P}_k = \{p : p = \sum_{\|\alpha\|_1 \leq k} c_\alpha x^\alpha\}$ or $\mathbb{Q}_k = \{p : p = \sum_{\|\alpha\|_\infty \leq k} c_\alpha x^\alpha\}$ in the standard multiindex notation. On an interior face F a function $v \in V_h$ is two-valued and its values v^- and v^+ are the restrictions from $T^-(F)$ and $T^+(F)$, respectively. For $F \in \mathcal{F}_h^i$ and $v \in V_h$ we introduce the jump and the weighted average

$$[\![v]\!]_F = v^- - v^+, \qquad \{v\}_\omega = \omega^- v^- + \omega^+ v^+,$$

with the weights satisfying $\omega^- + \omega^+ = 1$, $\omega^-, \omega^+ \geq 0$.

In the WIPG schemes the discrete solution $u_h \in V_h$ satisfies the variational equation

$$a_h(u_h, v) = l_h(v) \qquad \forall v \in V_h$$

with bilinear and linear forms defined as

$$a_h(u, v) = \sum_{T \in \mathcal{T}_h} (K \nabla u, \nabla v)_{0,T}$$

$$- \sum_{F \in \mathcal{F}_h^i} \left[(\{n_F^t K \nabla u\}_\omega, [\![v]\!])_{0,F} - \theta(\{n_F^t K \nabla v\}_\omega, [\![u]\!])_{0,F} - \gamma_F([\![u]\!], [\![v]\!])_{0,F} \right]$$

$$- \sum_{F \in \mathcal{F}_h^{\partial\Omega}} \left[(n_F^t K \nabla u, v)_{0,F} - \theta(n_F^t K \nabla v, u)_{0,F} - \gamma_F(u, v)_{0,F} \right],$$

$$l_h(u, v) = \sum_{T \in \mathcal{T}_h} (f, v)_{0,T} + \sum_{F \in \mathcal{F}_h^{\partial\Omega}} \left[\theta(n_F^t K \nabla v, g)_{0,F} + \gamma_F(g, v)_{0,F} \right].$$

For $\theta = -1$ we obtain the symmetric weighted interior penalty Galerkin method (SWIPG) used below for all tests.

The weights $\omega^\pm$ are defined with respect to the permeability as

$$\omega^- = \frac{\delta_{Kn}^+}{\delta_{Kn}^- + \delta_{Kn}^+}, \qquad \omega^+ = \frac{\delta_{Kn}^-}{\delta_{Kn}^- + \delta_{Kn}^+},$$

with $\delta_{Kn}^\pm = n_F^t K^\pm n_F$ for $F \in \mathcal{F}_h^i$ and $\delta_{Kn} = n_F^t K n_F$ for $F \in \mathcal{F}_h^{\partial\Omega}$.

The choice of the interior penalty parameter γ_F is crucial, as the scheme should be as independent of the problem and mesh parameters as possible. We use the following definition of the penalty parameter:

… Benchmark 3D: Symmetric Weighted Interior Penalty Discontinuous Galerkin Scheme

$$\forall F \in \mathscr{F}_h^i, \quad \gamma_F = \alpha \, \frac{2\delta_{Kn}^- \delta_{Kn}^+}{\delta_{Kn}^- + \delta_{Kn}^+} \, k(k+n-1) \, \frac{|F|}{\min(|T^-(F)|, |T^+(F)|)}, \quad (2a)$$

$$\forall F \in \mathscr{F}_h^{\partial\Omega}, \quad \gamma_F = \alpha \, \delta_{Kn} \, k(k+n-1) \, \frac{|F|}{|T^-(F)|}. \quad (2b)$$

with α a user-defined parameter. This choice is a combination of different papers: The harmonic average of "normal" permeabilities was introduced and analyzed in [6], the dependence on the polynomial degree was analyzed in [5] and the choice of the h-dependence is taken from [7]. The parameter α was chosen as follows:

Test	1	1	1	3	3	3	3	4	5
Mesh	tetra	kershaw	checkerboard	rand	rand	rand	rand	well	locraf
k	1-4	1-4	1-4	1	2	3	4	1-4	1-4
α	3.0	2.5	1.0	1000	2000	5000	10000	1000	0.7

Unfortunately, the choice of α is heuristic. It should be subject of future research to find a formula (2) that better takes into account the element shape as it was done in [5] for tetrahedral elements.

2 Numerical results

The L^2, H^1 and energy error are computed by numerical integration of order 12.

- **Test 1, Mild anisotropy, Tetrahedral meshes**

Table 1, $\mathbb{P}_1$

i	nu	nmat	umin	uemin	umax	uemax	normg
1	8012	150576	5.32E-02	2.03E-02	1.965	1.989	1.794
2	15592	297376	1.31E-02	6.84E-03	1.974	1.989	1.797
3	30844	593648	1.71E-02	9.13E-03	1.983	1.994	1.797
4	61064	1184192	1.05E-02	5.52E-03	1.992	1.997	1.797
5	121920	2379936	7.19E-03	1.49E-03	1.994	1.997	1.798
6	244208	4791872	3.69E-03	1.83E-03	1.994	1.997	1.798

Table 1, $\mathbb{P}_2$

i	nu	nmat	umin	uemin	umax	uemax	normg
1	20030	941100	2.11E-02	2.03E-02	1.989	1.989	1.799
2	38980	1858600	6.97E-03	6.84E-03	1.989	1.989	1.799
3	77110	3710300	9.19E-03	9.13E-03	1.993	1.994	1.799
4	152660	7401200	5.56E-03	5.52E-03	1.997	1.997	1.799
5	304800	14874600	1.51E-03	1.49E-03	1.997	1.997	1.799
6	610520	29949200	1.84E-03	1.83E-03	1.997	1.997	1.798

Table 1, $\mathbb{P}_3$

i	nu	nmat	umin	uemin	umax	uemax	normg
1	40060	3764400	2.04E-02	2.03E-02	1.989	1.989	1.798
2	77960	7434400	6.83E-03	6.84E-03	1.989	1.989	1.798
3	154220	14841200	9.14E-03	9.13E-03	1.993	1.994	1.798
4	305320	29604800	5.52E-03	5.52E-03	1.997	1.997	1.798
5	609600	59498400	1.49E-03	1.49E-03	1.997	1.997	1.798
6	1221040	119796800	1.83E-03	1.83E-03	1.997	1.997	1.798

Table 1, $\mathbb{P}_4$

i	nu	nmat	umin	uemin	umax	uemax	normg
1	70105	11528475	2.03E-02	2.03E-02	1.989	1.989	1.798
2	136430	22767850	6.84E-03	6.84E-03	1.989	1.989	1.798
3	269885	45451175	9.13E-03	9.13E-03	1.994	1.994	1.798
4	534310	90664700	5.52E-03	5.52E-03	1.997	1.997	1.798
5	1066800	182213850	1.49E-03	1.49E-03	1.997	1.997	1.798
6	2136820	366877700	1.83E-03	1.83E-03	1.997	1.997	1.798

Table 2, $\mathbb{P}_1$

i	nu	erl2	ratiol2	ergrad	ratiograd	ener	ratioener
1	8012	1.11E-02	—	2.28E-01	—	2.23E-01	—
2	15592	7.02E-03	2.084	1.82E-01	1.015	1.79E-01	1.004
3	30844	4.52E-03	1.934	1.46E-01	0.961	1.43E-01	0.988
4	61064	2.91E-03	1.931	1.16E-01	1.016	1.13E-01	1.022
5	121920	1.87E-03	1.925	9.23E-02	0.993	9.03E-02	0.979
6	244208	1.16E-03	2.068	7.28E-02	1.021	7.11E-02	1.034

Table 2, $\mathbb{P}_2$

i	nu	erl2	ratiol2	ergrad	ratiograd	ener	ratioener
1	20030	8.22E-04	—	2.40E-02	—	2.32E-02	—
2	38980	4.19E-04	3.034	1.54E-02	1.986	1.48E-02	2.033
3	77110	2.08E-04	3.079	9.51E-03	2.123	9.17E-03	2.104
4	152660	1.05E-04	3.016	6.05E-03	1.985	5.84E-03	1.980
5	304800	5.34E-05	2.925	3.85E-03	1.962	3.72E-03	1.960
6	610520	2.66E-05	3.015	2.42E-03	2.015	2.33E-03	2.022

Table 2, $\mathbb{P}_3$

i	nu	erl2	ratiol2	ergrad	ratiograd	ener	ratioener
1	40060	4.46E-05	—	1.71E-03	—	1.66E-03	—
2	77960	1.78E-05	4.152	8.58E-04	3.118	8.28E-04	3.125
3	154220	7.27E-06	3.928	4.38E-04	2.957	4.21E-04	2.969
4	305320	2.84E-06	4.135	2.17E-04	3.076	2.09E-04	3.074
5	609600	1.14E-06	3.969	1.09E-04	3.004	1.05E-04	3.006
6	1221040	4.43E-07	4.067	5.36E-05	3.052	5.15E-05	3.060

Table 2, $\mathbb{P}_4$

i	nu	erl2	ratiol2	ergrad	ratiograd	ener	ratioener
1	70105	2.31E-06	—	1.03E-04	—	9.93E-05	—
2	136430	7.36E-07	5.154	4.19E-05	4.068	3.97E-05	4.134
3	269885	2.30E-07	5.118	1.62E-05	4.186	1.54E-05	4.161
4	534310	7.00E-08	5.226	6.35E-06	4.110	6.06E-06	4.097
5	1066800	2.29E-08	4.850	2.59E-06	3.900	2.47E-06	3.896
6	2136820	7.08E-09	5.066	1.01E-06	4.047	9.64E-07	4.061

• Test 1, Mild anisotropy, Kershaw meshes

Table 1, $\mathbb{Q}_1$

i	nu	nmat	umin	uemin	umax	uemax	normg
1	4096	204800	9.58E-02	3.03E-02	1.850	1.958	1.771
2	32768	1736704	2.40E-02	1.06E-02	1.953	1.993	1.781
3	262144	14286848	5.39E-03	1.75E-03	1.987	1.997	1.786
4	2097152	115867648	1.71E-03	7.14E-04	1.997	1.999	1.791

Table 1, $\mathbb{Q}_2$

i	nu	nmat	umin	uemin	umax	uemax	normg
1	13824	2332800	3.12E-02	3.03E-02	1.944	1.958	1.796
2	110592	19782144	1.04E-02	1.06E-02	1.990	1.993	1.796
3	884736	162736128	1.71E-03	1.75E-03	1.997	1.997	1.798
4	7077888	1319804928	7.11E-04	7.14E-04	1.999	1.999	1.798

Table 1, $\mathbb{Q}_3$

i	nu	nmat	umin	uemin	umax	uemax	normg
1	32768	13107200	2.91E-02	3.03E-02	1.955	1.958	1.797
2	262144	111149056	1.05E-02	1.06E-02	1.992	1.993	1.798
3	2097152	914358272	1.75E-03	1.75E-03	1.997	1.997	1.798

Table 1, $\mathbb{Q}_4$

i	nu	nmat	umin	uemin	umax	uemax	normg
1	64000	50000000	3.02E-02	3.03E-02	1.958	1.958	1.798
2	512000	424000000	1.06E-02	1.06E-02	1.993	1.993	1.798
3	4096000	-806967296	1.75E-03	1.75E-03	1.997	1.997	1.798

Table 2, $\mathbb{Q}_1$

i	nu	erl2	ratiol2	ergrad	ratiograd	ener	ratioener
1	4096	6.65E-02	—	5.79E-01	—	5.52E-01	—
2	32768	4.57E-02	0.541	4.26E-01	0.441	4.00E-01	0.466
3	262144	2.53E-02	0.856	2.76E-01	0.627	2.59E-01	0.624
4	2097152	1.07E-02	1.246	1.57E-01	0.813	1.51E-01	0.778

Table 2, $\mathbb{Q}_2$

i	nu	erl2	ratiol2	ergrad	ratiograd	ener	ratioener
1	13824	2.95E-02	—	2.45E-01	—	2.17E-01	—
2	110592	7.51E-03	1.975	1.04E-01	1.236	9.80E-02	1.145
3	884736	9.75E-04	2.945	3.26E-02	1.675	3.22E-02	1.606
4	7077888	7.54E-05	3.693	8.76E-03	1.895	8.77E-03	1.877

Table 2, $\mathbb{Q}_3$

i	nu	erl2	ratiol2	ergrad	ratiograd	ener	ratioener
1	32768	5.72E-03	—	6.34E-02	—	5.57E-02	—
2	262144	7.05E-04	3.020	1.58E-02	2.002	1.50E-02	1.895
3	2097152	2.91E-05	4.598	2.45E-03	2.689	2.42E-03	2.629

Table 2, $\mathbb{Q}_4$

i	nu	erl2	ratiol2	ergrad	ratiograd	ener	ratioener
1	64000	1.61E-03	—	2.09E-02	—	1.82E-02	—
2	512000	5.64E-05	4.837	1.99E-03	3.389	1.89E-03	3.266
3	4096000	1.21E-06	5.540	1.51E-04	3.725	1.49E-04	3.664

- **Test 1, Mild anisotropy, Checkerboard meshes**

Table 1, $\mathbb{P}_1$

i	nu	nmat	umin	uemin	umax	uemax	normg
1	144	3648	2.35E-01	1.54E-01	1.784	1.846	1.550
2	1152	35328	6.71E-02	4.01E-02	1.931	1.960	1.667
3	9216	307200	1.47E-02	1.01E-02	1.985	1.990	1.751
4	73728	2555904	2.56E-03	2.54E-03	1.997	1.997	1.784
5	589824	20840448	6.36E-04	6.36E-04	1.999	1.999	1.795

Table 1, $\mathbb{P}_2$

i	nu	nmat	umin	uemin	umax	uemax	normg
1	360	22800	1.82E-01	1.54E-01	1.812	1.846	1.750
2	2880	220800	4.52E-02	4.01E-02	1.954	1.960	1.795
3	23040	1920000	1.05E-02	1.01E-02	1.989	1.990	1.799
4	184320	15974400	2.57E-03	2.54E-03	1.997	1.997	1.799
5	1474560	130252800	6.37E-04	6.36E-04	1.999	1.999	1.798

Table 1, $\mathbb{P}_3$

i	nu	nmat	umin	uemin	umax	uemax	normg
1	720	91200	1.61E-01	1.54E-01	1.839	1.846	1.783
2	5760	883200	4.02E-02	4.01E-02	1.960	1.960	1.798
3	46080	7680000	1.01E-02	1.01E-02	1.990	1.990	1.798
4	368640	63897600	2.54E-03	2.54E-03	1.997	1.997	1.798
5	2949120	521011200	6.36E-04	6.36E-04	1.999	1.999	1.798

Table 1, $\mathbb{P}_4$

i	nu	nmat	umin	uemin	umax	uemax	normg
1	1260	279300	1.55E-01	1.54E-01	1.845	1.846	1.797
2	10080	2704800	4.01E-02	4.01E-02	1.960	1.960	1.798
3	80640	23520000	1.01E-02	1.01E-02	1.990	1.990	1.798
4	645120	195686400	2.54E-03	2.54E-03	1.997	1.997	1.798
5	5160960	1595596800	6.36E-04	6.36E-04	1.999	1.999	1.798

Table 2, $\mathbb{P}_1$

i	nu	erl2	ratiol2	ergrad	ratiograd	ener	ratioener
1	144	9.89E-02	—	5.97E-01	—	5.71E-01	—
2	1152	3.13E-02	1.658	3.33E-01	0.842	3.35E-01	0.769
3	9216	9.17E-03	1.773	1.68E-01	0.984	1.68E-01	0.998
4	73728	2.51E-03	1.871	8.30E-02	1.022	8.22E-02	1.031
5	589824	6.47E-04	1.954	4.10E-02	1.015	4.05E-02	1.020

Table 2, $\mathbb{P}_2$

i	nu	erl2	ratiol2	ergrad	ratiograd	ener	ratioener
1	360	3.56E-02	—	2.70E-01	—	2.93E-01	—
2	2880	6.55E-03	2.442	8.25E-02	1.713	8.00E-02	1.874
3	23040	7.78E-04	3.075	2.10E-02	1.971	2.04E-02	1.970
4	184320	9.46E-05	3.039	5.26E-03	2.001	5.11E-03	1.998
5	1474560	1.17E-05	3.010	1.31E-03	2.001	1.28E-03	2.000

Table 2, $\mathbb{P}_3$

i	nu	erl2	ratiol2	ergrad	ratiograd	ener	ratioener
1	720	1.19E-02	—	1.01E-01	—	8.92E-02	—
2	5760	1.02E-03	3.544	1.54E-02	2.712	1.47E-02	2.602
3	46080	7.31E-05	3.803	2.08E-03	2.886	1.95E-03	2.912
4	368640	4.78E-06	3.933	2.67E-04	2.962	2.50E-04	2.968
5	2949120	3.04E-07	3.975	3.37E-05	2.986	3.15E-05	2.988

Table 2, $\mathbb{P}_4$

i	nu	erl2	ratiol2	ergrad	ratiograd	ener	ratioener
1	1260	2.72E-03	—	2.82E-02	—	3.04E-02	—
2	10080	1.19E-04	4.517	2.18E-03	3.693	2.07E-03	3.874
3	80640	4.29E-06	4.789	1.46E-04	3.903	1.37E-04	3.923
4	645120	1.44E-07	4.900	9.34E-06	3.964	8.68E-06	3.977
5	5160960	4.62E-09	4.958	5.90E-07	3.985	5.46E-07	3.992

- **Test 3, Flow on random meshes, Random meshes**

Table 1, $\mathbb{Q}_1$

i	nu	nmat	umin	uemin	umax	uemax	normg
1	512	22528	-4.34E-01	-7.59E-01	0.355	0.691	3.007
2	4096	204800	-8.45E-01	-9.39E-01	0.791	0.923	3.431
3	32768	1736704	-9.58E-01	-9.85E-01	0.946	0.982	3.565
4	262144	14286848	-9.90E-01	-9.96E-01	0.989	0.996	3.588

Table 1, $\mathbb{Q}_2$

i	nu	nmat	umin	uemin	umax	uemax	normg
1	1728	256608	-7.50E-01	-7.59E-01	0.676	0.691	3.637
2	13824	2332800	-9.40E-01	-9.39E-01	0.924	0.923	3.569
3	110592	19782144	-9.85E-01	-9.85E-01	0.982	0.982	3.597
4	884736	162736128	-9.96E-01	-9.96E-01	0.996	0.996	3.596

Table 1, $\mathbb{Q}_3$

i	nu	nmat	umin	uemin	umax	uemax	normg
1	4096	1441792	-7.53E-01	-7.59E-01	0.684	0.691	3.655
2	32768	13107200	-9.38E-01	-9.39E-01	0.922	0.923	3.570
3	262144	111149056	-9.85E-01	-9.85E-01	0.982	0.982	3.597
4	2097152	914358272	-9.96E-01	-9.96E-01	0.996	0.996	3.596

Table 1, $\mathbb{Q}_4$

i	nu	nmat	umin	uemin	umax	uemax	normg
1	8000	5500000	-7.59E-01	-7.59E-01	0.691	0.691	3.655
2	64000	50000000	-9.39E-01	-9.39E-01	0.923	0.923	3.570
3	512000	424000000	-9.85E-01	-9.85E-01	0.982	0.982	3.597

Table 2, $\mathbb{Q}_1$

i	nu	erl2	ratiol2	ergrad	ratiograd	ener	ratioener
1	512	3.04E-01	—	4.99E-01	—	5.00E-01	—
2	4096	8.38E-02	1.858	2.58E-01	0.952	2.53E-01	0.984
3	32768	2.16E-02	1.958	1.30E-01	0.988	1.25E-01	1.017
4	262144	5.77E-03	1.902	6.72E-02	0.953	6.30E-02	0.989

Table 2, $\mathbb{Q}_2$

i	nu	erl2	ratiol2	ergrad	ratiograd	ener	ratioener
1	1728	4.41E-02	—	1.25E-01	—	1.14E-01	—
2	13824	6.01E-03	2.874	2.98E-02	2.066	2.83E-02	2.014
3	110592	8.26E-04	2.864	7.77E-03	1.941	7.27E-03	1.962
4	884736	1.74E-04	2.248	2.60E-03	1.578	2.16E-03	1.750

Table 2, $\mathbb{Q}_3$

i	nu	erl2	ratiol2	ergrad	ratiograd	ener	ratioener
1	4096	5.29E-03	—	2.00E-02	—	1.85E-02	—
2	32768	3.86E-04	3.776	2.60E-03	2.943	2.30E-03	3.002
3	262144	6.22E-05	2.634	6.36E-04	2.032	3.72E-04	2.629
4	2097152	3.93E-05	0.661	5.81E-04	0.130	2.65E-04	0.489

Table 2, $\mathbb{Q}_4$

i	nu	erl2	ratiol2	ergrad	ratiograd	ener	ratioener
1	8000	5.44E-04	—	2.69E-03	—	2.26E-03	—
2	64000	4.30E-05	3.662	3.02E-04	3.159	1.76E-04	3.684
3	512000	2.15E-05	0.999	2.47E-04	0.291	9.24E-05	0.930

- **Test 4, Flow around a well, Well meshes**

Table 1, $\mathbb{Q}_1$

i	nu	nmat	umin	uemin	umax	uemax	normg
1	7120	356736	3.52E-01	4.14E-01	5.316	5.317	1686.482
2	17856	931328	2.23E-01	2.44E-01	5.328	5.328	1652.956
3	40128	2139904	1.46E-01	1.54E-01	5.329	5.329	1637.838
4	89760	4857216	1.13E-01	1.18E-01	5.330	5.330	1632.052
5	185680	10136320	8.72E-02	8.99E-02	5.339	5.339	1628.690
6	341064	18717760	7.06E-02	7.23E-02	5.345	5.345	1626.812
7	597432	32900416	5.55E-02	5.65E-02	5.361	5.361	1625.784

Table 1, $\mathbb{Q}_2$

i	nu	nmat	umin	uemin	umax	uemax	normg
1	24030	4063446	4.13E-01	4.14E-01	5.317	5.317	1625.412
2	60264	10608408	2.43E-01	2.44E-01	5.328	5.328	1623.883
3	135432	24374844	1.54E-01	1.54E-01	5.329	5.329	1623.603
4	302940	55326726	1.18E-01	1.18E-01	5.330	5.330	1623.529
5	626670	115459020	8.99E-02	8.99E-02	5.339	5.339	1623.506
6	1151091	213206985	7.23E-02	7.23E-02	5.345	5.345	1623.497
7	2016333	374756301	5.65E-02	5.65E-02	5.361	5.361	1623.491

Table 1, $\mathbb{Q}_3$

i	nu	nmat	umin	uemin	umax	uemax	normg
1	56960	22831104	4.15E-01	4.14E-01	5.317	5.317	1623.772
2	142848	59604992	2.44E-01	2.44E-01	5.328	5.328	1623.611
3	321024	136953856	1.54E-01	1.54E-01	5.329	5.329	1623.546
4	718080	310861824	1.18E-01	1.18E-01	5.330	5.330	1623.514
5	1485440	648724480	8.99E-02	8.99E-02	5.339	5.339	1623.500
6	2728512	1197936640	7.24E-02	7.23E-02	5.345	5.345	1623.493
7	4779456	2105626624	5.65E-02	5.65E-02	5.361	5.361	1623.489

Table 1, $\mathbb{Q}_4$

i	nu	nmat	umin	uemin	umax	uemax	normg
1	111250	87093750	4.14E-01	4.14E-01	5.317	5.317	1623.709
2	279000	227375000	2.44E-01	2.44E-01	5.328	5.328	1623.607
3	627000	522437500	1.54E-01	1.54E-01	5.329	5.329	1623.546
4	1402500	1185843750	1.18E-01	1.18E-01	5.330	5.330	1623.514
5	2901250	2474687500	8.99E-02	8.99E-02	5.339	5.339	1623.500

Table 2, $\mathbb{Q}_1$

i	nu	erl2	ratiol2	ergrad	ratiograd	ener	ratioener
1	7120	6.54E-03	—	2.50E-01	—	2.48E-01	—
2	17856	2.99E-03	2.551	1.70E-01	1.254	1.70E-01	1.239
3	40128	1.47E-03	2.622	1.19E-01	1.318	1.19E-01	1.310
4	89760	9.71E-04	1.555	9.08E-02	1.010	9.09E-02	1.008
5	185680	6.04E-04	1.959	7.07E-02	1.033	7.08E-02	1.031
6	341064	3.68E-04	2.440	5.71E-02	1.056	5.72E-02	1.056
7	597432	2.72E-04	1.629	4.77E-02	0.962	4.78E-02	0.959

Table 2, $\mathbb{Q}_2$

i	nu	erl2	ratiol2	ergrad	ratiograd	ener	ratioener
1	24030	3.77E-04	—	3.95E-02	—	3.95E-02	—
2	60264	1.17E-04	3.813	1.61E-02	2.932	1.61E-02	2.931
3	135432	4.96E-05	3.180	7.47E-03	2.843	7.48E-03	2.838
4	302940	3.33E-05	1.487	4.40E-03	1.976	4.40E-03	1.978
5	626670	1.86E-05	2.412	2.64E-03	2.099	2.65E-03	2.098
6	1151091	9.19E-06	3.469	1.69E-03	2.197	1.70E-03	2.194
7	2016333	6.09E-06	2.198	1.17E-03	1.966	1.17E-03	1.964

Table 2, $\mathbb{Q}_3$

i	nu	erl2	ratiol2	ergrad	ratiograd	ener	ratioener
1	56960	4.74E-05	—	8.16E-03	—	8.15E-03	—
2	142848	1.01E-05	5.050	2.25E-03	4.211	2.24E-03	4.210
3	321024	3.23E-06	4.225	7.44E-04	4.094	7.44E-04	4.090
4	718080	1.63E-06	2.552	3.22E-04	3.122	3.22E-04	3.123
5	1485440	7.41E-07	3.242	1.53E-04	3.070	1.53E-04	3.066
6	2728512	2.82E-07	4.766	8.03E-05	3.177	8.04E-05	3.177
7	4779456	2.18E-07	1.377	4.98E-05	2.560	4.98E-05	2.557

Table 2, $\mathbb{Q}_4$

i	nu	erl2	ratiol2	ergrad	ratiograd	ener	ratioener
1	111250	6.83E-06	—	1.51E-03	—	1.50E-03	—
2	279000	8.10E-07	6.958	2.41E-04	5.976	2.41E-04	5.975
3	627000	2.17E-07	4.874	5.24E-05	5.661	5.23E-05	5.659
4	1402500	9.17E-08	3.217	1.79E-05	4.011	1.78E-05	4.024
5	2901250	3.30E-08	4.213	6.69E-06	4.050	6.53E-06	4.131

- **Test 5, Discontinuous permeability, Locally refined meshes**

Table 1, $\mathbb{Q}_1$

i	nu	nmat	umin	uemin	umax	uemax	normg
1	176	7936	-5.46E+01	-1.00E+02	54.594	100.000	52.441
2	1408	71168	-3.09E+01	-3.54E+01	30.857	35.355	79.708
3	11264	600064	-7.05E+01	-7.89E+01	70.515	78.858	89.071
4	90112	4923392	-9.14E+01	-9.43E+01	91.442	94.346	96.089
5	720896	39878656	-9.78E+01	-9.86E+01	97.780	98.562	98.260

Table 1, $\mathbb{Q}_2$

i	nu	nmat	umin	uemin	umax	uemax	normg
1	594	90396	-1.18E+02	-1.00E+02	118.325	100.000	144.541
2	4752	810648	-3.83E+01	-3.54E+01	38.300	35.355	108.493
3	38016	6835104	-7.90E+01	-7.89E+01	78.962	78.858	100.680
4	304128	56080512	-9.44E+01	-9.43E+01	94.354	94.346	99.360
5	2433024	454242816	-9.86E+01	-9.86E+01	98.563	98.562	99.089

Table 1, $\mathbb{Q}_3$

i	nu	nmat	umin	uemin	umax	uemax	normg
1	1408	507904	-1.05E+02	-1.00E+02	104.586	100.000	123.703
2	11264	4554752	-3.55E+01	-3.54E+01	35.484	35.355	100.084
3	90112	38404096	-7.88E+01	-7.89E+01	78.828	78.858	99.078
4	720896	315097088	-9.43E+01	-9.43E+01	94.343	94.346	99.013
5	5767168	2552233984	-9.86E+01	-9.86E+01	98.562	98.562	99.010

Table 2, $\mathbb{Q}_1$

i	nu	erl2	ratiol2	ergrad	ratiograd	ener	ratioener
1	176	1.31E+00	—	1.12E+00	—	1.29E+00	—
2	1408	2.71E-01	2.277	5.30E-01	1.074	5.89E-01	1.126
3	11264	6.42E-02	2.079	2.46E-01	1.109	2.62E-01	1.169
4	90112	1.61E-02	1.992	1.16E-01	1.080	1.18E-01	1.152
5	720896	4.05E-03	1.993	5.70E-02	1.027	5.71E-02	1.044

Table 2, $\mathbb{Q}_2$

i	nu	erl2	ratiol2	ergrad	ratiograd	ener	ratioener
1	594	3.58E-01	—	6.29E-01	—	8.35E-01	—
2	4752	6.14E-02	2.543	1.88E-01	1.740	2.53E-01	1.720
3	38016	6.35E-03	3.274	3.82E-02	2.302	4.72E-02	2.426
4	304128	6.54E-04	3.278	8.34E-03	2.196	9.44E-03	2.321
5	2433024	6.72E-05	3.284	1.88E-03	2.148	2.01E-03	2.228

Table 2, $\mathbb{Q}_3$

i	nu	erl2	ratiol2	ergrad	ratiograd	ener	ratioener
1	1408	2.37E-01	—	5.81E-01	—	8.64E-01	—
2	11264	7.67E-03	4.951	3.58E-02	4.023	5.19E-02	4.059
3	90112	3.04E-04	4.656	2.82E-03	3.665	3.51E-03	3.884
4	720896	1.37E-05	4.474	2.35E-04	3.583	2.49E-04	3.816
5	5767168	8.04E-07	4.088	2.50E-05	3.236	2.50E-05	3.316

3 Comments

Tests 1 and 5 were uncritical on all meshes. The penalty parameter can be chosen small and the corresponding symmetric and positive definite systems are easily solved. Either $\mathbb{P}_k$ or $\mathbb{Q}_k$ can be chosen, with $\mathbb{Q}_k$ being slightly more efficient in terms of error with respect to degrees of freedom. Tests 3 and 4 are much more difficult with two consequences: First, only $\mathbb{Q}_k$ did work on these meshes. Secondly, the global penalty parameter α has to be chosen large in order to obtain optimal convergence rates. In test 3 it has to be increased with polynomial degree (and even with these values the convergence rate breaks down on the finest mesh with $k = 4$). Large penalty parameters lead to very ill conditioned matrices which take a large number of iterations to solve. Note that the standard continuous Galerkin finite element method has no difficulties at all with tests 3 and 4. All numerical tests have been performed with the DUNE software framework[1, 2, 4], using the `dune-pdelab` discretization framework described in [3].

Acknowledgements I would like to thank Robert Klöfkorn for providing all the meshes in DUNE grid format and lots of discussions.

References

1. P. Bastian, M. Blatt, A. Dedner, C. Engwer, R. Klöfkorn, R. Kornhuber, M. Ohlberger, and O. Sander. A generic grid interface for parallel and adaptive scientific computing. part II: implementation and tests in DUNE. *Computing*, 82(2-3):121–138, 2008.
2. P. Bastian, M. Blatt, A. Dedner, C. Engwer, R. Klöfkorn, M. Ohlberger, and O. Sander. A generic grid interface for parallel and adaptive scientific computing. part I: abstract framework. *Computing*, 82(2-3):103–119, 2008.
3. P. Bastian, F. Heimann, and S. Marnach. Generic implementation of finite element methods in the distributed and unified numerics environment (dune). *Kybernetika*, 46(2):294–315, 2010.
4. M. Blatt and P. Bastian. The iterative solver template library. In B. Kagström, E. Elmroth, J. Dongarra, and J. Wasniewski, editors, *Applied Parallel Computing. State of the Art in Scientific Computing*, number 4699 in Lecture Notes in Scientific Computing, pages 666–675, 2007.
5. Y. Epshteyn and B. Rivière. Estimation of penalty parameters for symmetric interior penalty Galerkin methods. *Journal of Computational and Applied Mathematics*, 206:843–872, 2007.
6. A. Ern, A. F. Stephansen, and P. Zunino. A discontinuous Galerkin method with weighted averages for advection-diffusion equations with locally small and anisotropic diffusivity. *IMA Journal of Numerical Analysis*, 29:235–256, 2009. doi:10.1093/imanum/drm050.
7. P. Houston and R. Hartmann. An optimal order interior penalty discontinuous Galerkin discretization of the compressible Navier-Stokes equations. *J. Comp. Phys.*, 227:9670–9685, 2008.
8. B. Rivière. *Discontinuous Galerkin methods for solving elliptic and parabolic equations*. Frontiers in Applied Mathematics. SIAM, 2008.

The paper is in final form and no similar paper has been or is being submitted elsewhere.

Benchmark 3D: A Mimetic Finite Difference Method

Peter Bastian, Olaf Ippisch, and Sven Marnach

1 Presentation of the scheme

In the two-dimensional discretisation benchmark session at the FVCA5 conference, we participated with a Mimetic Finite Difference (MFD) method [7]. In this paper, we present results for the three-dimensional case using the same method. Since the previous conference, the equivalence of MFD, Hybrid Finite Volume and Mixed Finite Volume methods has been demonstrated in [6]. Our outline of the method as used in our computations follows the exposition in [5].

First, the diffusion problem is restated as a system of two first order PDEs

$$\begin{aligned} \text{div}\mathbf{v} &= f \quad \text{in } \Omega, \\ \mathbf{v} &= -\mathbf{K}\nabla u \quad \text{in } \Omega, \\ u &= \bar{u} \quad \text{on } \Gamma_D, \\ \mathbf{K}\nabla u \cdot n &= g \quad \text{on } \Gamma_N. \end{aligned} \qquad (1)$$

Our aim will be the definition of discrete analogues of the divergence operator div and the flux operator $-\mathbf{K}\nabla$. To this end, we first define the spaces of discrete scalar and vector functions. Let $\mathcal{T}_h$ denote a conforming triangulation of the domain Ω. The elements $E \in \mathcal{T}_h$ are assumed to be polyhedra. For details on the requirements of $\mathcal{T}_h$, see [4].

A *discrete scalar function* u is assumed to be constant on the elements E of the triangulation. The value of u in the element E is denoted by u_E. The dimension of the space Q_h of discrete scalar functions is equal to the number of elements of $\mathcal{T}_h$.

Peter Bastian, Olaf Ippisch, and Sven Marnach
Interdisciplinary Center for Scientific Computing, University of Heidelberg.
Corresponding author Sven Marnach, e-mail: sven.marnach@iwr.uni-heidelberg.de

A *discrete vector function* **v** is given by assigning a real number $\mathbf{v}_E^e$ to each face e of each element E. These numbers are regarded as the normal components of the vector function with respect to the outer normal $\mathbf{n}_E^e$ on the face e. For a face e that is shared by elements E_1 and E_2, we require the compatibility of the normal components

$$\mathbf{v}_{E_1}^e = -\mathbf{v}_{E_2}^e. \tag{2}$$

Thus the dimension of the space X_h of discrete vector functions is equal to the number of faces of $\mathscr{T}_h$.

We now introduce the discrete differential operators. Again, see [4] for the details. The *discrete divergence operator* $\mathrm{div}_h : X_h \to Q_h$ is defined to comply with the divergence theorem on each element,

$$(\mathrm{div}_h \mathbf{v})_E = \frac{1}{|E|} \sum_{e \subset \partial E} |e|\, v_E^e, \tag{3}$$

where the sum is over all faces e of the element E. For the definition of the *discrete flux operator*, we introduce additional scalar unknowns on each face e of the triangulation denoted by u^e, and define $(-\mathbf{K}\nabla)_h : Q_h \to X_h$ by

$$\bigl((-\mathbf{K}\nabla)_h u\bigr)_E^e = \sum_{f \subset \partial E} \mathbb{W}_E^{e,f} |f| (u_E - u^f), \tag{4}$$

where $\mathbb{W}_E$ is a symmetric and positive definite matrix defined below. The scalar unknowns on the faces can be eliminated by compatibility requirement (2) on the inner faces and by boundary conditions (1) on the outer faces.

Now we can give the whole linear system discretising the linear diffusion problem. For each element E, we have the equation

$$\mathrm{div}_h (-\mathbf{K}\nabla)_h u = \frac{1}{|E|} \sum_{e \subset \partial E} |e| \sum_{f \subset \partial E} \mathbb{W}_E^{e,f} |f| (u_E - u^f) = q_E. \tag{5}$$

The equation for an inner face e shared by the elements E_1 and E_2 reads

$$\bigl((-\mathbf{K}\nabla)_h u\bigr)_{E_1}^e + \bigl((-\mathbf{K}\nabla)_h u\bigr)_{E_2}^e$$
$$= \sum_{f \subset \partial E_1} \mathbb{W}_{E_1}^{e,f} |f| (u_{E_1} - u^f) + \sum_{f \subset \partial E_2} \mathbb{W}_{E_2}^{e,f} |f| (u_{E_2} - u^f) = 0. \tag{6}$$

For each face e on the Neumann boundary Γ_N, we get

$$\bigl((-\mathbf{K}\nabla)_h u\bigr)_E^e = \sum_{f \subset \partial E} \mathbb{W}_E^{e,f} |f| (u_E - u^f) = g^e. \tag{7}$$

Finally, a face e on the Dirichlet boundary provides us with the trivial equation

$$u^e = \bar{u}^e. \tag{8}$$

To obtain a symmetric and positive definite stiffness matrix, we first eliminate the unknowns on the Dirichlet boundaries. Then, we scale (5) by $|E|$ and (6) as well as (7) by $-|e|$, see [8].

We finally give the definition of the matrix $\mathbb{W}_E$ for an element E. Let k_E denote the number of faces of E. Define the $k_E \times 3$ matrices $\mathbb{R}$ and $\mathbb{N}$ by

$$\mathbb{R}_{e,i} = \int_e (x_i - x_{E,i}), \quad \mathbb{N}_{e,i} = \mathbf{e}_i \cdot n_E^e, \tag{9}$$

where e ranges over the faces of E, $i = 1, 2, 3$, x_i is the i-th coordinate function, x_E is the "centre" of E and $\mathbf{e}_i$ is the unit vector in the direction of the i-th axis. The centre point x_E can be chosen on each cell individually (subject to some restrictions). A good choice is to use the centre of mass, which we used for the tetrahedral mesh. For the hexahedral meshes, we used the image of the centre of the unit cube under the usual trilinear coordinate mappings.

We construct a $k_E \times k_E$ matrix $\mathbb{W}_E$ according to algorithm 1 in [5]. In short, that means the following:

1. Orthonormalise the columns of the matrix $\mathbb{R}$ using the Gram–Schmidt algorithm and call the resulting matrix $\tilde{\mathbb{R}}$.
2. Set $\mathbb{D} = \mathbb{I} - \tilde{\mathbb{R}}\tilde{\mathbb{R}}^T$, where $\mathbb{I}$ denotes the $k_E \times k_E$ unit matrix.
3. Define $\mathbb{W}_E$ by

$$\mathbb{W}_E = \frac{1}{|E|} \mathbb{N} \mathbf{K} \mathbb{N}^T + \omega \mathbb{D}, \tag{10}$$

where ω is an arbitrary positive real number and $\mathbf{K}$ is simply evaluated at the cell centre x_E.

We used the common choice for ω

$$\omega = \frac{\text{trace } \mathbf{K}}{|E|}, \tag{11}$$

which was suggested in [5].

2 Numerical results

For estimating the L_2 error, we compared the approximate solution u_E on a cell E with the average

$$u_{E,\text{exact}} = \frac{1}{|E|} \int_E u(x) dx$$

of the exact solution over the cell,

$$\text{erl2}^2 = \frac{\sum_E |E|(u_E - u_{E,\text{exact}})^2}{\sum_E |E|(u_{E,\text{exact}})^2}. \tag{12}$$

The MFD method provides values for the fluxes on the faces, but does not allow the direct computation of approximate gradients of the solution. In some cases it would be possible to get a reconstruction of the gradients in the interior of a cell using the Piola transformation, but this fails, for example, for cells with hanging nodes on some faces. To circumvent these problems, we substituted the H_1 semi-norm by the somewhat unnatural "flux norm"

$$\text{ergrad}^2 = \frac{\sum_e |e|(\mathbf{v}^e - \mathbf{v}^e_{\text{exact}})^2}{\sum_e |e|(\mathbf{v}^e_{\text{exact}})^2}, \tag{13}$$

where the exact average flux over the face e is given by

$$\mathbf{v}^e_{\text{exact}} = \frac{1}{|e|} \int_e \mathbf{n}(x) \cdot A(x) \nabla u(x) dx$$

We did not provide any values for the energy norm E. Though it is possible to give an approximation of the energy norm using the formulation

$$E = \int_\Omega \mathbf{K} \nabla u \cdot \nabla u$$
$$= \int_\Omega \mathbf{K}^{-1}(-\mathbf{K}\nabla u) \cdot (-\mathbf{K}\nabla u)$$
$$= \sum_{E \in \mathcal{T}_h} \int_E \mathbf{K}^{-1}(-\mathbf{K}\nabla u) \cdot (-\mathbf{K}\nabla u)$$

and the scalar product on X_h, this would not provide much information, since E would coincide with the exact energy norm up to the accuracy of the linear solver by construction of the method.

All numerical tests have been performed with the DUNE software framework [1,3] using the `dune-pdelab` discretisation framework described in [2].

- **Test 1 Mild anisotropy,** $u(x,y,z) = 1 + \sin(\pi x)\sin\left(\pi\left(y+\frac{1}{2}\right)\right)\sin\left(\pi\left(z+\frac{1}{3}\right)\right)$
min = 0, max = 2, **Tetrahedral meshes**

i	nu	nmat	umin	uemin	umax	uemax	normg
1	6311	46371	2.26E-02	2.03E-02	1.986	1.989	0.000
2	12146	90106	5.50E-03	6.84E-03	1.989	1.989	0.000
3	23859	178079	8.50E-03	9.13E-03	1.994	1.994	0.000
4	46957	352277	5.10E-03	5.52E-03	1.997	1.997	0.000
5	93267	702867	1.91E-03	1.49E-03	1.996	1.997	0.000
6	186040	1407080	1.75E-03	1.83E-03	1.997	1.997	0.000

i	nu	erl2	ratiol2	ergrad	ratiograd	ener	ratioener
1	6311	4.55E-03	0.000	1.41E-01	0.000	0.00E+00	0.000
2	12146	2.88E-03	2.102	1.05E-01	1.376	0.00E+00	0.000
3	23859	1.82E-03	2.030	9.73E-02	0.328	0.00E+00	0.000
4	46957	1.20E-03	1.859	7.03E-02	1.440	0.00E+00	0.000
5	93267	7.38E-04	2.120	6.13E-02	0.602	0.00E+00	0.000
6	186040	4.65E-04	2.004	4.75E-02	1.102	0.00E+00	0.000

- **Test 1 Mild anisotropy,** $u(x,y,z) = 1 + \sin(\pi x)\sin\left(\pi\left(y+\frac{1}{2}\right)\right)\sin\left(\pi\left(z+\frac{1}{3}\right)\right)$
min = 0, max = 2, **Kershaw meshes**

i	nu	nmat	umin	uemin	umax	uemax	normg
1	2240	23744	-6.03E-01	3.03E-02	2.100	1.958	0.000
2	17152	189184	-5.83E-03	1.06E-02	2.008	1.993	0.000
3	134144	1510400	-1.11E-03	1.75E-03	2.000	1.997	0.000
4	1060864	12070912	1.65E-04	7.14E-04	2.000	1.999	0.000

i	nu	erl2	ratiol2	ergrad	ratiograd	ener	ratioener
1	2240	3.27E-01	0.000	1.19E+01	0.000	0.00E+00	0.000
2	17152	5.28E-02	2.685	3.37E+00	1.859	0.00E+00	0.000
3	134144	8.89E-03	2.600	5.26E-01	2.709	0.00E+00	0.000
4	1060864	2.02E-03	2.146	1.12E-01	2.245	0.00E+00	0.000

- **Test 1 Mild anisotropy**, $u(x, y, z) = 1 + \sin(\pi x) \sin\left(\pi \left(y + \frac{1}{2}\right)\right) \sin\left(\pi \left(z + \frac{1}{3}\right)\right)$ min $= 0$, max $= 2$, **Checkerboard meshes**

i	nu	nmat	umin	uemin	umax	uemax	normg
1	192	2496	1.27E-01	1.54E-01	1.883	1.846	0.000
2	1488	25248	-1.32E-01	4.01E-02	2.150	1.960	0.000
3	11712	225696	-5.37E-02	1.01E-02	2.053	1.990	0.000
4	92928	1905600	-1.40E-02	2.54E-03	2.014	1.997	0.000
5	740352	15655296	-3.52E-03	6.36E-04	2.004	1.999	0.000

i	nu	erl2	ratiol2	ergrad	ratiograd	ener	ratioener
1	192	2.81E-01	0.000	4.00E-01	0.000	0.00E+00	0.000
2	1488	1.19E-01	1.263	2.15E-01	0.906	0.00E+00	0.000
3	11712	3.44E-02	1.801	1.13E-01	0.936	0.00E+00	0.000
4	92928	9.39E-03	1.879	5.71E-02	0.992	0.00E+00	0.000
5	740352	2.46E-03	1.936	2.86E-02	1.001	0.00E+00	0.000

- **Test 3 Flow on random meshes**, $u(x, y, z) = \sin(2\pi x) \sin(2\pi y) \sin(2\pi z)$, min $= -1$, max $= 1$, **Random meshes**

i	nu	nmat	umin	uemin	umax	uemax	normg
1	304	2992	-1.02E+00	-7.59E-01	1.045	0.691	0.000
2	2240	23744	-9.79E-01	-9.39E-01	1.019	0.923	0.000
3	17152	189184	-1.02E+00	-9.85E-01	1.008	0.982	0.000
4	134144	1510400	-1.00E+00	-9.96E-01	1.000	0.996	0.000

i	nu	erl2	ratiol2	ergrad	ratiograd	ener	ratioener
1	304	8.38E-01	0.000	9.62E-01	0.000	0.00E+00	0.000
2	2240	1.58E-01	2.502	3.14E-01	1.679	0.00E+00	0.000
3	17152	3.91E-02	2.060	1.29E-01	1.314	0.00E+00	0.000
4	134144	1.15E-02	1.788	6.61E-02	0.975	0.00E+00	0.000

Benchmark 3D: A Mimetic Finite Difference Method 967

- **Test 4 Flow around a well, Well meshes,** min = 0, max = 5.415

i	nu	nmat	umin	uemin	umax	uemax	normg
1	3888	41268	5.74E-01	4.14E-01	5.317	5.317	0.000
2	9464	103208	2.96E-01	2.44E-01	5.328	5.328	0.000
3	20902	231574	1.75E-01	1.54E-01	5.329	5.329	0.000
4	46203	517443	1.30E-01	1.18E-01	5.330	5.330	0.000
5	94885	1069705	9.66E-02	8.99E-02	5.339	5.339	0.000
6	173515	1964101	7.67E-02	7.23E-02	5.345	5.345	0.000
7	303058	3439576	5.91E-02	5.65E-02	5.361	5.361	0.000

i	nu	erl2	ratiol2	ergrad	ratiograd	ener	ratioener
1	3888	5.53E-03	0.000	2.70E-01	0.000	0.00E+00	0.000
2	9464	1.64E-03	4.090	1.08E-01	3.089	0.00E+00	0.000
3	20902	8.43E-04	2.530	5.04E-02	2.888	0.00E+00	0.000
4	46203	8.27E-04	0.074	2.84E-02	2.173	0.00E+00	0.000
5	94885	6.83E-04	0.796	1.70E-02	2.142	0.00E+00	0.000
6	173515	4.84E-04	1.709	1.12E-02	2.060	0.00E+00	0.000
7	303058	4.07E-04	0.932	7.78E-03	1.965	0.00E+00	0.000

- **Test 5 Discontinuous permeability,** $u(x, y, z) = a_i \sin(2\pi x) \sin(2\pi y) \sin(2\pi z)$, min = -100, max = 100, **Locally refined meshes**

i	nu	nmat	umin	uemin	umax	uemax	normg
1	115	1231	-2.51E+02	-1.00E+02	250.808	100.000	0.000
2	812	8972	-4.44E+01	-3.54E+01	44.367	35.355	0.000
3	6064	68272	-8.32E+01	-7.89E+01	83.205	78.858	0.000
4	46784	532160	-9.56E+01	-9.43E+01	95.600	94.346	0.000
5	367360	4201216	-9.89E+01	-9.86E+01	98.887	98.562	0.000

i	nu	erl2	ratiol2	ergrad	ratiograd	ener	ratioener
1	115	8.77E+00	0.000	2.92E+00	0.000	0.00E+00	0.000
2	812	7.09E-01	3.860	7.88E-01	2.012	0.00E+00	0.000
3	6064	1.41E-01	2.413	3.15E-01	1.367	0.00E+00	0.000
4	46784	3.33E-02	2.116	2.13E-01	0.576	0.00E+00	0.000
5	367360	8.21E-03	2.038	1.55E-01	0.464	0.00E+00	0.000

References

1. Bastian, P., Blatt, M., Dedner, A., Engwer, C., Klöfkorn, R., Kornhuber, R., Ohlberger, M., Sander, O.: A generic grid interface for parallel and adaptive scientific computing. part ii: Implementation and tests in dune. Computing **82**(2–3), 121–138 (2007)
2. Bastian, P., Heimann, F., Marnach, S.: Generic software components for finite elements (2009). To appear in the proceedings of Algorithmy 2009
3. Blatt, M., Bastian, P.: The iterative solver template library. In: B. Kagström, E. Elmroth, J. Dongarra, J. Wasniewski (eds.) Applied Parallel Computing. State of the Art in Scientific Computing, *Lecture Notes in Scientific Computing*, vol. 4699, pp. 666–675. Springer (2007)
4. Brezzi, F., Lipnikov, K., Shashkov, M.: Convergence of the mimetic finite difference method for diffusion problems on polyhedral meshes. SIAM Journal on Numerical Analysis **43**(5), 1872–1896 (2005)
5. Brezzi, F., Lipnikov, K., Simoncini, V.: A family of mimetic finite difference methods on polygonal and polyhedral meshes. Mathematical Models and Methods in Applied Sciences **15**(10), 1533–1551 (2005)
6. Droniou, J., Eymard, R., Gallouët, T., Herbin, R.: A unified approach to mimetic finite difference, hybrid finite volume and mixed finite volume methods. Mathematical Models and Methods in Applied Sciences **20**(2), 265 – 295 (2010)
7. Marnach, S.: Benchmark on anisotropic problems – a mimetic finite difference method. In: J.M.H. Robert Eymard (ed.) Finite Volumes for Complex Applications V – Problems and Perspectives. ISTE Ltd and John Wiley & Sons Inc (2008)
8. Morel, J.E., Roberts, R.M., Shashkov, M.J.: A local support-operators diffusion discretization scheme for quadrilateral r-z meshes. Journal of Computational Physics **144**, 17–51 (1998)

The paper is in final form and no similar paper has been or is being submitted elsewhere.

Benchmark 3D: A Composite Hexahedral Mixed Finite Element

Ibtihel Ben Gharbia, Jérôme Jaffré, N. Suresh Kumar, and Jean E. Roberts

1 The Numerical Scheme

The numerical method used here (see [6]) is a mixed finite element method based on the weak formulation of the problem:

Find $(p, \boldsymbol{u}) \in L^2(\Omega) \times H(\text{div}; \Omega)$ such that

$$\int_\Omega \boldsymbol{K}^{-1}\boldsymbol{u} \cdot \boldsymbol{v} - \int_\Omega p \, \text{div}\boldsymbol{v} = -\int_{\Gamma_D} \bar{p} \, \boldsymbol{v} \cdot \boldsymbol{n} \quad \forall \boldsymbol{v} \in H(\text{div}; \Omega) \quad (1)$$

$$\int_\Omega \text{div}\boldsymbol{v} \, q = \int_\Omega fq \quad \forall q \in L^2(\Omega).$$

Straightforward extensions of the Raviart-Thomas-Nédélec mixed finite elements [3, 5] to hexahedral meshes do not converge. Therefore in [6] a composite mixed finite element was introduced and analyzed.

Given a discretization $\mathcal{T}_h$ of Ω into hexahedra (with planar faces) we solve the following system:

Ibtihel Ben Gharbia, Jérôme Jaffré, and Jean E. Roberts
INRIA Paris-Rocquencourt, 78153 LeChesnay, France,
e-mail: ibtihel.ben-gharbia@inria.fr, jerome.jaffre@inria.fr, jean.roberts@inria.fr

N. Suresh Kumar
Department of Mathematics, National Institute of Technology Calicut, India,
e-mail: sureshknsk@gmail.com

Find $(p_h, \boldsymbol{u}_h) \in M_h \times \boldsymbol{W}_h$ such that

$$\int_\Omega K^{-1} \boldsymbol{u}_h \cdot \boldsymbol{v}_h - \int_\Omega p_h \mathrm{div} \boldsymbol{v}_h = -\int_{\Gamma_D} \bar{p} \boldsymbol{v}_h \cdot \boldsymbol{n} \quad \forall \boldsymbol{v}_h \in \boldsymbol{W}_h, \qquad (2)$$

$$\int_\Omega \mathrm{div} \boldsymbol{v}_h \, q_h = \int_\Omega f q_h \quad \forall q_h \in M_h,$$

where $M_h \subset L^2(\Omega)$ is the space of piecewise constant functions (just as in the lowest order Raviart-Thomas-Nedelec spaces for tetrahedra or for rectangular solids). The space $\boldsymbol{W}_h \subset H(\mathrm{div}; \Omega)$ is constructed following ideas of Kuznetsov and Repin see [4]. It is a space of composite elements satisfying the following 4 conditions (all of which are satisfied by the Raviart-Thomas-Nédelec elements when the underlying spatial discretization is made up of tetrahedra and/or rectangular solids):

- $\boldsymbol{W}_h \subset H(\mathrm{div}; \Omega)$; i.e. elements of $\boldsymbol{W}_h$ are locally in $H(\mathrm{div}; T)$; $\forall T \in \mathcal{T}_h$, and normal components of elements of $\boldsymbol{W}_h$ are continuous across edges of the hexahedra in $\mathcal{T}_h$.
- normal components of elements of $\boldsymbol{W}_h$ are constant on each face of an element of $\mathcal{T}_h$.
- $\mathrm{div} \boldsymbol{W}_h \subset M_h$; i.e. the divergence of an element of $\boldsymbol{W}_h$ is constant on each hexahedron of $\mathcal{T}_h$.
- an element of $\boldsymbol{W}_h$ is uniquely determined by its flux through the faces of elements of $\mathcal{T}_h$; i.e. $\boldsymbol{W}_h$ has a basis of functions $\{\boldsymbol{v}_F : F \in \mathcal{F}_h\}$, where $\mathcal{F}_h$ is the set of all faces of hexahedra in $\mathcal{T}_h$, not lying on Γ_N, and for $F \in \mathcal{F}_h$, $\boldsymbol{v}_F$ is the unique function in $\boldsymbol{W}_h$ having normal component with flux across the face $E \in \mathcal{F}_h$ equal to $\delta_{E,F}$.

The space $\boldsymbol{W}_h$ is constructed element by element: for an element $T \in \mathcal{T}_h$ we define the space $\boldsymbol{W}_T$ of functions on T, and then $\boldsymbol{W}_h$ is defined to be the subspace of $H(\mathrm{div}; \Omega)$ consisting of those functions whose restriction to T is in $\boldsymbol{W}_T$ for each $T \in \mathcal{T}_h$. To construct $\boldsymbol{W}_T$ for an element $T \in \mathcal{T}_h$, T is subdivided into 5 tetrahedra as follows: starting from any vertex V_1 of T there are 3 vertices (say V_2, V_4, and V_5) of T that can be joined to V_1 by an edge of T, there are 3 other vertices (say V_3, V_6, and V_8) that lie on a face with V_1 (but not on an edge with V_1). The remaining vertex V_7 together with V_2, V_4, and V_5 forms a tetrahedron S_0 having no face lying on the boundary of T. Then $T \setminus S_0$ is made up of 4 tetrahedra S_1, S_2, S_3 and S_4, each of which has 3 faces lying on the surface of T and one face in common with S_0; see Fig. 1.

The collection of tetrahedra $\mathcal{T}_T = \{S_i : i = 0, 1, \cdots, 4\}$ is a discretization of T by tetrahedra, and we denote by $\widetilde{\boldsymbol{W}}_T$ the Raviart-Thomas-Nédelec space of lowest order associated with $\mathcal{T}_T$. We let $\widetilde{M}_T$ denote the set of functions constant on each of the five tetrahedra in $\mathcal{T}_T$, let $\widetilde{\boldsymbol{W}}_{T,0} \subset \widetilde{\boldsymbol{W}}_T$ denote the set of functions in $\widetilde{\boldsymbol{W}}_T$ whose normal traces vanish on all of ∂T, and let $|T|$ denote the volume of T. For each face F of T, letting $|F|$ denote the area of F and letting $\widetilde{\boldsymbol{W}}_{T,F} \subset \widetilde{\boldsymbol{W}}_T$ denote the set of functions in $\widetilde{\boldsymbol{W}}_T$ whose normal traces vanish on all of $\partial T \setminus F$ and are identically

Fig. 1 A partition of the reference hexahedron into 5 tetrahedra: one tetrahedron lies in the interior of T and is determined by the vertices V_2, V_4, V_5, V_7. The four other tetrahedra have each three faces on the surface of T and each contains one of the vertices V_1, V_3, V_6, V_8. There are two possible such constructions depending on which vertex is chosen as V_1

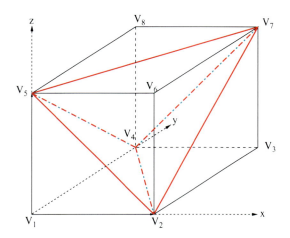

equal to $\frac{1}{|F|}$ on F, we define $\boldsymbol{v}_F$ to be the second component of the solution of the problem

$$\text{Find}(q_F, \boldsymbol{v}_F) \in \widetilde{M}_T \times \widetilde{\boldsymbol{W}}_{T,F} \text{ such that}$$

$$\int_T \boldsymbol{v}_F \cdot \boldsymbol{v}_h - \int_T q_F \text{div}\boldsymbol{v}_h = 0, \quad \forall \boldsymbol{v}_h \in \widetilde{\boldsymbol{W}}_{T,0}, \tag{3}$$

$$\int_T \text{div}\boldsymbol{v}_F \, q_h = \frac{1}{|T|} \int_T q_h \quad \forall q_h \in \widetilde{M}_T.$$

The pure Neumann problem (3) has a solution since the compatibility condition - that the integral over ∂T of the Neumann data function be equal to the integral over T of the source term - is satisfied. The second component $\boldsymbol{v}_F$ of the solution is uniquely determined: in the algebraic system associated with problem (3), the four equations corresponding to the four exterior tetrahedra, $S_1, \cdots, S_4$, determine $\boldsymbol{v}_F$, the equation associated with S_0 is redundant but is not a problem since the compatibility condition is satisfied. (The four equations associated with the internal faces, the four faces of S_0, imply that q_F is constant on all of T, but do not determine the value of the constant, but this is not needed here.) Then $\boldsymbol{W}_T \subset \widetilde{\boldsymbol{W}}_T$ is defined to be simply the six-dimensional subspace generated by the basis elements $\{\boldsymbol{v}_F : F \text{ is a face of } T\}$. Now defining $\boldsymbol{W}_h$ by

$$\boldsymbol{W}_h = \{\boldsymbol{v} \in H(\text{div}; \Omega) : \boldsymbol{v}_{|T} \in \boldsymbol{W}_T, \quad \forall T \in \mathscr{T}_h\},$$

one can easily check that $\boldsymbol{W}_h$ satisfies the four conditions listed above.

Remark 1. We point out that there are two possible choices for $\mathscr{T}_T$ (and thus for $\boldsymbol{W}_T$) depending on whether (in the notation used above) vertices $\{V_2, V_4, V_5, V_7\}$ or the vertices $\{V_1, V_3, V_6, V_8\}$ are used to form the interior tetrahedron. Also it is not always possible to choose the sets $\mathcal{T}_T$ is such a way that $\cup_{T\in\mathcal{T}_h}\mathcal{T}_T$ forms a finite element decomposition of Ω into tetrahedra.

Remark 2. This method is not appropriate for meshes containing deformed cubes which are not true hexahedra; i.e. for meshes containing deformed cubes with nonplanar "faces". In applications nonplanar "faces" arise when a cube is deformed in such a way that four vertices defining a face of the cube are moved so that they are no longer planar. However any three of the vertices remain planar so for either choice of the decomposition into five tetrahedra the nonplanar "face" is divided into two (planar) triangles so that one obtains a polyhedron (with planar faces) of from seven to twelve sides depending on how many nonplanar "faces" the original "hexahedron" had. Thus the new polyhedron is divided into five tetrahedra and one could generalize the method used here to include this case. However, as mentioned above it may not be possible to choose the divisions of the hexahedra into five tetrahedra in such a way that the resulting collection of tetrahedra forms a finite element mesh; i. e. in such a way that the resulting division of the quadrilateral interior faces into two triangles is compatible for each pair of adjacent "hexahedra". The resulting pair of adjacent polyhedra may then either overlap or leave a void space between the two polyhedra. A new composite mixed finite element is now under development to treat the case of nonplanar faces.

Remark 3. One could in a perhaps more natural way divide each of the hexahedra into 6 tetrahedra (all of equal volume for the reference hexahedron) by adding a central edge between any single pair of vertices not belonging to a common face. The six tetrahedra would all have this edge in common and each would have two internal faces and two external faces. One could form a system similar to (2) for each of the six faces of T. The dimension of $\widetilde{M}_T$ would then be 6 instead of 5 and that of $\widetilde{W}_{T,0}$ would be 6 instead of 4 as there would be 6 interior faces. The six equations of the linear system corresponding to one of the six tetrahedra would each only give a relation between the fluxes through the internal faces of the tetrahedron, so the second component of the solution would be determined only up to a (divergence free) flow going around the central edge. One would then need to impose a condition to make the macro elements rotational free (as are the Raviart-Thomas-Nédelec elements on tetrahedra and on rectangular solids as well as are those defined above on hexahedra using a decomposition into five tetrahedra). We have not further investigated this possibility.

Error estimates

In this paragraph we briefly recall the error estimates obtained in [6]. Following [1] we define the notion of shape regularity for a family of meshes of hexahedra.

Definition 1. For S a tetrahedron let ρ_S and h_S denote respectively the radius of the inscribed sphere of S and the diameter of S. Then for a hexahedron T, as seen earlier, there are two possible ways of decomposing T into five tetrahedra. Let ρ_T

be the smallest of the ρ_S's for these 10 tetrahedra, let h_T be the diameter of T and let $\sigma_T = h_T/\rho_T$ be the shape constant of T. For a mesh $\mathcal{T}_h$ of hexahedra, the shape constant of $\mathcal{T}_h$ is the largest σ_T for $T \in \mathcal{T}_h$. A family $\{\mathcal{T}_h : h \in \mathcal{H}\}$ of meshes $\mathcal{T}_h$ made up of hexahedra is said to be *shape regular* if the shape constants for the meshes can be uniformly bounded.

In [6] it is shown that if $(p, u) \in L^2(\Omega) \times H(\text{div}; \Omega)$ is the solution of problem (1) and $(p_h, u_h) \in M_h \times W_h$ is the solution of problem (2) and the family $\{\mathcal{T}_h : h \in \mathcal{H}\}$ of meshes $\mathcal{T}_h$ made up of hexahedra is shape regular then there is a constant C independent of h such that

$$\|p_h - p\|^2_{L^2(\Omega)} + \|u_h - u\|^2_{H(\text{div};\Omega)} \le C h^2 \left(|p|^2_{H^1(\Omega)} + \|u\|^2_{H^1(\Omega)} + \|\text{div} u\|^2_{H^1(\Omega)}\right),$$

provided that p and u are sufficiently regular for the righthand side to be defined.

Mixed-hybrid finite elements and solution of the linear problem

As with the Raviart-Thomas-Nédelec elements for tetrahedra and rectangular solids, the solution (u_h, p_h) is sought in a subspace $M_h \times W_h$ of $L^2(\Omega) \times H(\text{div}; \Omega)$ in which the degrees of freedom are the average values of the pressure over the hexahedra of the grid and the fluxes through the faces of the grid. The resulting linear system then has exactly the same form as that for the Raviart-Thomas-Nédelec elements for grids of rectangular solids (when the problem has full tensor coefficients). As in [2] we can relax the condition that the approximate solution be sought in a subspace of $H(\text{div}; \Omega)$ and enforce this condition using Lagrange multipliers. We then define the approximation space W_h^* by

$$W_h^* = \{v \in (L^2(\Omega))^3 : v_{|T} \in W_T, \quad \forall T \in \mathcal{T}_h\},$$

and introduce a space of Lagrange multipliers $\Lambda_h = \{\lambda_h = \{\lambda_F\}_{F \in \mathcal{F}_h} \in R^{n_F}\}$ where n_F is the number of faces in $\mathcal{F}_h$. Then the following problem has a unique solution:

Find $(p_h^*, u_h^*, \lambda_h) \in M_h \times W_h^* \times \Lambda_h$ such that

$$\sum_{T \in \mathcal{T}_h} \int_T K^{-1} u_h^* \cdot v_h - \sum_{T \in \mathcal{T}_h} \int_T p_h^* \text{div} v_h - \sum_{F \in \mathcal{F}_h} \int_F \lambda_F [v_h \cdot n_F] =$$
$$- \int_{\Gamma_D} \bar{p} v_h \cdot n \quad \forall v_h \in W_h^*,$$

$$\sum_{T \in \mathcal{T}_h} \text{div} u_h^* \, q_h = \int_\Omega f q_h \quad \forall q_h \in M_h,$$

$$\sum_{F \in \mathcal{F}_h} \int_F [u_h^* \cdot n_F] \mu_F = 0, \quad \forall \mu_h \in \Lambda_h,$$

where for $F \in \mathscr{F}_h$, $\boldsymbol{n}_F$ is a unit vector normal to F and for $\boldsymbol{v}_h \in \boldsymbol{W}_h^*$, $[\boldsymbol{v}_h \cdot \boldsymbol{n}_F]$ denotes the jump across F of $\boldsymbol{v}_h \cdot \boldsymbol{n}_F$ in the direction of $\boldsymbol{n}_F$. As with the Raviart-Thomas-Nédelec method it is now easy to eliminate first $\boldsymbol{u}_h^*$ and then p_h^* from the linear system and thus obtain a symmetric positive definite system with λ_h as the only unknown. For $F \in \mathscr{F}_h$ the multiplier λ_F enforcing continuity of $\boldsymbol{u}_h^* \cdot \boldsymbol{n}_F$ across F is in fact an approximation of the trace of the pressure p on F.

Once λ_h is found one can recover $\boldsymbol{u}_h^*$ and p_h^* through local calculations given by the first two equations of system (4). One shows easily that $\boldsymbol{u}_h^*$ is in fact in $\boldsymbol{W}_h$ and is equal to $\boldsymbol{u}_h$ and that $p_h^* = p_h$.

2 Numerical experiments

The data are provided by the *FVCA6 3D anisotropic benchmark*. We have chosen to do the first test case with mild anisotropy and Kershaw grids. Table 1 gives results obtained for the 4 Kershaw meshes which were proposed in the benchmark. The index i, i = 1, 2, 3, 4, denotes the mesh index for the $8 \times 8 \times 8$, the $16 \times 16 \times 16$, the $32 \times 32 \times 32$, and the $64 \times 64 \times 64$ Kershaw meshes respectively. As mentioned earlier, the matrix of the linear system associated with the mixed-hybrid finite element after elimination of p_h^* and $\boldsymbol{u}_h^*$ is symmetric and positive definite, and the unknowns are the Lagrangian multipliers λ_h which are approximations of the averages of the trace of the scalar variable (pressure) over the faces. From λ_h local calculations yield the cell pressure unknowns of p_h and the fluxes across the faces of the velocity $\boldsymbol{u}_h$.

In Table 1 nu, the number of unknowns of the linear system, is the number of degrees of freedom for λ_h which is the number of faces. The number of matrix nonzeros, nmat, is given in the table for the full matrix (not the upper or lower halves). umin, uemin, λmin (resp. umax, uemax, λmax) the minimum (resp. maximum) of p_h, p and λ_h.

The function p_h is constant inside each hexahedral cell, so the L^2 error erl2 between p and p_h is calculated as

$$\text{erl2} = \frac{\sqrt{\int_\Omega (p - p_h)^2}}{\sqrt{\int_\Omega p^2}}$$

where the integrals in the numerator are calculated using on each cell an integration formula exact for polynomials of degree 2 in 3D.

The mixed finite element method calculates also the velocity $\boldsymbol{u}_h$ approximating the vector unknown $\boldsymbol{u} = -\boldsymbol{K}\nabla p$ as a piecewise polynomial vector function. The usual error calculated with the mixed method is the L^2 error for $\boldsymbol{u}_h$ in addition to the L^2 error for p_h. However in this benchmark the errors for p_h in the H^1 seminorm and in the energy norm are asked for. These norms are actually equivalent to the

Benchmark 3D: a Composite Hexahedral Mixed Finite Element

Table 1 Results obtained for a composite hexahedral mixed finite element on a sequence of Kershaw meshes

i	nu	nmat	umin	uemin	umax	uemax	normg
1	576	2496	-0.03255	0.	1.94685	2.	1.84064
2	4352	32512	-0.04618	0.	1.99488	2.	1.85063
3	33792	310272	-0.03621	0.	2.00028	2.	1.85242
4	266240	2682880	-0.00837	0.	2.00061	2.	1.84036

i	nu	erl2	ratiol2	ergrad	ratiograd	ener	ratioener
1	576	0.063751		1.63849		1.49726	
2	4352	0.038971	0.73	0.96309	0.79	0.86755	0.81
3	33792	0.019424	1.02	0.51181	0.92	0.44853	0.97
4	266240	0.009148	1.09	0.25421	1.02	0.21819	1.05

L^2 norm of $\boldsymbol{u}$. Indeed we have $|\nabla p|^2 = |\boldsymbol{K}^{-1}\boldsymbol{u}|^2$ and $(\boldsymbol{K}\nabla p) \cdot \nabla p = (\boldsymbol{K}^{-1}\boldsymbol{u}) \cdot \boldsymbol{u}$. Therefore we calculate the error for the gradient and the error in the energy norm with the formula

$$\text{ergrad} = \frac{\sqrt{\int_\Omega |\boldsymbol{K}^{-1}(\boldsymbol{u}-\boldsymbol{u}_h)|^2}}{\sqrt{\int_\Omega |\boldsymbol{K}^{-1}\boldsymbol{u}|^2}}, \quad \text{ener} = \frac{\sqrt{\int_\Omega (\boldsymbol{K}^{-1}(\boldsymbol{u}-\boldsymbol{u}_h)) \cdot (\boldsymbol{u}-\boldsymbol{u}_h)}}{\sqrt{\int_\Omega (\boldsymbol{K}^{-1}\boldsymbol{u}) \cdot \boldsymbol{u}}}$$

where again the integrals in the numerator were calculated with an integration formula exact for polynomials of degree 2 inside each cell.

Similarly the L^1 norm of the gradient of p_h was calculated as

$$\text{normgrad} = \int_\Omega |\boldsymbol{K}^{-1}\boldsymbol{u}_h|.$$

The rates of convergence ratiol2, ratioener and ratiograd are calculated as required by the benchmark by comparing the errors erl2, ergrad and ener obtained on meshes i and i-1 using the formula

$$\text{ratio}(i) = -3\frac{\log(\text{error}(i)/\text{error}(i-1))}{\log(\text{nu}(i)/\text{nu}(i-1))}.$$

All errors behave as predicted by the theory and show an asymptotic rate of convergence of order 1. The exact solution is such that $0 \leq p \leq 2$ and the calculated solution has small undershoots which become smaller as the meshes are refined.

3 Conclusion

In spite of the bad aspect ratios of some of the hexahedra in the Kershaw meshes, the proposed composite hexahedral mixed finite element shows first order convergence for the pressure as well as for the velocity, as it was predicted by the analysis of the method.

Acknowledgements This work was partially supported by the GNR MoMaS (PACEN/CNRS, ANDRA, BRGM, CEA, EDF, IRSN).

References

1. D. N. Arnold, D. Boffi, and R. S. Falk. Quadrilateral h(div) finite elements. *SIAM J. Numer. Anal.*, 42:2429–2451, 2005.
2. D. N. Arnold and F. Brezzi. Mixed and nonconforming finite element methods : implementation, postprocessing and error estimates. *M2AN*, 19:7–32, 1985.
3. F. Brezzi and M. Fortin. *Mixed and Hybrid Finite Element Methods*. Springer-Verlag, New York, 1991.
4. Yu. Kuznetzov and S. Repin. Convergence analysis and error estimates for mixed finite element method on distorted meshes. *Russ. J. Numer. Anal. Math. Modelling J. Numer. Math.*, 13:33–51, 2005.
5. J. E. Roberts and J.-M. Thomas. Mixed and hybrid methods. In P. G. Ciarlet and J. L. Lions, editors, *Handbook of Numerical Analysis Vol.II*, pages 523–639. North Holland, Amsterdam, 1991.
6. A. Sboui, J. Jaffré, and J. E. Roberts. A composite mixed finite element for hexahedral grids. *SIAM J. Sci. Comput.*, 31(3):2623–2645, 2009.

The paper is in final form and no similar paper has been or is being submitted elsewhere.

Benchmark 3D: CeVeFE-DDFV, a discrete duality scheme with cell/vertex/face+edge unknowns

Yves Coudière, Florence Hubert, and Gianmarco Manzini

1 Presentation of the scheme

The method that we investigate in this contribution was proposed by Y. Coudière and F. Hubert in [1] as a three-dimensional (3D) extension of the finite volume scheme previously studied by F. Hermeline in [4] and K. Domelevo and P. Omnès in [3]. This method belongs to the family of Discrete Duality Finite Volume (DDFV) methods, which can naturally handle anisotropic or non-linear problems on general distorted meshes.

In this benchmark paper, we present the results obtained by using the formulation in [1] and the variant for discontinuous permeabilities that is presented in the proceeding paper [2].

The DDFV method that we consider herein makes use of three polyhedral meshes for the solution approximation, denoted by $\mathcal{M}$, $\mathcal{N}$, $\mathcal{FE}$, and the mesh of diamonds for the solution gradient approximation, denoted by $\mathcal{D}$.

We denote the control volumes of the primal mesh $\mathcal{M}$ by K and L, and with every primal cell we associate an internal point, e.g., $x_K \in K$. Different choices are possible, which give rise to different versions of the same scheme, such as the barycenters or the arithmetic average of the position vector of cell vertices (also called "iso-barycenters"). For the results shown here, we used the second choice, but apparently there is no significant difference between the two choices mentioned above as far as accuracy and convergence behavior are concerned. The vertices, the

---

Yves Coudière
Laboratoire de Mathématiques Jean Leray, Nantes, FRANCE, e-mail: Yves.Coudiere@univ-nantes.fr

Florence Hubert
LATP, Université de Provence, Marseille, FRANCE, e-mail: fhubert@cmi.univ-mrs.fr

Gianmarco Manzini
IMATI and CESNA-IUSS, Pavia, ITALY, e-mail: gm.manzini@gmail.com

edges, and the faces of mesh $\mathscr{M}$ are denoted by x_A, E and F, respectively. Also, we denote the midpoint of E by x_E and the barycenter of F by x_F. We associate a degree of freedom (the scheme unknowns) with each one of these points; hence, the unknown scalar variable takes the form:

$$u^{\mathscr{T}} = \left((u_K)_{K\in\mathscr{M}}, (u_A)_{A\in\mathscr{N}}, (u_E)_{E\in\mathscr{E}}, (u_F)_{F\in\mathscr{F}}\right).$$

We denote the collections of the boundary items (vertices, edges and faces) by $\partial\mathscr{N}$, $\partial\mathscr{F}\mathscr{E}$ and we introduce the set of boundary cells $\partial\mathscr{M}$ which is composed by the boundary faces here considered as degenerated control volumes. Dirichlet boundary conditions are easily introduced into the scheme through the set of boundary data

$$\delta u^{\mathscr{T}} = \left((u_K)_{x_K\in\partial\mathscr{M}}, (u_A)_{x_A\in\partial\mathscr{N}}, (u_E)_{x_E\in\partial\mathscr{F}\mathscr{E}}, (u_F)_{x_F\in\partial\mathscr{F}\mathscr{E}}\right).$$

The scalar solution field u is approximated by the degrees of freedom $(u^{\mathscr{T}}, \delta u^{\mathscr{T}})$.

The gradient formula is given on each diamond cell $D \in \mathscr{D}$, which is the convex hull of the points K, L, x_A, x_B, x_F, x_E, by

$$\nabla^D_{\delta u} u^{\mathscr{T}} = \frac{1}{3|D|}\left((u_L - u_K)N_{KL} + (u_B - u_A)N_{AB} + (u_F - u_E)N_{EF}\right) \quad (1)$$

using the normal vectors $N_{KL} = \frac{1}{2}(x_B - x_A) \times (x_F - x_E)$, $N_{AB} = \frac{1}{2}(x_F - x_E) \times (x_L - x_K)$ and $N_{EF} = \frac{1}{2}(x_L - x_K) \times (x_B - x_A)$. Gradient formula (1) allows us to define the numerical flux through each interface of the control volumes of the three meshes $\mathscr{M}$, $\mathscr{N}$ and $\mathscr{F}\mathscr{E}$. Let $\mathbf{Q}$ be the linear space of piecewise constant vector fields defined on the mesh of diamonds $\mathscr{D}$ and X be the linear space of triples of piecewise constant scalar fields defined on the three meshes $\mathscr{M}$, $\mathscr{N}$ and $\mathscr{F}\mathscr{E}$. Three finite volume schemes are written by using a discrete divergence operator that maps each vector field in $\mathbf{Q}$ to a triple of scalar functions in X. Formally, we introduce the operator

$$\text{div}^{\mathscr{T}} : \xi = (\xi_D)_{D\in\mathscr{D}} \in \mathbf{Q} \mapsto (\text{div}^{\mathscr{M}}\xi, \text{div}^{\mathscr{N}}\xi, \text{div}^{\mathscr{F}\mathscr{E}}\xi) \in X$$

whose components

$$\text{div}^{\mathscr{M}}\xi = (\text{div}_K\xi)_K, \text{div}^{\mathscr{N}}\xi = (\text{div}_A\xi)_A \text{ and } \text{div}^{\mathscr{F}\mathscr{E}}\xi = \{(\text{div}_E\xi)_E, (\text{div}_F\xi)_F\}$$

are given by

$$|K|\text{div}_K\xi = \sum_{D\in D_K} \xi_D \cdot N_{KL}, \quad |A|\text{div}_A\xi = \sum_{D\in D_A} \xi_D \cdot N_{AB}, \quad (2)$$

$$|E|\text{div}_E\xi = \sum_{D\in D_E} \xi_D \cdot N_{EF}, \quad |F|\text{div}_F\xi = \sum_{D\in D_F} \xi_D \cdot (-N_{EF}). \quad (3)$$

In the previous statements, the symbols D_K, D_A, D_E, D_F refer to the diamond cells which overlap the cells labeled by the corresponding subscripted indices K, A, E, and L.

Since each of the $\text{div}_C \xi$ approximates $\frac{1}{|C|} \int_C \text{div} \xi$ (for $C = K, A, E, F$), the right hand side of the discrete problem is given by the piecewise constant projection of the function f onto the space X, $\pi^{\mathcal{T}} f = \{(f_K)_{K \in \mathcal{M}}, (f_A)_{A \in \mathcal{N}}, (f_E, f_F)_{E \in \mathcal{E}, F \in \mathcal{F}}\}$ with $f_C = \frac{1}{|C|} \int_C f(x) dx$ for any cell $C = K \in \mathcal{M}$ or $A \in \mathcal{N}$ or F or $E \in \mathcal{FE}$.

The CeVeFE-DDFV scheme reads:

$$-\text{div}^{\mathcal{T}}(\mathbf{K}_D \nabla^D_{\delta u} u^{\mathcal{T}}) = \pi^{\mathcal{T}} f, \qquad (4)$$

where $\mathbf{K}_D = \frac{1}{|D|} \int_D \mathbf{K}(x) dx$ is a piecewise constant tensor field on the mesh of the diamond cells. The scheme in (4) originates a symmetric and positive-definite linear system of equations (see [1] for a thourough discussion of the other properties). The case of the discontinuous permeability tensor of test 5 deserves a special treatment that is thouroughly discussed in [2].

Mesure on the error

To put the discrete and the exact solutions "at the same level", we use the projection $\pi^{\mathcal{T}} u_e$ of the exact solution and the associated discrete gradient reconstruction $\nabla^{\mathcal{T}} \pi^{\mathcal{T}} u_e$. Approximation errors are evaluated through the following norms:

$$\text{erl2} = \|e^{\mathcal{T}}\|_{L^2} / \|\pi^{\mathcal{T}} u_e\|_{L^2} \text{ with } \|e^{\mathcal{T}}\|^2_{L^2} = \frac{1}{3} \sum_{C \in \mathcal{M} \cup \mathcal{N} \cup \mathcal{FE}} |C||e_C|^2$$

$$\text{ergrad} = \|\nabla^{\mathcal{T}} e^{\mathcal{T}}\|_{L^2} / \|\nabla^{\mathcal{T}} \pi^{\mathcal{T}} u_e\|_{L^2} \text{ with } \|\nabla^{\mathcal{T}} e^{\mathcal{T}}\|^2 = \sum_{D \in \mathcal{D}} |D| |\nabla^D e^{\mathcal{T}}|^2$$

$$\text{ener} = (\mathbf{K}^{\mathcal{D}} \nabla^{\mathcal{T}} e^{\mathcal{T}}, \nabla^{\mathcal{T}} e^{\mathcal{T}})_{L^2} / (\mathbf{K}^{\mathcal{D}} \nabla^{\mathcal{T}} \pi^{\mathcal{T}} u_e, \nabla^{\mathcal{T}} \pi^{\mathcal{T}} u_e)_{L^2}$$

$$\text{with } (\mathbf{K}^{\mathcal{D}} \nabla^{\mathcal{T}} e^{\mathcal{T}}, \nabla^{\mathcal{T}} e^{\mathcal{T}})_{L^2} = \sum_{D \in \mathcal{D}} |D| (\mathbf{K}_D \nabla^D e^{\mathcal{T}}, \nabla^D e^{\mathcal{T}})$$

In the case of the discontinuous tensor of test 5, the diamond cell D is divided in two subdiamond cells, namely, D_K and D_L. The gradient $\nabla^D u$ is constant on D_K (respectively, D_L) with value $\nabla^D_K u$ (respectively, $\nabla^D_L u$). The quantities $\|\nabla^{\mathcal{T}} e^{\mathcal{T}}\|^2_{L^2}$ and $(\mathbf{K}^{\mathcal{D}} \nabla^{\mathcal{T}} e^{\mathcal{T}}, \nabla^{\mathcal{T}} e^{\mathcal{T}})_{L^2}$ become

$$\|\nabla^{\mathcal{T}} e^{\mathcal{T}}\|^2_{L^2} = \sum_{D \in \mathcal{D}} \left(|D_K| |\nabla_{D_K} e^{\mathcal{T}}|^2 + |D_L| |\nabla_{D_L} e^{\mathcal{T}}|^2 \right)$$

and

$$(\mathbf{K}^{\mathcal{D}} \nabla^{\mathcal{T}} e^{\mathcal{T}}, \nabla^{\mathcal{T}} e^{\mathcal{T}})_{L^2} = \sum_{D \in \mathcal{D}} \left(|D_K| (\mathbf{K}_{D_K} \nabla_{D_K} e^{\mathcal{T}}, \nabla_{D_K} e^{\mathcal{T}}) + |D_L| (\mathbf{K}_{D_L} \nabla_{D_L} e^{\mathcal{T}}, \nabla_{D_L} e^{\mathcal{T}}) \right).$$

2 Numerical results

The following results were obtained by using a BiCG-stab solver with ILU(0) preconditioner (routine MI26 of HSL implementation).

- **Test 1** Mild anisotropy, $u(x,y,z) = 1 + \sin(\pi x)\sin\left(\pi\left(y+\frac{1}{2}\right)\right)\sin\left(\pi\left(z+\frac{1}{3}\right)\right)$ min = 0, max = 2, **Tetrahedral meshes**

i	nu	nmat	umin	uemin	umax	uemax	normg
1	7777	100569	6.09E-03	1.05E-02	1.988	1.980	1.790
2	15495	208527	7.48E-03	9.35E-03	1.995	1.994	1.793
3	31139	431667	3.19E-03	5.93E-03	1.993	1.993	1.795
4	62419	885735	1.48E-03	2.98E-03	1.996	1.996	1.796
5	125993	1823199	1.56E-03	2.28E-03	2.000	1.999	1.797
6	254657	3746829	1.93E-03	2.70E-03	1.999	1.998	1.798

i	nu	erl2	ratiol2	ergrad	ratiograd	ener	ratioener
1	7777	0.228E-02	-	0.562E-01	-	0.528E-01	-
2	15495	0.147E-02	1.904	0.441E-01	1.051	0.415E-01	1.054
3	31139	0.916E-03	2.036	0.349E-01	1.011	0.327E-01	1.021
4	62419	0.573E-03	2.025	0.276E-01	1.006	0.258E-01	1.022
5	125993	0.374E-03	1.819	0.219E-01	0.994	0.206E-01	0.969
6	254657	0.231E-03	2.067	0.174E-01	0.983	0.163E-01	0.990

- **Test 1** Mild anisotropy, $u(x,y,z) = 1 + \sin(\pi x)\sin\left(\pi\left(y+\frac{1}{2}\right)\right)\sin\left(\pi\left(z+\frac{1}{3}\right)\right)$ min = 0, max = 2, **Voronoi meshes**

i	nu	nmat	umin	uemin	umax	uemax	normg
1	345	4559	7.93E-02	1.51E-01	1.875	1.844	1.719
2	933	12811	4.79E-02	4.74E-02	1.989	1.982	1.785
3	2075	29291	5.46E-02	5.64E-02	1.987	1.978	1.794
4	3963	56947	3.25E-02	3.23E-02	2.000	1.996	1.795
5	6909	101229	1.28E-02	3.17E-02	2.000	1.996	1.797

i	nu	erl2	ratiol2	ergrad	ratiograd	ener	ratioener
1	345	0.274E-01	-	0.179E+00	-	0.162E+00	-
2	933	0.223E-01	0.622	0.149E+00	0.556	0.139E+00	0.458
3	2075	0.119E-01	2.364	0.102E+00	1.409	0.964E-01	1.373
4	3963	0.819E-02	1.724	0.835E-01	0.933	0.782E-01	0.972
5	6909	0.599E-02	1.694	0.691E-01	1.021	0.655E-01	0.953

- **Test 1 Mild anisotropy,** $u(x,y,z) = 1 + \sin(\pi x)\sin\left(\pi\left(y+\frac{1}{2}\right)\right)\sin\left(\pi\left(z+\frac{1}{3}\right)\right)$
min = 0, max = 2, **Kershaw meshes**

i	nu	nmat	umin	uemin	umax	uemax	normg
1	3375	49071	5.67E-02	3.43E-02	1.940	1.974	1.767
2	29791	455895	9.19E-03	7.33E-03	1.988	1.991	1.782
3	250047	3916359	2.42E-03	1.59E-03	1.999	1.998	1.793
4	2048383	32446751	6.52E-04	6.17E-04	2.000	1.999	1.797

i	nu	erl2	ratiol2	ergrad	ratiograd	ener	ratioener
1	3375	0.287E-01	-	0.481E+00	-	0.589E+00	-
2	29791	0.113E-01	1.289	0.218E+00	1.088	0.233E+00	1.277
3	250047	0.330E-02	1.730	0.904E-01	1.243	0.953E-01	1.260
4	2048383	0.859E-03	1.922	0.395E-01	1.180	0.422E-01	1.161

- **Test 1 Mild anisotropy,** $u(x,y,z) = 1 + \sin(\pi x)\sin\left(\pi\left(y+\frac{1}{2}\right)\right)\sin\left(\pi\left(z+\frac{1}{3}\right)\right)$
min = 0, max = 2, **Checkerboard meshes**

i	nu	nmat	umin	uemin	umax	uemax	normg
1	239	2871	8.58E-02	8.40E-02	1.903	1.916	1.795
2	2543	34927	2.90E-02	2.13E-02	1.971	1.979	1.804
3	23135	336735	4.68E-03	5.35E-03	1.995	1.995	1.800
4	196799	2943487	1.69E-03	1.34E-03	1.998	1.999	1.799
5	1622399	24588351	2.88E-04	3.35E-04	2.000	2.000	1.799

i	nu	erl2	ratiol2	ergrad	ratiograd	ener	ratioener
1	239	0.307E-01	-	0.141E+00	-	0.139E+00	-
2	2543	0.120E-01	1.190	0.104E+00	0.384	0.101E+00	0.405
3	23135	0.323E-02	1.786	0.571E-01	0.814	0.550E-01	0.827
4	196799	0.830E-03	1.905	0.298E-01	0.909	0.285E-01	0.920
5	1622399	0.210E-03	1.955	0.154E-01	0.937	0.147E-01	0.945

- **Test 2 Heterogeneous anisotropy,** $u(x,y,z) = x^3 y^2 z + x\sin(2\pi xz)\sin(2\pi xy)\sin(2\pi z)$, min = -0.862, max = 1.0487, **Prism meshes**

i	nu	nmat	umin	uemin	umax	uemax	normg
1	12179	188089	-8.55E-01	-8.46E-01	1.014	1.009	1.693
2	96759	1545215	-8.55E-01	-8.57E-01	1.026	1.031	1.706
3	325739	5259545	-8.61E-01	-8.59E-01	1.037	1.035	1.708
4	771119	12518433	-8.60E-01	-8.60E-01	1.040	1.041	1.709

i	nu	erl2	ratiol2	ergrad	ratiograd	ener	ratioener
1	12179	0.392E-01	-	0.811E-01	-	0.803E-01	-
2	96759	0.109E-01	1.854	0.392E-01	1.054	0.397E-01	1.019
3	325739	0.502E-02	1.917	0.256E-01	1.051	0.261E-01	1.040
4	771119	0.287E-02	1.942	0.190E-01	1.039	0.194E-01	1.034

- **Test 3 Flow on random meshes,** $u(x, y, z) = \sin(\pi x) \sin(\pi y) \sin(\pi z)$, min $= 0$, max $= 1$, **Random meshes**

i	nu	nmat	umin	uemin	umax	uemax	normg
1	343	4447	-4.25E+01	-9.78E-01	49.169	0.931	38.139
2	3375	49855	-2.22E+01	-9.94E-01	21.970	0.982	21.514
3	29791	466111	-6.96E+00	-9.95E-01	7.051	0.993	12.536
4	250047	4019647	-2.67E+00	-9.98E-01	2.725	0.998	7.541

i	nu	erl2	ratiol2	ergrad	ratiograd	ener	ratioener
1	343	0.147E+03	-	0.238E+02	-	0.162E+01	-
2	3375	0.956E+01	3.589	0.121E+02	0.892	0.888E+00	0.787
3	29791	0.681E+00	3.640	0.632E+01	0.891	0.459E+00	0.909
4	250047	0.447E-01	3.840	0.314E+01	0.988	0.229E+00	0.979

- **Test 4 Flow around a well, Well meshes,** min $= 0$, max $= 5.415$

i	nu	nmat	umin	uemin	umax	uemax	normg
1	5868	86728	3.83E-01	4.13E-01	5.317	5.317	1596.292
2	15776	243104	2.37E-01	2.43E-01	5.328	5.328	1611.158
3	36846	580244	1.54E-01	1.54E-01	5.329	5.329	1617.452
4	84546	1350382	1.17E-01	1.18E-01	5.330	5.330	1620.143
5	177590	2860258	8.96E-02	8.98E-02	5.339	5.339	1621.406
6	329236	5329338	7.22E-02	7.23E-02	5.345	5.345	1622.053
7	580190	9422104	5.66E-02	5.64E-02	5.361	5.361	1622.472

i	nu	erl2	ratiol2	ergrad	ratiograd	ener	ratioener
1	5868	0.141E-04	-	0.128E+00	-	0.116E+00	-
2	15776	0.476E-05	3.290	0.877E-01	1.144	0.781E-01	1.212
3	36846	0.208E-05	2.924	0.610E-01	1.283	0.542E-01	1.287
4	84546	0.141E-05	1.411	0.466E-01	0.975	0.408E-01	1.033
5	177590	0.914E-06	1.747	0.362E-01	1.021	0.316E-01	1.023
6	329236	0.609E-06	1.976	0.293E-01	1.026	0.258E-01	0.998
7	580190	0.422E-06	1.941	0.244E-01	0.964	0.214E-01	0.971

- **Test 5 Discontinuous permeability,** $u(x, y, z) = \sin(\pi x)\sin(\pi y)\sin(\pi z)$, min = 0, max = 1, **Locally refined meshes**

i	nu	nmat	umin	uemin	umax	uemax	normg
1	131	1017	-6.34E+01	-1.00E+02	64.462	100.000	85.763
2	1215	8303	-3.10E+02	-1.00E+02	309.886	100.000	192.379
3	10463	65007	-1.34E+02	-1.00E+02	134.323	100.000	139.345
4	86847	509567	-1.09E+02	-1.00E+02	109.373	100.000	114.251
5	707711	4024287	-1.02E+02	-1.00E+02	102.394	100.000	104.279

i	nu	erl2	ratiol2	ergrad	ratiograd	ener	ratioener
1	131	0.218E+01	-	0.450E+00	-	0.406E+00	-
2	1215	0.193E+01	0.159	0.187E+01	-1.917	0.623E+00	-0.578
3	10463	0.862E-01	4.334	0.828E+00	1.134	0.297E+00	1.033
4	86847	0.517E-02	3.989	0.407E+00	1.006	0.147E+00	0.995
5	707711	0.326E-03	3.953	0.203E+00	0.994	0.734E-01	0.994

3 Comments

This finite volume method assigns one degree of freedom to any mesh item (cells, faces, edges, and vertices). For this reason, the scheme has a large number of degrees of freedom if compared to other finite volume methods or similar discretization techniques (such as mimetic finite differences). Nonetheless, the method was proved very effective both for two and three dimensional problems with strong anisotropic coefficients and using meshes with strongly distorted cells. Among the other advantages offered by the method, we mention the coercivity of the method that eases the convergence analysis and the fact that this finite volume method generally shows second order of accuracy in all numerical experiments where the exact solution is sufficiently regular. The results shown in the tables of the previous section confirm this general behavior.

All linear systems were solved efficiently by standard preconditioned Krylov methods as BiCG-stab or GMRES. Direct solvers for general asymmetric systems (UMFPACK) can also be used, but they normally require a huge memory storage, in particular for the biggest problems. In Table 1-2, we see an example of the performance of the different solvers offered by the benchmark site when solving Test 1 on the checkerboard meshes $8 \times 8 \times 8$ and $16 \times 16 \times 16$. The comparison reveals that PETSc implementation of the CG solver is the fastest one, in particular, when combined with the diagonal preconditioner (Jacobi). A good performance in terms of CPU costs is also provided by the ISTL-BiCGstab implementation using Jacobi or ILU(0) preconditioners. CPU times are usually smaller than those obtained by using the direct solver UMFPACK, which is also available in the benchmark site. For example, in the case of $8 \times 8 \times 8$-size mesh we note that UMFPACK requires a CPU time of 3.180 seconds.

Table 1 CeVeFe-DDFV method, test 1 using checkerboard mesh, grid resolution $8 \times 8 \times 8$; CPU times are measured in seconds

solver	precond	CPU time	# iters	Rel. resid.
PETSc-CG	Jacobi	0.209	202	6.368e-11
PETSc-CG	none	0.243	242	1.715e-10
ISTL-BiCGstab	ILU(0)	0.404	53	6.832e-11
ISTL-BiCGstab	none	0.563	167	3.656e-11
ISTL-BiCGstab	Jacobi	0.680	120	4.415e-11
ISTL-GMRES	ILU(0)	0.683	152	4.241e-11

Table 2 CeVeFe-DDFV method, test 1 using checkerboard mesh, grid resolution $16 \times 16 \times 16$; CPU times are measured in seconds

solver	precond	CPU time	# iters	Rel. resid.
PETSc-CG	Jacobi	3.946	369	4.248e-11
PETSc-CG	none	4.989	471	8.038e-11
ISTL-CG	ILU(0)	5.540	166	4.041e-11
ISTL-CG	none	7.319	471	8.038e-11
ISTL-BiCGstab	ILU(0)	8.681	107	3.873e-11
ISTL-CG	Jacobi	10.989	368	4.281e-11

Acknowledgements The work of the first and second author was supported by Groupement de Recherche MOMAS. The work of the third author was partially supported by the Italian MIUR through the program PRIN2008.

References

1. Y. Coudière and F. Hubert. A 3D discrete duality finite volume method for nonlinear elliptic equation. Preprint HAL, URL: http://hal.archives-ouvertes.fr/hal-00456837/fr, 2010.
2. Y. Coudière, F. Hubert, and G. Manzini. A cevefe-ddfv scheme for discontinuous permeability tensors. In *Finite Volume For Complex Applications, Problems And Perspectives. 6th International Conference (this volume)*, 2011.
3. K. Domelevo and P. Omnès. A finite volume method for the laplace equation on almost arbitrary two-dimensional grids. *M2AN, Math. Model. Numer. Anal.*, 39(6):1203–1249, 2005.
4. F. Hermeline. Approximation of diffusion operators with discontinuous tensor coefficients on distorted meshes. *Comput. Methods Appl. Mech. Engrg.*, 192(16):1939–1959, 2003.

The paper is in final form and no similar paper has been or is being submitted elsewhere.

Benchmark 3D: The Cell-Centered Finite Volume Method Using Least Squares Vertex Reconstruction ("Diamond Scheme")

Yves Coudière and Gianmarco Manzini

1 Presentation of the scheme

We consider, for this contribution, the cell-centered finite volume method based on least squares vertex reconstruction. This method, which is also popularly known as *"the diamond scheme"*, was originally presented for the advection-diffusion equation in two-dimensions and then extended in 3-D. The discretization of the diffusive term in 2-D and 3-D is found in in [1, 3–5]. The scalar solution of the diffusion problem u is numerically approximated by a piecewise constant function u_T on the cells K of mesh T. The numerical approximation u_T is defined as $u_T(x) = \sum_{K \in T} u_K \chi_K(x)$ ($\chi_K(x)$ being the characteristic function of cell K) through the values $(u_K)_{K \in T}$. To define the numerical diffusive flux through the interface f of the mesh, a polyhedral cell is built around this interface. This polyhedral cell, which has a quadrilateral shape in two dimensions, is named after its shape as *"diamond cell"*, which also motivates the name of the method. Specifically, let $x_K \in K$ be the center of gravity of the cell K of mesh T. The diamond cell D associated to the interface f between two cells K and L in T is the convex hull $D = \text{hull}(f, x_K, x_L)$. If f is a boundary face, thus defined by $f = \partial K \cap \partial \Omega$ where $\partial \Omega$ is the boundary of the computational domain Ω, then the diamond cell associated to f is the convex hull $D = \text{hull}(f, x_K)$.

The numerical diffusive flux is built by using a constant approximation of the solution gradient on each diamond cell. Let $(x_1, x_2, \ldots x_m)$ denote the vertices of face f, x_K and x_L the centers of gravity of the cells K and L that share this face, and D the convex hull of these points. For any function $u \in H^1(D)$, the Green-Gauss formula yields the relation $\int_D \nabla u(x) dx = \int_{\partial D} u(x) n(x) d\sigma(x)$, where n is

---

Yves Coudière
LMJL, Université de Nantes, France, e-mail: Yves.Coudiere@univ-nantes.fr

Gianmarco Manzini
IMATI-CNR and CESNA-IUSS, Pavia, Italy, e-mail: marco.manzini@imati.cnr.it

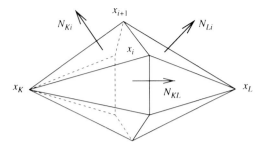

Fig. 1 Geometry of the *diamond* cell

the unit vector orthogonal to ∂D and pointing out of D. If the restriction of u to the face f of ∂D is an affine function, the boundary integral only depends on the values of u at the vertices of D. In this case, the Green-Gauss divergence theorem yields

$$\frac{1}{|D|}\int_D \nabla u(x)dx = \frac{1}{|D|}\int_{\partial D} u(x)n(x)dx = \frac{1}{|D|}\sum_{f\in\partial D}\int_f u(x)n(x)dx$$

$$= \frac{1}{3|D|}\sum_{i=1}^m \left(N_{Ki}(u(x_i) + u(x_{i+1}) + u(x_K)) + N_{Li}(u(x_i) + u(x_{i+1}) + u(x_L))\right)$$

$$= \frac{1}{3|D|}\left((u(x_L) - u(x_K))N_{KL} + \sum_{i=1}^m u(x_i)N_i\right),$$

where N_{Ki} and N_{Li} are the vectors orthogonal to the triangular facets hull(x_K, x_i, x_{i+1}) and hull(x_L, x_i, x_{i+1}), respectively, and having lengths equal to the measure of the facets; specifically, $N_{Ki} = \frac{1}{2}(x_i - x_K) \times (x_{i+1} - x_K)$ and $N_{Li} = -\frac{1}{2}(x_i - x_L) \times (x_{i+1} - x_L)$, see Fig. 1. The vectors N_{KL} and N_i actually used during the computations are

$$N_{KL} = \sum_{i=1}^m N_{Li} = -\sum_{i=1}^m N_{Ki} \quad \text{and} \quad N_i = N_{Ki} + N_{Li} + N_{Ki-1} + N_{Li-1}.$$

To derive the gradient formula, we consider the right-hand side of the above equality with the unknown u_K and u_L replacing $u(x_K)$ and $u(x_L)$ and some values u_i replacing the values $u(x_i)$. These values, $u_1, u_2, \ldots u_m$, are linearly interpolated from the values $(u_K)_{K\in T}$ as follows. For any vertex x_i of the mesh T, we consider $u_i = \sum_{K\in T_i} \omega_{iK} u_K$ where $T_i = \{K : x_i \in K\}$ denotes the subset of the mesh cells which share the vertex x_i. The interpolation weights ω_{iK} are assumed to verify the consistency relations [4]:

$$\sum_{K\in T_i} \omega_{iK} = 1 \quad \text{and} \quad \sum_{K\in T_i} \omega_{iK}(x_i - x_K) = 0.$$

The interpolation weights ω_{iK} are obtained by solving the reconstruction problem that approximates the cell-averaged data set $\{(x_K, u_K) \text{ for } K \in T_i\}$ by the affine function
$$\tilde{u}_i(x) = \alpha + \beta \cdot (x - x_i) \text{ for } x \in \mathcal{V}_i$$
on the co-volume $\mathcal{V}_i = \bigcup_{K \in T_i} K$ and in a least square sense, cf. [2, 4]. The reconstructed value at vertex x_i is now given by taking $u_i = \tilde{u}_i(x_i) = \alpha$. The coefficients $(\alpha, \beta)^T$ are the minimizers of the least squares functional
$$\mathcal{J}(\alpha, \beta^T) = \sum_{K: x_i \in K} \left(\alpha + \beta \cdot (x - x_i) - u_K\right)^2.$$

Imposing the zero gradient condition, i.e., $\nabla_{\alpha,\beta} \mathcal{J}(\alpha, \beta^T) = 0$, yields a linear system for the coefficients (α, β), whose solution returns the interpolation weights. The values u_i at the vertices $x_i \in \partial\Omega$ on the Dirichlet boundary are constrained to the boundary data, for instance $u_i = 0$ for a homogeneous condition. Other kinds of boundary conditions, e.g., Neumann or Robin, can be taken into account by extending to the 3-D case the technique investigated in [2]. Finally, the scheme reads as
$$\forall K \in T, \quad -\sum_{f \subset \partial K} \Lambda_f \nabla_D u_T \cdot N_{KL} = f_K |K| := \int_K f(x) dx,$$
where Λ_f is an arithmetic average of the diffusion tensor Λ over the diamond cell located around face f and N_{KL} is exactly the normal from above.

2 Numerical results

- **Test 1 Mild anisotropy,** $u(x, y, z) = 1 + \sin(\pi x) \sin\left(\pi \left(y + \frac{1}{2}\right)\right) \sin\left(\pi \left(z + \frac{1}{3}\right)\right)$ min $= 0$, max $= 2$, **Tetrahedral meshes**

i	nu	nmat	umin	uemin	umax	uemax	normg
0	215	6985	3.02E-02	3.15E-02	1.949	1.948	1.627
1	2003	107331	2.03E-02	1.13E-02	1.989	1.995	1.730
2	3898	227618	6.84E-03	4.21E-03	1.989	1.990	1.750
3	7711	476645	9.13E-03	8.18E-03	1.994	1.995	1.767
4	15266	994892	5.52E-03	4.10E-03	1.997	1.997	1.776
5	30480	2072944	1.49E-03	2.57E-04	1.997	1.999	1.784
6	61052	4292073	1.83E-03	1.20E-03	1.997	1.998	1.789

i	nu	erl2	ratiol2	ergrad	ratiograd	ener	ratioener
0	215	3.750E-02	-	3.503E-01	-	2.636E-01	-
1	2003	9.173E-03	1.892	1.568E-01	1.081	1.071E-01	1.210
2	3898	5.897E-03	1.991	1.215E-01	1.149	8.159E-02	1.225
3	7711	3.551E-03	2.230	9.410E-02	1.122	6.016E-02	1.339
4	15266	2.255E-03	1.994	7.387E-02	1.063	4.648E-02	1.132
5	30480	1.412E-03	2.032	5.768E-02	1.073	3.565E-02	1.152
6	61052	8.882E-04	2.001	4.502E-02	1.070	2.733E-02	1.147

Name of the solver: BiCG-stab with Jacobi preconditioner (in-house implementation).

- **Test 1 Mild anisotropy,** $u(x,y,z) = 1+\sin(\pi x)\sin\left(\pi\left(y+\frac{1}{2}\right)\right)\sin\left(\pi\left(z+\frac{1}{3}\right)\right)$ min $= 0$, max $= 2$, **Voronoi meshes**

i	nu	nmat	umin	uemin	umax	uemax	normg
1	29	257	8.51E-02	-6.18E+00	1.870	7.968	20.435
2	66	660	1.43E-01	2.06E-01	1.854	1.846	1.887
3	130	1410	3.85E-02	7.00E-03	1.925	1.941	1.855
4	228	2620	1.74E-02	2.37E-02	1.914	1.920	2.067
5	356	4424	2.84E-03	-2.18E+00	1.979	3.546	3.274

i	nu	erl2	ratiol2	ergrad	ratiograd	ener	ratioener
1	29	2.639E+00	-	3.631E+01	-	9.728E+00	-
2	66	9.077E-02	12.292	9.457E-01	13.307	3.751E-01	11.876
3	130	5.508E-02	2.210	7.434E-01	1.065	3.055E-01	0.907
4	228	6.650E-02	-1.006	1.163E+00	-2.391	3.224E-01	-0.287
5	356	3.674E-01	-11.507	5.071E+00	-9.912	1.287E+00	-9.321

Name of the solver: BiCG-stab with Jacobi preconditioner (in-house implementation).

- **Test 1 Mild anisotropy,** $u(x,y,z) = 1+\sin(\pi x)\sin\left(\pi\left(y+\frac{1}{2}\right)\right)\sin\left(\pi\left(z+\frac{1}{3}\right)\right)$ min $= 0$, max $= 2$, **Kershaw meshes**

i	nu	nmat	umin	uemin	umax	uemax	normg
1	512	10648	3.03E-02	8.74E-02	1.958	1.916	1.768
2	4096	97336	1.06E-02	3.00E-02	1.993	1.973	1.700
3	32768	830584	1.75E-03	5.87E-03	1.997	1.991	1.726
4	262144	6859000	7.14E-04	9.88E-04	1.999	1.998	1.765

Benchmark 3D: The Cell-Centered Finite Volume Method

i	nu	erl2	ratiol2	ergrad	ratiograd	ener	ratioener
1	512	6.846E-02	-	6.798E-01	-	4.901E-01	-
2	4096	4.715E-02	0.537	3.403E-01	0.998	2.715E-01	0.851
3	32768	2.866E-02	0.718	1.831E-01	0.894	1.532E-01	0.825
4	262144	1.315E-02	1.123	8.289E-02	1.143	6.942E-02	1.142

Name of the solver: BiCG-stab with Jacobi preconditioner (in-house implementation).

- **Test 1 Mild anisotropy,** $u(x, y, z) = 1 + \sin(\pi x) \sin\left(\pi \left(y + \frac{1}{2}\right)\right) \sin\left(\pi \left(z + \frac{1}{3}\right)\right)$ min $= 0$, max $= 2$, **Checkerboard meshes**

i	nu	nmat	umin	uemin	umax	uemax	normg
1	36	424	1.54E-01	1.33E-01	1.846	1.833	1.588
2	288	4528	4.01E-02	3.47E-02	1.960	1.958	1.721
3	2304	41896	1.01E-02	8.74E-03	1.990	1.990	1.773
4	18432	360280	2.54E-03	1.98E-03	1.997	1.998	1.791
5	147456	2987704	6.36E-04	5.26E-04	1.999	1.999	1.796

i	nu	erl2	ratiol2	ergrad	ratiograd	ener	ratioener
1	36	1.356E-01	-	2.488E-01	-	3.406E-01	-
2	288	4.427E-02	1.615	1.471E-01	0.758	1.346E-01	1.339
3	2304	1.191E-02	1.894	7.031E-02	1.065	4.678E-02	1.524
4	18432	3.112E-03	1.936	3.410E-02	1.043	1.687E-02	1.471
5	147456	7.976E-04	1.964	1.695E-02	1.008	7.003E-03	1.268

Name of the solver: BiCG-stab with Jacobi preconditioner (in-house implementation).

- **Test 2 Heterogeneous anisotropy,** $u(x, y, z) = x^3 y^2 z + x \sin(2\pi xz) \sin(2\pi xy) \sin(2\pi z)$, min $= -0.862$, max $= 1.0487$, **Prism meshes**

i	nu	nmat	umin	uemin	umax	uemax	normg
1	1210	21308	-8.42E-01	-8.57E-01	0.978	0.977	1.481
2	8820	169418	-8.38E-01	-8.41E-01	1.010	1.011	1.638
3	28830	570328	-8.58E-01	-8.60E-01	1.032	1.033	1.676
4	67240	1350038	-8.57E-01	-8.58E-01	1.033	1.034	1.690

i	nu	erl2	ratiol2	ergrad	ratiograd	ener	ratioener
1	1210	9.551E-02	-	2.356E-01	-	2.404E-01	-
2	8820	2.403E-02	2.084	8.174E-02	1.598	7.974E-02	1.666
3	28830	1.067E-02	2.057	4.167E-02	1.706	3.947E-02	1.781
4	67240	6.013E-03	2.030	2.562E-02	1.722	2.371E-02	1.805

Name of the solver: BiCG-stab with Jacobi preconditioner (in-house implementation).

- **Test 3 Flow on random meshes,** $u(x, y, z) = \sin(2\pi x) \sin(2\pi y) \sin(2\pi z)$, min = 0, max = 1, **Random meshes**

i	nu	nmat	umin	uemin	umax	uemax	normg
1	64	1000	-7.56E-01	-7.11E-01	0.711	0.525	1.650
2	512	10648	-9.39E-01	-8.32E-01	0.926	0.933	2.674
3	4096	97336	-9.86E-01	-9.77E-01	0.982	0.978	3.330
4	32768	830584	-9.96E-01	-9.92E-01	0.996	0.990	3.527

i	nu	erl2	ratiol2	ergrad	ratiograd	ener	ratioener
1	64	5.548E-01	-	7.060E-01	-	7.651E-01	-
2	512	1.427E-01	1.958	3.067E-01	1.202	3.264E-01	1.229
3	4096	2.967E-02	2.266	9.532E-02	1.685	9.569E-02	1.770
4	32768	7.166E-03	2.049	3.253E-02	1.551	2.529E-02	1.919

Name of the solver: BiCG-stab with Jacobi preconditioner (in-house implementation).

- **Test 4 Flow around a well, Well meshes,** min = 0, max = 5.415

i	nu	nmat	umin	uemin	umax	uemax	normg
1	890	18876	4.57E-01	5.26E-01	5.317	5.318	1573.020
2	2232	51800	2.61E-01	2.89E-01	5.329	5.329	1600.780
3	5016	121584	1.62E-01	1.73E-01	5.329	5.329	1613.840
4	11220	280868	1.23E-01	1.29E-01	5.330	5.330	1619.520
5	23210	592448	9.28E-02	9.66E-02	5.339	5.339	1620.960
6	42633	1100865	7.42E-02	7.67E-02	5.345	5.345	1621.200
7	74679	1942619	5.75E-02	5.91E-02	5.361	5.361	1621.930

Benchmark 3D: The Cell-Centered Finite Volume Method

i	nu	erl2	ratiol2	ergrad	ratiograd	ener	ratioener
1	890	9.562E-03	-	8.767E-02	-	5.372E-02	-
2	2232	3.699E-03	3.098	3.903E-02	2.640	2.305E-02	2.761
3	5016	1.676E-03	2.932	1.916E-02	2.636	1.104E-02	2.727
4	11220	1.190E-03	1.275	1.270E-02	1.531	7.205E-03	1.589
5	23210	7.545E-04	1.882	8.919E-03	1.458	5.053E-03	1.463
6	42633	4.601E-04	2.439	6.148E-03	1.835	3.522E-03	1.781
7	74679	3.402E-04	1.616	5.400E-03	0.693	3.174E-03	0.556

Name of the solver: BiCG-stab with Jacobi preconditioner (in-house implementation).

- **Test 5 Discontinuous permeability**, $u(x, y, z) = \sin(\pi x) \sin(\pi y) \sin(\pi z)$, min = 0, max = 1, **Locally refined meshes**

i	nu	nmat	umin	uemin	umax	uemax	normg
1	22	252	-1.00E+02	-5.24E+01	100.000	52.359	58.097
2	176	3220	-3.54E+01	-2.62E+01	35.355	26.180	43.055
3	1408	31524	-7.89E+01	-7.30E+01	78.858	73.021	76.757
4	11264	277396	-9.43E+01	-9.25E+01	94.346	92.545	92.422
5	90112	2324532	-9.86E+01	-9.81E+01	98.562	98.089	97.247

i	nu	erl2	ratiol2	ergrad	ratiograd	ener	ratioener
1	22	9.831E-01	-	7.196E-01	-	3.176E+02	-
2	176	5.072E-01	0.954	7.376E-01	-0.035	8.184E-01	8.600
3	1408	1.376E-01	1.882	3.770E-01	0.968	6.058E-01	0.433
4	11264	3.347E-02	2.039	1.874E-01	1.008	4.685E-01	0.370
5	90112	9.731E-03	1.782	1.159E-01	0.693	3.448E-01	0.442

Name of the solver: BiCG-stab with Jacobi preconditioner (in-house implementation).

3 Comments

This finite volume method is truly cell-centered and, for this reason, it has a relatively small number of degrees of freedom with respect to other finite volume discretizations which introduce face unknowns to approximate the scalar variable. The coercivity was proved only for simple cases (see [3] for details), so very few can be said from a theoretical standpoint about the convergence properties of this scheme and the literature misses a general convergence analysis. Despite this fact, the resulting finite volume method generally show second order of accuracy in all numerical experiments where the exact solution is enough regular and on

Table 1 LS-FVM method, test 1 using Kershaw mesh with grid resolution $32 \times 32 \times 32$; CPU times are measured in seconds.

solver	precond	CPU time	# iters	Rel. resid.
UMFPACK	none	28.712	0	7.287e-15
ISTL-BiCGstab	Jacobi	36.048	2990	1.126e-10
ISTL-GMRES	Jacobi	64.775	9186	3.931e-10
ISTL-BiCGstab	none	78.880	11417	3.431e-10
ISTL-BiCGstab	ILU(4)	1888.48	2	1.191e-12
ISTL-GMRES	ILU(4)	1692.49	4	9.103e-10

"reasonable" meshes. Moreover, it can be easily applied to complex, distorted meshes and anisotropic permeabilities for which it provides a reliable numerical approximation. It is also generally robust even if a locking phenomenon for the convergence has been reported in the literature [6].

The linear system for the cell-centered unknowns that is originated by this scheme on a general polyhedral mesh leads to an *asymmetric sparse matrix*. Therefore, this system can be solved efficiently by standard preconditioned Krylov methods (BiCG-stab or GMRES) or by direct solvers for general asymmetric systems (UMFPACK). An example of a typical behavior is reported in Table 1 for a subset of the combinations solvers and preconditioners available on the benchmark site. The comparison among these results reveals that the BiCG-stab solver using a diagonal Jacobi preconditioner seems to be the more efficient choice in most of the cases. The performance is usually comparable with that offered by the direct solver (**UMFPACK**), but the memory storage required by this latter may be from 2 to 60 times greater.

References

1. Bertolazzi, E., Manzini, G.: A cell-centered second-order accurate finite volume method for convection-diffusion problems on unstructured meshes. Math. Models Methods Appl. Sci. **8**, 1235–1260 (2004)
2. Bertolazzi, E., Manzini, G.: On vertex reconstructions for cell-centered finite volume approximations of 2-D anisotropic diffusion problems. Math. Models Methods Appl. Sci. **17**(1), 1–32 (2007)
3. Coudière, Y.: Analyse de schémas volumes finis sur maillages non structurés pour des problèmes linéaires hyperboliques et elliptiques. Ph.D. thesis, Université "P. Sabatier" de Toulouse, Toulouse III, Toulouse, France (1999)
4. Coudière, Y., Vila, J.P., Villedieu, P.: Convergence rate of a finite volume scheme for a two-dimensional diffusion convection problem. M2AN, Math. Model. Numer. Anal. **33**(3), 493–516 (1999)
5. Coudière, Y., Villedieu, P.: Convergence of a finite volume scheme for the linear convection-diffusion equation on locally refined meshes. M2AN, Math. Model. Numer. Anal. **34**(6), 1123–1149 (2000)
6. Manzini, G., Putti, M.: Mesh locking effects in the finite volume solution of 2-D anisotropic diffusion equations. J. Comput. Phys. **220**(2), 751–771 (2007)

The paper is in final form and no similar paper has been or is being submitted elsewhere.

Benchmark 3D: A Monotone Nonlinear Finite Volume Method for Diffusion Equations on Polyhedral Meshes

Alexander Danilov and Yuri Vassilevski

1 Presentation of the scheme

We propose a new monotone FV method based on a nonlinear two-point flux approximation scheme. The original idea belongs to C. LePotier [2] who proposed a monotone FV scheme for the discretization of parabolic equations on triangular meshes, which was extended to steady-state diffusion problems with full anisotropic tensors on triangulations or scalar diffusion coefficients on shape regular polygonal meshes [3]. Later a new interpolation-free monotone cell-centered FV method with nonlinear two-point flux approximation was proposed for full diffusion tensors and unstructured conformal polygonal 2D meshes [4]. In this paper, we extend the last approach to the case of 3D conformal polyhedral meshes [1].

Let Ω be a three-dimensional polyhedral domain with boundary Γ. We consider a model diffusion problem for unknown concentration u:

$$\begin{aligned} -\mathrm{div}(\mathbb{K}\nabla u) &= g \quad \text{in} \quad \Omega \\ u &= g_D \quad \text{on} \quad \Gamma \end{aligned} \quad (1)$$

where $\mathbb{K}(\mathbf{x}) = \mathbb{K}^T(\mathbf{x}) > 0$ is an anisotropic diffusion tensor, and g is a source term.

We consider a conformal polyhedral mesh $\mathscr{T}$ composed of shape-regular cells with planar faces. We assume that each cell is a star-shaped 3D domain with respect to its barycenter, and each face is a star-shaped 2D domain with respect to face's barycenter. Let $N_\mathscr{T}$ be the number of polyhedral cells and $N_\mathscr{B}$ be the number of boundary faces. The tensor function $\mathbb{K}(\mathbf{x})$ is assumed to be smooth for the sake of simplicity of presentation; however the original method is designed for discontinuous tensor function, which may jump across mesh faces as well as may change orientation of principal directions [1].

We denote by $\mathscr{F}_I$, $\mathscr{F}_B$ disjoint sets of interior and boundary faces, respectively. The cardinality of set $\mathscr{F}_*$ is denoted by $N_{\mathscr{F}_*}$. Let $\mathscr{F}_T$ and $\mathscr{E}_T$ denote the sets of faces and edges of polyhedron T, respectively.

Let $\mathbf{q} = -\mathbb{K}\nabla u$ denote the flux which satisfies the mass balance equation:

$$\operatorname{div} \mathbf{q} = g \quad \text{in} \quad \Omega. \tag{2}$$

We derive a FV scheme with a nonlinear two-point flux approximation. Integrating equation (2) over a polyhedron T and using the Green's formula we get:

$$\int_{\partial T} \mathbf{q} \cdot \mathbf{n}_T \, ds = \int_T g \, dx, \tag{3}$$

where $\mathbf{n}_T$ denotes the external unit normal to ∂T. Let f denote a face of cell T and $\mathbf{n}_f$ be the corresponding normal vector. It will be convenient to assume that $|\mathbf{n}_f| = |f|$ where $|f|$ denotes the area of face f. The equation (3) becomes

$$\sum_{f \in \partial T} \mathbf{q}_f \cdot \mathbf{n}_f = \int_T g \, dx, \tag{4}$$

where $\mathbf{q}_f$ is the average flux density for face f.

For each cell T, we assign one degree of freedom, U_T, for concentration u. Let U be the vector of all unknown concentrations. If two cells T_+ and T_- have a common face f, the two-point flux approximation is as follows:

$$\mathbf{q}_f^h \cdot \mathbf{n}_f = M_f^+ U_{T_+} - M_f^- U_{T_-}, \tag{5}$$

where M_f^+ and M_f^- are some coefficients. In a linear FV method, these coefficients are equal and fixed. In the nonlinear FV method, they may be different and depend on concentrations in surrounding cells.

For every cell T in $\mathscr{T}$, we define the collocation point $\mathbf{x}_T$ at the barycenter of T. For every face $f \in \mathscr{F}_B$, we denote the face barycenter by $\mathbf{x}_f$ and associate a collocation point with $\mathbf{x}_f$ for $f \in \mathscr{F}_B$.

We shall refer to collocation points on faces as the *auxiliary* collocation points. They are introduced for mathematical convenience and will not enter the final algebraic system although will affect system coefficients. In contrast, we shall refer to the other collocation points as the *primary* collocation points whose discrete concentrations form the unknown vector in the algebraic system.

For every cell T we define a set Σ_T of nearby collocation points. We assume that for every cell-face pair $T \in \mathscr{T}$, $f \in \mathscr{F}_T$, there exist three points $\mathbf{x}_{f,1}$, $\mathbf{x}_{f,2}$, and $\mathbf{x}_{f,3}$ in set Σ_T such that the following condition holds:

The co-normal vector $\boldsymbol{\ell}_f = \mathbb{K}(\mathbf{x}_f)\mathbf{n}_f$ started from $\mathbf{x}_T$ belongs to the trihedral corner formed by vectors

$$\mathbf{t}_{f,1} = \mathbf{x}_{f,1} - \mathbf{x}_T, \quad \mathbf{t}_{f,2} = \mathbf{x}_{f,2} - \mathbf{x}_T, \quad \mathbf{t}_{f,3} = \mathbf{x}_{f,3} - \mathbf{x}_T, \tag{6}$$

and

$$\frac{1}{|\boldsymbol{\ell}_f|}\boldsymbol{\ell}_f = \frac{\alpha_f}{|\mathbf{t}_{f,1}|}\mathbf{t}_{f,1} + \frac{\beta_f}{|\mathbf{t}_{f,2}|}\mathbf{t}_{f,2} + \frac{\gamma_f}{|\mathbf{t}_{f,3}|}\mathbf{t}_{f,3}, \tag{7}$$

where $\alpha_f \geq 0$, $\beta_f \geq 0$, $\gamma_f \geq 0$.

Let f be an internal face. We denote by T_+ and T_- the cells that share f and assume that $\mathbf{n}_f$ is outward for T_+. Let $\mathbf{x}_\pm$ be the collocation point of $T_\pm$. Let $U_\pm$ be the discrete concentrations in $T_\pm$.

Let $T = T_+$ and $\mathbb{K}_f = \mathbb{K}(\mathbf{x}_f)$. Using the above notations, definition of the directional derivative,

$$\frac{\partial u}{\partial \boldsymbol{\ell}_f}|\boldsymbol{\ell}_f| = \nabla u \cdot (\mathbb{K}_f \mathbf{n}_f),$$

and assumption (7), we write

$$\mathbf{q}_f \cdot \mathbf{n}_f = -\frac{|\boldsymbol{\ell}_f|}{|f|} \int_f \frac{\partial u}{\partial \boldsymbol{\ell}_f} ds = -\frac{|\boldsymbol{\ell}_f|}{|f|} \int_f \left(\alpha_f \frac{\partial u}{\partial \mathbf{t}_{f,1}} + \beta_f \frac{\partial u}{\partial \mathbf{t}_{f,2}} + \gamma_f \frac{\partial u}{\partial \mathbf{t}_{f,3}} \right) ds. \tag{8}$$

Replacing directional derivatives by finite differences, we get

$$\int_f \frac{\partial u}{\partial \mathbf{t}_{f,i}} ds = \frac{U_{f,i} - U_T}{|\mathbf{x}_{f,i} - \mathbf{x}_T|}|f| + O(h_T^2), \quad i = 1, 2, 3, \tag{9}$$

where h_T is the diameter of cell T. Using the finite difference approximations (9), we transform formula (8) to

$$\mathbf{q}_f^h \cdot \mathbf{n}_f = -|\boldsymbol{\ell}_f| \left(\frac{\alpha_f}{|\mathbf{t}_{f,1}|}(U_{f,1} - U_T) + \frac{\beta_f}{|\mathbf{t}_{f,2}|}(U_{f,2} - U_T) + \frac{\gamma_f}{|\mathbf{t}_{f,3}|}(U_{f,3} - U_T) \right). \tag{10}$$

At the moment, this flux involves four rather than two concentrations. To derive a two-point flux approximation, we consider the cell T_- and derive another approximation of flux through face f. To distinguish between T_+ and T_-, we add subscripts $\pm$ and omit subscript f. Since $\mathbf{n}_f$ is the internal normal vector for T_-, we have to change sign of the right hand side:

$$\mathbf{q}_\pm^h \cdot \mathbf{n}_f = \mp|\boldsymbol{\ell}_f| \left(\frac{\alpha_\pm}{|\mathbf{t}_{\pm,1}|}(U_{\pm,1} - U_\pm) + \frac{\beta_\pm}{|\mathbf{t}_{\pm,2}|}(U_{\pm,2} - U_\pm) + \frac{\gamma_\pm}{|\mathbf{t}_{\pm,3}|}(U_{\pm,3} - U_\pm) \right), \tag{11}$$

where $\alpha_\pm, \beta_\pm$ and $\gamma_\pm$ are given by (7) and $U_{\pm,i}$ denote concentrations at points $\mathbf{x}_{\pm,i}$ from $\Sigma_{T_\pm}$. We define a new discrete flux as a linear combination of $\mathbf{q}_\pm^h \cdot \mathbf{n}_f$ with non-negative weights $\mu_\pm$:

$$\begin{aligned}
\mathbf{q}_f^h \cdot \mathbf{n}_f &= \mu_+ \mathbf{q}_+^h \cdot \mathbf{n}_f + \mu_- \mathbf{q}_-^h \cdot \mathbf{n}_f \\
&= \mu_+ |\ell_f| \left(\frac{\alpha_+}{|\mathbf{t}_{+,1}|} + \frac{\beta_+}{|\mathbf{t}_{+,2}|} + \frac{\gamma_+}{|\mathbf{t}_{+,3}|} \right) U_+ \\
&\quad - \mu_- |\ell_f| \left(\frac{\alpha_-}{|\mathbf{t}_{-,1}|} + \frac{\beta_-}{|\mathbf{t}_{-,2}|} + \frac{\gamma_-}{|\mathbf{t}_{-,3}|} \right) U_- \\
&\quad - \mu_+ |\ell_f| \left(\frac{\alpha_+}{|\mathbf{t}_{+,1}|} U_{+,1} + \frac{\beta_+}{|\mathbf{t}_{+,2}|} U_{+,2} + \frac{\gamma_+}{|\mathbf{t}_{+,3}|} U_{+,3} \right) \\
&\quad + \mu_- |\ell_f| \left(\frac{\alpha_-}{|\mathbf{t}_{-,1}|} U_{-,1} + \frac{\beta_-}{|\mathbf{t}_{-,2}|} U_{-,2} + \frac{\gamma_-}{|\mathbf{t}_{-,3}|} U_{-,3} \right).
\end{aligned} \qquad (12)$$

The obvious requirement for the weights is to cancel the terms in the last two rows of (12) which results in a two-point flux formula. The second requirement is to approximate the true flux. These requirements lead us to the following system

$$\begin{cases} -\mu_+ d_+ + \mu_- d_- = 0, \\ \mu_+ + \mu_- = 1, \end{cases} \qquad (13)$$

where

$$d_\pm = |\ell_f| \left(\frac{\alpha_\pm}{|\mathbf{t}_{\pm,1}|} U_{\pm,1} + \frac{\beta_\pm}{|\mathbf{t}_{\pm,2}|} U_{\pm,2} + \frac{\gamma_\pm}{|\mathbf{t}_{\pm,3}|} U_{\pm,3} \right).$$

Since coefficients $d_\pm$ depend on both geometry and concentration, the weights $\mu_\pm$ do as well. Thus, the resulting two-point flux approximation is *nonlinear*.

It may happen that concentration $U_{+,i}$, $(U_{-,i})$ $i = 1, 2, 3$, is defined at the same collocation point as U_- (U_+). In this case the terms to be cancelled are changed so that they do not incorporate $U_\pm$. By doing so, for the Laplace operator we recover the classical linear scheme with the 6-1-1-1-1-1 stencil on uniform cubic meshes.

The solution of (13) can be written explicitly. In all cases $d_\pm \geq 0$ if $U \geq 0$. If $d_\pm = 0$, we set $\mu_+ = \mu_- = \frac{1}{2}$. Otherwise,

$$\mu_+ = \frac{d_-}{d_- + d_+} \quad \text{and} \quad \mu_- = \frac{d_+}{d_- + d_+}.$$

This implies that the weights $\mu_\pm$ are non-negative. Substituting this into (12), we get the two-point flux formula (5) with coefficients

$$M_f^\pm = \mu_\pm |\ell_f| (\alpha_\pm/|\mathbf{t}_{\pm,1}| + \beta_\pm/|\mathbf{t}_{\pm,2}| + \gamma_\pm/|\mathbf{t}_{\pm,3}|). \qquad (14)$$

Now we consider the case of Dirichlet boundary face $f \in \mathscr{F}_B$ where we define

$$U_f = \bar{g}_{D,f} = \frac{1}{|f|} \int_f g_D \, ds. \qquad (15)$$

It may be convenient to think about f as the ghost cell with zero volume. Let T be the cell with face f. Replacing U_+ and U_- with U_T and U_f, and Σ_{T_+}, Σ_{T_-} with Σ_T, $\Sigma_{f,T}$ respectively, we get

$$\mathbf{q}_f^h \cdot \mathbf{n}_f = M_f^+ U_T - M_f^- U_f, \tag{16}$$

where coefficients $M_f^{\pm}$ are given by (14).

For every T in $\mathscr{T}$, the cell equation (4) is

$$\sum_{f \in \mathscr{F}_T} \chi(T, f) \mathbf{q}_f^h \cdot \mathbf{n}_f = \int_T f \, dx, \tag{17}$$

where $\chi(T, f) = sign(\mathbf{n}_f \cdot \mathbf{n}_T(\mathbf{x}_f))$. Substituting two-point flux formula (5) with non-negative coefficients given by (14) into (17), and using equations (15) and (16) to eliminate concentrations at boundary faces, we get a nonlinear system of $N_\mathscr{T}$ equations

$$\mathbb{M}(U)U = G(U). \tag{18}$$

The right hand side vector $G(U)$ is generated by the source and the boundary data:

$$G_T(U) = \int_T g \, dx + \sum_{f \in \mathscr{F}_B \cap \mathscr{F}_T} M_f^-(U) \bar{g}_{D,f}, \quad \forall T \in \mathscr{T}. \tag{19}$$

For data functions $g \geq 0$ and $g_D \geq 0$ the components of vector G are non-negative. We use the Picard iterations to solve the nonlinear system (18).

The details of the presented scheme, algorithms, modifications of the scheme for Neumann boundary conditions and discontinuous diffusion tensor coefficients, as well as monotonicity analysis of the scheme are presented in [1].

2 Numerical results

We use discrete L_2-norm to evaluate discretization errors for the concentration u:

$$\text{erl2} = \left[\frac{\sum_{T \in \mathscr{T}} (u(\mathbf{x}_T) - U_T)^2 |T|}{\sum_{T \in \mathscr{T}} (u(\mathbf{x}_T))^2 |T|} \right]^{1/2}.$$

For each cell T we derive the value of ∇u from the linear reconstruction of the concentration over T introduced in [5]:

$$\mathscr{R}_T(\mathbf{x}) = U_T + \nabla U_T(\mathbf{x} - \mathbf{x}_T).$$

This reconstruction minimizes the deviation of $\mathscr{R}_T(\mathbf{x}_k)$ from targeted values U_k in nearby cells. We denote the approximate gradient in cell T as ∇U_T.

We use the following estimate for L_1-norm of the euclidean norm of the approximate gradient:
$$\text{normg} = \sum_{T \in \mathscr{T}} ||\nabla U_T|| \, |T|.$$

The discrete H^1 and energy norms of the error are defined as follows:

$$\text{ergrad} = \left[\frac{\sum_{T \in \mathscr{T}} ||\nabla u(\mathbf{x}_T) - \nabla U_T||^2 |T|}{\sum_{T \in \mathscr{T}} ||\nabla u(\mathbf{x}_T)||^2 |T|} \right]^{1/2}$$

$$\text{ener} = \left[\frac{\sum_{T \in \mathscr{T}} \mathbb{K}(\nabla u(\mathbf{x}_T) - \nabla U_T) \cdot (\nabla u(\mathbf{x}_T) - \nabla U_T) |T|}{\sum_{T \in \mathscr{T}} K \nabla u(\mathbf{x}_T) \cdot \nabla u(\mathbf{x}_T) |T|} \right]^{1/2}$$

The proposed method is designed for non-negative solutions, and may behave unexpectedly if the values of solution go below zero. Since several test cases have negative values of exact solution, we added positive constants to exact solutions to force their positivity. We added $+1$ in tests 2 and 3, and $+100$ in test 5. We subtracted back the positive constants at the end of test runs. We denote minimum and maximum values for discrete solution as umin and umax respectively, and exact values of u at the cell centers as uemin and uemax respectively.

We use Picard method to solve the nonlinear system (18). The values nu and nmat correspond to the number of unknowns and number of non-zero terms in linearized system.

- **Test 1 Mild anisotropy,** $u(x, y, z) = 1 + \sin(\pi x) \sin\left(\pi \left(y + \frac{1}{2}\right)\right) \sin\left(\pi \left(z + \frac{1}{3}\right)\right)$ min $= 0$, max $= 2$, **Tetrahedral meshes**

i	nu	nmat	umin	uemin	umax	uemax	normg
1	2003	9411	0.028	0.020	1.997	1.989	1.790
2	3898	18586	0.014	0.007	1.992	1.989	1.794
3	7711	37103	0.014	0.009	1.997	1.994	1.795
4	15266	74012	0.008	0.006	1.998	1.997	1.797
5	30480	148746	0.004	0.001	1.999	1.997	1.797
6	61052	299492	0.003	0.002	1.998	1.997	1.798

i	nu	erl2	ratiol2	ergrad	ratiograd	ener	ratioener
1	2003	5.31e-03		1.30e-01		1.26e-01	
2	3898	4.00e-03	1.281	1.07e-01	0.879	1.04e-01	0.858
3	7711	2.44e-03	2.167	8.47e-02	1.039	8.16e-02	1.057
4	15266	1.70e-03	1.592	6.78e-02	0.979	6.52e-02	0.986
5	30480	9.57e-04	2.490	5.36e-02	1.016	5.20e-02	0.986
6	61052	6.42e-04	1.726	4.23e-02	1.026	4.09e-02	1.037

- **Test 1 Mild anisotropy,** $u(x,y,z) = 1 + \sin(\pi x) \sin\left(\pi\left(y + \frac{1}{2}\right)\right) \sin\left(\pi\left(z + \frac{1}{3}\right)\right)$
  min = 0, max = 2, **Voronoi meshes**

i	nu	nmat	umin	uemin	umax	uemax	normg
1	29	257	0.011	0.085	1.991	1.870	1.303
2	66	660	0.107	0.143	1.902	1.854	1.614
3	130	1410	0.038	0.038	1.963	1.925	1.609
4	228	2620	0.021	0.017	1.941	1.914	1.685
5	356	4424	0.002	0.003	2.004	1.979	1.686

i	nu	erl2	ratiol2	ergrad	ratiograd	ener	ratioener
1	29	7.49e-02		6.70e-01		6.81e-01	
2	66	6.16e-02	0.710	4.96e-01	1.099	4.66e-01	1.384
3	130	3.44e-02	2.579	3.65e-01	1.351	3.70e-01	1.023
4	228	2.32e-02	2.098	2.78e-01	1.470	2.72e-01	1.636
5	356	1.73e-02	1.988	2.27e-01	1.364	2.23e-01	1.341

- **Test 1 Mild anisotropy,** $u(x,y,z) = 1 + \sin(\pi x) \sin\left(\pi\left(y + \frac{1}{2}\right)\right) \sin\left(\pi\left(z + \frac{1}{3}\right)\right)$
  min = 0, max = 2, **Kershaw meshes**

i	nu	nmat	umin	uemin	umax	uemax	normg
1	512	3200	0.112	0.030	1.942	1.958	1.695
2	4096	27136	0.037	0.011	1.977	1.993	1.763
3	32768	223232	0.011	0.002	1.989	1.997	1.749
4	262144	1810432	0.003	0.001	1.997	1.999	1.761

i	nu	erl2	ratiol2	ergrad	ratiograd	ener	ratioener
1	512	6.02e-02		4.53e-01		4.27e-01	
2	4096	4.98e-02	0.276	5.55e-01	-0.294	6.25e-01	-0.548
3	32768	3.70e-02	0.428	3.45e-01	0.687	3.74e-01	0.740
4	262144	2.22e-02	0.737	1.83e-01	0.910	1.89e-01	0.988

- **Test 1 Mild anisotropy,** $u(x,y,z) = 1 + \sin(\pi x)\sin\left(\pi\left(y+\frac{1}{2}\right)\right)\sin\left(\pi\left(z+\frac{1}{3}\right)\right)$
  min = 0, max = 2, **Checkerboard meshes**

i	nu	nmat	umin	uemin	umax	uemax	normg
1	36	228	0.122	0.154	1.905	1.846	1.769
2	288	2208	0.053	0.040	1.966	1.960	1.735
3	2304	19200	0.014	0.010	1.992	1.990	1.772
4	18432	159744	0.005	0.003	1.998	1.997	1.790
5	147456	1302528	0.001	0.001	2.000	1.999	1.796

i	nu	erl2	ratiol2	ergrad	ratiograd	ener	ratioener
1	36	7.24e-02		5.48e-01		5.21e-01	
2	288	2.83e-02	1.356	2.77e-01	0.988	2.68e-01	0.959
3	2304	7.82e-03	1.854	1.21e-01	1.192	1.17e-01	1.190
4	18432	2.20e-03	1.829	5.48e-02	1.142	5.34e-02	1.139
5	147456	6.49e-04	1.761	2.59e-02	1.085	2.52e-02	1.084

- **Test 2 Heterogeneous anisotropy,** $u(x,y,z) = x^3 y^2 z + x\sin(2\pi xz)\sin(2\pi xy)\sin(2\pi z)$, min = −0.862, max = 1.0487, **Prism meshes**

i	nu	nmat	umin	uemin	umax	uemax	normg
1	1210	9788	-0.854	-0.842	1.002	0.978	1.579
2	8820	75178	-0.840	-0.838	1.014	1.010	1.669
3	28830	250168	-0.859	-0.858	1.034	1.032	1.689
4	67240	588758	-0.858	-0.857	1.034	1.033	1.698

i	nu	erl2	ratiol2	ergrad	ratiograd	ener	ratioener
1	1210	6.07e-02		2.11e-01		2.11e-01	
2	8820	1.69e-02	1.928	8.51e-02	1.368	8.62e-02	1.352
3	28830	7.96e-03	1.911	4.63e-02	1.543	4.74e-02	1.517
4	67240	4.62e-03	1.933	2.93e-02	1.626	3.02e-02	1.597

- **Test 3 Flow on random meshes,** $u(x,y,z) = \sin(2\pi x)\sin(2\pi y)\sin(2\pi z)$, min = −1, max = 1, **Random meshes**

i	nu	nmat	umin	uemin	umax	uemax	normg
1	64	352	-0.905	-0.778	0.759	0.702	2.241
2	512	3200	-0.928	-0.937	0.959	0.930	3.167
3	4096	27136	-1.005	-0.985	0.996	0.982	3.492
4	32768	223232	-0.989	-0.996	1.001	0.996	3.568

i	nu	erl2	ratiol2	ergrad	ratiograd	ener	ratioener
1	64	2.83e-01		5.55e-01		5.50e-01	
2	512	8.74e-02	1.698	1.81e-01	1.618	1.63e-01	1.756
3	4096	2.71e-02	1.688	7.01e-02	1.366	5.61e-02	1.537
4	32768	7.58e-03	1.839	3.23e-02	1.118	2.39e-02	1.230

- **Test 4 Flow around a well, Well meshes,** min $= 0$, max $= 5.415$

i	nu	nmat	umin	uemin	umax	uemax	normg
1	890	5574	0.518	0.458	5.318	5.317	1484.035
2	2232	14552	0.287	0.262	5.329	5.329	1541.433
3	5016	33436	0.173	0.162	5.329	5.329	1577.828
4	11220	75894	0.129	0.123	5.330	5.330	1596.743
5	23210	158380	0.096	0.093	5.339	5.339	1606.920
6	42633	292465	0.077	0.074	5.345	5.345	1611.935
7	74679	514069	0.059	0.058	5.361	5.361	1615.163

i	nu	erl2	ratiol2	ergrad	ratiograd	ener	ratioener
1	890	9.82e-03		1.88e-01		1.86e-01	
2	2232	4.07e-03	2.871	1.05e-01	1.892	1.04e-01	1.883
3	5016	1.77e-03	3.081	5.95e-02	2.120	5.90e-02	2.116
4	11220	1.09e-03	1.813	3.86e-02	1.611	3.80e-02	1.642
5	23210	6.44e-04	2.169	2.58e-02	1.656	2.54e-02	1.653
6	42633	4.58e-04	1.680	1.78e-02	1.854	1.73e-02	1.894
7	74679	3.20e-04	1.923	1.37e-02	1.373	1.33e-02	1.409

- **Test 5 Discontinuous permeability,** $u(x, y, z) = a_i \sin(2\pi x) \sin(2\pi y) \sin(2\pi z)$, min $= -100$, max $= 100$, **Locally refined meshes**

i	nu	nmat	umin	uemin	umax	uemax	normg
1	22	124	-246.736	-100.000	246.736	100.000	342.699
2	176	1112	-43.618	-35.355	43.618	35.355	68.108
3	1408	9376	-83.040	-78.858	83.040	78.858	92.094
4	11264	76928	-95.567	-94.346	95.567	94.346	97.550
5	90112	623104	-98.880	-98.562	98.880	98.562	98.676
6	720896	5015552	-99.719	-99.639	99.719	99.639	98.928

i	nu	erl2	ratiol2	ergrad	ratiograd	ener	ratioener
1	22	1.47e+00		2.14e+03		6.60e+03	
2	176	2.34e-01	2.651	6.29e-01	11.733	1.16e+00	12.475
3	1408	5.30e-02	2.140	2.40e-01	1.390	7.21e-01	0.684
4	11264	1.30e-02	2.034	1.43e-01	0.750	4.99e-01	0.532
5	90112	3.22e-03	2.008	9.78e-02	0.547	3.51e-01	0.507
6	720896	8.04e-04	2.002	6.87e-02	0.510	2.48e-01	0.502

3 Comments

In our experiments the linear systems in Picard method with the non-symmetric matrices were solved by the Bi-Conjugate Gradient Stabilized (BiCGStab) method with the ILU0 preconditioner. The nonlinear iterations are terminated when the relative norm of the residual norm becomes smaller then $\varepsilon_{non} = 10^{-9}$. The convergence tolerance for the linear solver is set to $\varepsilon_{lin} = 10^{-12}$. The number of Picard iterations for different test cases are presented in the table (Test 1 Mild anisotropy: 1B – Tetrahedral meshes, 1C – Voronoi meshes, 1D – Kershaw meshes, 1I – Checkerboard meshes; Test 2 Heterogeneous anisotropy: 2F – Prism meshes; Test 3 Flow on random meshes: 3AA – Random meshes; Test 4 Flow around a well: 4BB – Well meshes; Test 5 Discontinuous permeability: 5H – Locally refined meshes).

i	1B	1C	1D	1I	2F	3AA	4BB	5H
1	37	10	43	14	23	15	18	14
2	47	13	112	27	35	28	18	12
3	41	14	190	37	41	52	20	12
4	50	17	351	41	45	92	21	11
5	57	19		40			23	11
6	58						24	12
7							24	

Acknowledgements This work has been supported in part by RFBR grants 09-01-00115-a, 11-01-00971-a and the federal program "Scientific and scientific-pedagogical personnel of innovative Russia"

References

1. Danilov A., Vassilevski Yu. A monotone nonlinear finite volume method for diffusion equations on conformal polyhedral meshes. *Russian J. Numer. Anal. Math. Modelling*, No.24, pp.207-227, 2009.

2. Le Potier C. Schema volumes finis monotone pour des operateurs de diffusion fortement anisotropes sur des maillages de triangle non structures. *C.R.Acad. Sci. Paris*, Ser. I 341, pp.787-792, 2005.
3. Lipnikov K., Svyatskiy D., Shashkov M., Vassilevski Yu. Monotone finite volume schemes for diffusion equations on unstructured triangular and shape-regular polygonal meshes. *J. Comp. Phys.* Vol.227, pp.492-512, 2007.
4. Lipnikov K., Svyatskiy D., Vassilevski Yu. Interpolation-free monotone finite volume method for diffusion equations on polygonal meshes. *J. Comp. Phys.* Vol.228, No.3, pp.703-716, 2009.
5. Nikitin K., Vassilevski Yu. A monotone nonlinear finite volume method for advection-diffusion equations on unstructured polyhedral meshes in 3D. *Russian J. Numer. Anal. Math. Modelling*, Vol.25, pp.335-358, 2010.

The paper is in final form and no similar paper has been or is being submitted elsewhere.

Benchmark 3D: the SUSHI Scheme

Robert Eymard, Thierry Gallouët, and Raphaèle Herbin

1 Presentation of the scheme

We present the SUSHI scheme [2] in the case of a general heterogeneous and anisotropic diffusion problem with homogeneous Dirichlet boundary conditions. Let Ω be a bounded open domain of $\mathbb{R}^d$, with $d \in \mathbb{N}^*$, let $f \in L^2(\Omega)$ and let Λ be a measurable function from Ω to the set $\mathcal{M}_d(\mathbb{R})$ of $d \times d$ matrices, such that for a.e. $x \in \Omega$, $\Lambda(x)$ is symmetric, and such that the set of its eigenvalues is included in $[\underline{\lambda}, \overline{\lambda}]$, where $0 < \underline{\lambda} \leq \overline{\lambda}$. We wish to approximate the function u solution of

$$u \in H_0^1(\Omega) \text{ and } \forall v \in H_0^1(\Omega), \int_\Omega \Lambda(x)\nabla u(x) \cdot \nabla v(x) \mathrm{d}x = \int_\Omega f(x)v(x)\mathrm{d}x,$$

by the following scheme:

$$U \in X_{\mathcal{D}}, \ \forall V \in X_{\mathcal{D}}, \int_\Omega \Lambda(x)\nabla_{\mathcal{D}} U(x) \cdot \nabla_{\mathcal{D}} V(x) \mathrm{d}x = \int_\Omega f(x)\Pi_{\mathcal{D}} V(x) \mathrm{d}x,$$

where the reconstruction operator $\Pi_{\mathcal{D}}$ and the discrete gradient operator $\nabla_{\mathcal{D}}$ acting on the discrete functional space $X_{\mathcal{D}}$, depending on the discretization $\mathcal{D}$, are now defined, along with some notations:

1. $\mathcal{M}$ is the set of grid cells, that are disjoint open subsets of Ω such that $\bigcup_{K \in \mathcal{M}} \overline{K} = \overline{\Omega}$, $\mathcal{F}$ is the set of the faces of the mesh; note that each non-planar face is decomposed into planar faces without increasing the cost of the method. We assume that Λ is constant on all $K \in \mathcal{M}$, and we denote by Λ_K its value in K; a point x_K is chosen in K such that K is star-shaped with respect to x_K;

R. Eymard
Université Paris-Est, France, e-mail: robert.eymard@univ-mlv.fr

T. Gallouët and R. Herbin
Université Aix-Marseille, France, e-mail: Thierry.Gallouet@latp.univ-mrs.fr,
Raphaele.Herbin@latp.univ-mrs.fr

2. the set of discrete unknowns $X_\mathcal{D}$ is the finite dimensional vector space on $\mathbb{R}$, containing all real families $U = (u_K)_{K \in \mathcal{M}}$;
3. the space step $h_\mathcal{D} \in (0, +\infty)$ is the maximum diameter of all control volumes;
4. the mapping $\Pi_\mathcal{D} : X_\mathcal{D} \to L^2(\Omega)$ is the reconstruction of the approximate function defined by the value u_K in each $K \in \mathcal{M}$;
5. the mapping $\nabla_\mathcal{D} : X_\mathcal{D} \to L^2(\Omega)^d$ is the reconstruction of the gradient of the function, defined below.

The construction of $\nabla_\mathcal{D}$ involves the following steps.

1. for all exterior faces $\sigma \in \mathcal{F}_{\text{ext}}$, a value u_σ is given at the barycentre $\boldsymbol{x}_\sigma$ of σ
2. for each face $\sigma \in \mathcal{F}_{\text{int}}$ and $U = (u_K)_{K \in \mathcal{M}}$, a value u_σ, meant to approximate u at the barycentre $\boldsymbol{x}_\sigma$ of σ, is computed such that

$$u_\sigma = \sum_{K \in \mathcal{M}} \alpha_\sigma^K u_K + \sum_{\tau \in \mathcal{F}_{\text{ext}}} \beta_\sigma^\tau u_\tau, \text{ with } \sum_{K \in \mathcal{M}} \alpha_\sigma^K + \sum_{\tau \in \mathcal{F}_{\text{ext}}} \beta_\sigma^\tau = 1,$$

where the coefficients α_σ^K and β_σ^τ are chosen as explained below. Note that the interior faces of the mesh cannot be defined as $\partial K \cap \partial L$ for two neighbouring control volumes K and L, since there may exist more than one common face between K and L, in particular, if non-planar faces are split in triangular faces. Indeed, the definition of $\nabla_K U$ below is exact for affine functions only in the case of planar faces (see the comments on the results in the last section of this paper).
3. Denoting by $\mathcal{F}_K$ the subset of $\mathcal{F}$ containing all the faces of $K \in \mathcal{M}$ and, for $\sigma \in \mathcal{F}_K$, by $\boldsymbol{n}_{K,\sigma}$ the unit normal vector to σ outward to K, one defines

$$\nabla_K U = \frac{1}{|K|} \sum_{\sigma \in \mathcal{F}_K} |\sigma| (u_\sigma - u_K) \boldsymbol{n}_{K,\sigma},$$

and, for all $\sigma \in \mathcal{F}_K$, denoting by $d_{K,\sigma}$ the orthogonal distance between $\boldsymbol{x}_K$ and $\sigma \in \mathcal{F}_K$

$$\nabla_{K,\sigma} U = \nabla_K U + \frac{\sqrt{3}}{d_{K,\sigma}} (u_\sigma - u_K - \nabla_K U \cdot (\boldsymbol{x}_\sigma - \boldsymbol{x}_K)) \boldsymbol{n}_{K,\sigma},$$

4. $\nabla_\mathcal{D} U$ is given by the constant value $\nabla_{K,\sigma} U$ in the cone with vertex $\boldsymbol{x}_K$ and basis σ.

Let us now turn to the computation of α_σ^K and β_σ^τ. Let K and L be two grid cells separated by a common face σ. Let τ be a face of K or L, which is not common to K and L. We first compute a value w_τ at some point $\boldsymbol{y}_\tau$ by the following method:

1. if $\tau \in \mathcal{F}_{\text{ext}}$, then $\boldsymbol{y}_\tau = \boldsymbol{x}_\tau$ and $w_\tau = u_\tau$;
2. if $\tau \in \mathcal{F}_{\text{int}}$ is a common face to grid cells M and N (with one and one only of them being equal to K or L), then we define

$$\boldsymbol{y}_\tau = \frac{\lambda_N d_{M,\tau} \boldsymbol{y}_N + \lambda_M d_{N,\tau} \boldsymbol{y}_M + d_{M,\tau} d_{N,\tau} (\lambda_N^\sigma - \lambda_M^\sigma)}{\lambda_N d_{M,\tau} + \lambda_M d_{N,\tau}},$$

where, denoting by $\mathscr{P}(\boldsymbol{x}, \tau)$ the orthogonal projection on τ of any point $\boldsymbol{x}$ and by $\boldsymbol{n}_{MN}$ the unit normal vector, orthogonal to τ, oriented from M to N, we set

$$\begin{aligned} \boldsymbol{y}_M &= \mathscr{P}(\boldsymbol{x}_M, \tau), & \lambda_M &= \boldsymbol{n}_{MN} \cdot \Lambda_M \boldsymbol{n}_{MN}, & \boldsymbol{\lambda}_M^\tau &= \Lambda_M \boldsymbol{n}_{MN} - \lambda_M \boldsymbol{n}_{MN}, \\ \boldsymbol{y}_N &= \mathscr{P}(\boldsymbol{x}_N, \tau), & \lambda_N &= \boldsymbol{n}_{MN} \cdot \Lambda_N \boldsymbol{n}_{MN}, & \boldsymbol{\lambda}_N^\tau &= \Lambda_N \boldsymbol{n}_{MN} - \lambda_N \boldsymbol{n}_{MN}; \end{aligned}$$

then the following averaging formula is used to define the values w_τ as linear combinations of u_M and u_N:

$$w_\tau = \frac{\lambda_N d_{M,\tau} u_N + \lambda_M d_{N,\tau} u_M}{\lambda_N d_{M,\tau} + \lambda_M d_{N,\tau}};$$

3. two faces $\tau \in \mathscr{F}_K$ and $\tau' \in \mathscr{F}_L$ are then selected so that there exists a unique function w, affine in K and in L, continuous on σ, such that $\Lambda_K(\nabla w)_K \cdot \boldsymbol{n}_{KL} = \Lambda_L(\nabla w)_L \cdot \boldsymbol{n}_{KL}$ and such that $u_K = w(\boldsymbol{x}_K), u_L = w(\boldsymbol{x}_L), w_\tau = w(\boldsymbol{y}_\tau)$ and $w_{\tau'} = w(\boldsymbol{y}_{\tau'})$; we then set $u_\sigma = w(\boldsymbol{x}_\sigma)$, hence defining u_σ as a linear combination of u_K, u_L, w_τ and $w_{\tau'}$; the choice of τ and τ' is done thanks to an invertibility criterion of the 4×4 local linear systems thus obtained.

We refer to [1] for the complete presentation of the mathematical properties of the scheme, which are obtained for a slightly different choice of the coefficients α_σ^K and β_σ^τ from the one presented here. These mathematical properties remain valid in the case of the coefficients chosen here, which present the advantage of preserving exact affine solutions even in the heterogeneous case.

2 Numerical results

In this section, denoting by $|\cdot|$ the Euclidean norm, the norms have been computed by the following formula:

$$\mathrm{normg} = \sum_{K \in \mathscr{M}} |K| \, |\nabla_K U|,$$

$$\mathrm{erl2} = \left(\left(\sum_{K \in \mathscr{M}} |K| \, (u_K - u(\boldsymbol{x}_K))^2 \right) \Big/ \left(\sum_{K \in \mathscr{M}} |K| \, u(\boldsymbol{x}_K)^2 \right) \right)^{1/2},$$

$$\mathrm{ergrad} = \left(\left(\sum_{K \in \mathscr{M}} |K| \, |\nabla_K U - \nabla u(\boldsymbol{x}_K)|^2 \right) \Big/ \left(\sum_{K \in \mathscr{M}} |K| \, |\nabla u(\boldsymbol{x}_K)|^2 \right) \right)^{1/2},$$

$$\mathrm{ener} = \left(\left(\sum_{K \in \mathscr{M}} |K| \, |\nabla_K U - \nabla u(\boldsymbol{x}_K)|_\Lambda^2 \right) \Big/ \left(\sum_{K \in \mathscr{M}} |K| \, |\nabla u(\boldsymbol{x}_K)|_\Lambda^2 \right) \right)^{1/2},$$

setting, for any $K \in \mathscr{M}$, $|\xi|_\Lambda^2 = \Lambda_K \xi \cdot \xi$ for all $\xi \in \mathbb{R}^3$.

- **Test 1 Mild anisotropy,** $u(x, y, z) = 1 + \sin(\pi x) \sin\left(\pi \left(y + \frac{1}{2}\right)\right) \sin\left(\pi \left(z + \frac{1}{3}\right)\right)$
min = 0, max = 2, **Tetrahedral meshes**

i	nu	nmat	umin	uemin	umax	uemax	normg
1	2003	59943	3.21E-02	2.03E-02	1.98E+00	1.99E+00	1.77E+00
2	3898	122098	1.29E-02	6.84E-03	1.98E+00	1.99E+00	1.77E+00
3	7711	249457	1.30E-02	9.13E-03	1.99E+00	1.99E+00	1.78E+00
4	15266	504716	4.66E-03	5.52E-03	1.99E+00	2.00E+00	1.79E+00
5	30480	1029682	4.03E-03	1.49E-03	2.00E+00	2.00E+00	1.79E+00
6	61052	2102030	1.74E-03	1.83E-03	2.00E+00	2.00E+00	1.79E+00

i	nu	erl2	ratiol2	ergrad	ratiograd	ener	ratioener
1	2003	8.39E-03	-	1.63E-01	-	1.55E-01	-
2	3898	6.28E-03	1.31E+00	1.32E-01	9.50E-01	1.25E-01	9.41E-01
3	7711	3.98E-03	2.01E+00	1.04E-01	1.03E+00	9.90E-02	1.04E+00
4	15266	2.63E-03	1.83E+00	8.34E-02	9.77E-01	7.87E-02	1.01E+00
5	30480	1.67E-03	1.96E+00	6.59E-02	1.02E+00	6.26E-02	9.92E-01
6	61052	1.04E-03	2.03E+00	5.21E-02	1.02E+00	4.93E-02	1.03E+00

- **Test 1 Mild anisotropy,** $u(x, y, z) = 1 + \sin(\pi x) \sin\left(\pi \left(y + \frac{1}{2}\right)\right) \sin\left(\pi \left(z + \frac{1}{3}\right)\right)$
min = 0, max = 2, **Voronoi meshes**

i	nu	nmat	umin	uemin	umax	uemax	normg
1	29	765	1.11E-01	1.56E-01	1.90E+00	1.86E+00	1.36E+00
2	66	2934	1.29E-01	1.79E-01	1.85E+00	1.81E+00	1.56E+00
3	130	7598	2.67E-02	2.67E-02	1.94E+00	1.93E+00	1.63E+00
4	228	16210	9.62E-03	1.20E-02	1.93E+00	1.91E+00	1.67E+00
5	356	29820	1.02E-02	3.85E-03	2.00E+00	1.97E+00	1.69E+00

i	nu	erl2	ratiol2	ergrad	ratiograd	ener	ratioener
1	29	7.37E-02	-	3.97E-01	-	4.01E-01	-
2	66	6.41E-02	5.08E-01	3.12E-01	8.80E-01	2.89E-01	1.19E+00
3	130	4.05E-02	2.03E+00	2.64E-01	7.42E-01	2.48E-01	6.83E-01
4	228	2.81E-02	1.95E+00	2.11E-01	1.20E+00	1.97E-01	1.22E+00
5	356	1.86E-02	2.79E+00	1.85E-01	8.87E-01	1.78E-01	6.86E-01

- **Test 1 Mild anisotropy**, $u(x, y, z) = 1 + \sin(\pi x) \sin\left(\pi \left(y + \frac{1}{2}\right)\right) \sin\left(\pi \left(z + \frac{1}{3}\right)\right)$
min = 0, max = 2, **Kershaw meshes**

i	nu	nmat	umin	uemin	umax	uemax	normg
1	512	21422	-2.14E-03	3.03E-02	1.91E+00	1.96E+00	1.67E+00
2	4096	192664	1.58E-02	1.06E-02	1.96E+00	1.99E+00	1.73E+00
3	32768	1618164	4.90E-03	1.75E-03	1.99E+00	2.00E+00	1.74E+00
4	262144	13109746	8.51E-04	7.14E-04	2.00E+00	2.00E+00	1.76E+00

i	nu	erl2	ratiol2	ergrad	ratiograd	ener	ratioener
1	512	7.45E-02	-	4.84E-01	-	4.40E-01	-
2	4096	6.48E-02	2.00E-01	4.43E-01	1.28E-01	3.85E-01	1.92E-01
3	32768	4.32E-02	5.84E-01	3.02E-01	5.50E-01	2.56E-01	5.92E-01
4	262144	2.31E-02	9.02E-01	1.66E-01	8.62E-01	1.40E-01	8.74E-01

- **Test 1 Mild anisotropy**, $u(x, y, z) = 1 + \sin(\pi x) \sin\left(\pi \left(y + \frac{1}{2}\right)\right) \sin\left(\pi \left(z + \frac{1}{3}\right)\right)$
min = 0, max = 2, **Checkerboard meshes**

i	nu	nmat	umin	uemin	umax	uemax	normg
1	36	836	1.05E-01	1.54E-01	1.87E+00	1.85E+00	1.60E+00
2	288	15848	3.67E-02	4.01E-02	1.96E+00	1.96E+00	1.71E+00
3	2304	173048	5.91E-03	1.01E-02	1.99E+00	1.99E+00	1.77E+00
4	18432	1560014	1.71E-03	2.54E-03	2.00E+00	2.00E+00	1.79E+00
5	147456	13339482	3.83E-04	6.36E-04	2.00E+00	2.00E+00	1.80E+00

i	nu	erl2	ratiol2	ergrad	ratiograd	ener	ratioener
1	36	1.11E-01	-	2.77E-01	-	2.49E-01	-
2	288	3.34E-02	1.74E+00	1.50E-01	8.86E-01	1.40E-01	8.27E-01
3	2304	8.76E-03	1.93E+00	6.88E-02	1.13E+00	6.63E-02	1.08E+00
4	18432	2.33E-03	1.91E+00	3.34E-02	1.04E+00	3.33E-02	9.95E-01
5	147456	5.82E-04	2.00E+00	1.55E-02	1.10E+00	1.54E-02	1.12E+00

- **Test 2 Heterogeneous anisotropy**, $u(x, y, z) = x^3 y^2 z + x \sin(2\pi xz) \sin(2\pi xy) \sin(2\pi z)$, min = -0.862, max = 1.0487, **Prism meshes**

i	nu	nmat	umin	uemin	umax	uemax	normg
1	1210	65648	-8.22E-01	-8.41E-01	9.82E-01	9.84E-01	1.50E+00
2	8820	553442	-8.33E-01	-8.39E-01	1.00E+00	1.01E+00	1.64E+00
3	28830	1935862	-8.55E-01	-8.59E-01	1.03E+00	1.03E+00	1.68E+00
4	67240	4710944	-8.55E-01	-8.57E-01	1.03E+00	1.03E+00	1.69E+00

i	nu	erl2	ratiol2	ergrad	ratiograd	ener	ratioener
1	1210	5.95E-02	-	1.88E-01	-	1.91E-01	-
2	8820	1.85E-02	1.76E+00	6.67E-02	1.56E+00	6.75E-02	1.57E+00
3	28830	8.94E-03	1.85E+00	3.46E-02	1.66E+00	3.50E-02	1.66E+00
4	67240	5.37E-03	1.80E+00	2.23E-02	1.55E+00	2.25E-02	1.56E+00

- **Test 3 Flow on random meshes,** $u(x, y, z) = \sin(2\pi x) \sin(2\pi y) \sin(2\pi z)$, $\min = -1$, $\max = 1$, **Random meshes**

i	nu	nmat	umin	uemin	umax	uemax	normg
1	64	2306	-7.51E-01	-7.59E-01	7.58E-01	6.91E-01	1.43E+00
2	512	31576	-8.36E-01	-9.39E-01	8.64E-01	9.23E-01	2.58E+00
3	4096	317246	-9.69E-01	-9.85E-01	9.58E-01	9.82E-01	3.28E+00
4	32768	2819464	-9.90E-01	-9.96E-01	9.89E-01	9.96E-01	3.51E+00

i	nu	erl2	ratiol2	ergrad	ratiograd	ener	ratioener
1	64	2.00E-01	-	6.47E-01	-	6.64E-01	-
2	512	1.28E-01	6.48E-01	3.00E-01	1.11E+00	2.98E-01	1.16E+00
3	4096	4.67E-02	1.45E+00	1.06E-01	1.50E+00	1.05E-01	1.50E+00
4	32768	1.32E-02	1.82E+00	3.85E-02	1.46E+00	3.72E-02	1.50E+00

- **Test 4 Flow around a well, Well meshes,** $\min = 0$, $\max = 5.415$

i	nu	nmat	umin	uemin	umax	uemax	normg
1	890	56952	4.26E-01	4.14E-01	5.32E+00	5.32E+00	1.58E+03
2	2232	164566	2.58E-01	2.44E-01	5.33E+00	5.33E+00	1.58E+03
3	5016	394986	1.61E-01	1.54E-01	5.33E+00	5.33E+00	1.60E+03
4	11220	927684	1.23E-01	1.18E-01	5.33E+00	5.33E+00	1.61E+03
5	23210	1980998	9.28E-02	8.99E-02	5.34E+00	5.34E+00	1.62E+03
6	42633	3702759	7.41E-02	7.23E-02	5.35E+00	5.35E+00	1.62E+03
7	74679	6573107	5.78E-02	5.65E-02	5.36E+00	5.36E+00	1.62E+03

i	nu	erl2	ratiol2	ergrad	ratiograd	ener	ratioener
1	890	3.79E-03	-	9.69E-02	-	9.56E-02	-
2	2232	3.07E-03	6.86E-01	5.21E-02	2.03E+00	4.96E-02	2.14E+00
3	5016	1.60E-03	2.42E+00	2.81E-02	2.29E+00	2.66E-02	2.31E+00
4	11220	1.10E-03	1.38E+00	2.10E-02	1.08E+00	1.95E-02	1.16E+00
5	23210	7.77E-04	1.45E+00	1.57E-02	1.19E+00	1.45E-02	1.21E+00
6	42633	4.78E-04	2.39E+00	1.07E-02	1.89E+00	1.01E-02	1.81E+00
7	74679	4.56E-04	2.59E-01	9.98E-03	3.72E-01	9.27E-03	4.42E-01

- **Test 5 Discontinuous permeability,** $u(x, y, z) = \sin(\pi x) \sin(\pi y) \sin(\pi z)$, min = 0, max = 1, **Locally refined meshes**

i	nu	nmat	umin	uemin	umax	uemax	normg
1	22	358	-2.49E+02	-1.00E+02	2.49E+02	1.00E+02	1.03E+02
2	176	4570	-4.63E+01	-3.54E+01	4.63E+01	3.54E+01	9.21E+01
3	1408	37730	-8.35E+01	-7.89E+01	8.35E+01	7.89E+01	9.64E+01
4	11264	293666	-9.56E+01	-9.43E+01	9.56E+01	9.43E+01	9.85E+01
5	90112	2309882	-9.89E+01	-9.86E+01	9.89E+01	9.86E+01	9.91E+01

i	nu	erl2	ratiol2	ergrad	ratiograd	ener	ratioener
1	22	1.55E+00	-	1.02E+01	-	4.09E+01	-
2	176	3.00E-01	2.37E+00	2.42E-01	5.39E+00	2.35E-01	7.44E+00
3	1408	6.57E-02	2.19E+00	6.08E-02	2.00E+00	6.20E-02	1.92E+00
4	11264	1.63E-02	2.01E+00	1.90E-02	1.68E+00	1.69E-02	1.88E+00
5	90112	4.62E-03	1.82E+00	8.18E-03	1.22E+00	5.10E-03	1.73E+00

3 Comments on the results

All the linear solvers could be solved using the conjugate gradient solver of the PETSC library with ILU(2) preconditioning with tolerance (or reduction factor) set to 10^{-10}. The following results have been obtained:

1. Using the conjugate gradient solver of the PETSC library, the ILU(0) seems to be the fastest preconditioning on some cases. For example, using the fourth Kershaw mesh, for test 1, we obtain the following CPU times: for ILU(2), 178s, for ILU(1), 60s, for ILU(0), 14s and for Jacobi, 20s. We systematically used ILU(2) in order to prevent from any possible failure.
2. The computing times, using the conjugate gradient solver of the PETSC library with ILU(2) preconditioning are the following, for tetrahedral meshes 2 to 6 on test 1: 1.06, 2.45, 5.73, 6.07 and 23.65s. For the conjugate gradient solver of the ISTL library, with ILU(0), we obtain 0.05, 0.10, 0.29, 0.73 and 1.83s, which seems to show that the computing time which can be expected on full scale studies will be acceptable.

A second remark concerns the treatment of non-planar faces. In the above results, we used the possibility to decompose the non-planar faces in triangles, in particular in test3, "Flow on random meshes", the results which are obtained without using this possibility are the following:

i	nu	nmat	umin	uemin	umax	uemax	normg
1	64	1774	-9.69E-01	-7.55E-01	8.61E-01	6.98E-01	2.57E+00
2	512	21812	-9.30E-01	-9.39E-01	9.97E-01	9.24E-01	3.22E+00
3	4096	210534	-1.02E+00	-9.85E-01	1.01E+00	9.82E-01	3.51E+00
4	32768	1839254	-1.00E+00	-9.96E-01	1.01E+00	9.96E-01	3.58E+00

i	nu	erl2	ratiol2	ergrad	ratiograd	ener	ratioener
1	64	2.96E-01	-	3.89E-01	-	3.54E-01	-
2	512	9.59E-02	1.62E+00	1.87E-01	1.06E+00	1.57E-01	1.17E+00
3	4096	3.89E-02	1.30E+00	1.23E-01	6.00E-01	6.93E-02	1.18E+00
4	32768	1.92E-02	1.02E+00	1.13E-01	1.20E-01	4.94E-02	4.87E-01

They show a clear loss of accuracy of the scheme (the order of convergence being around 1 and not 2).

Acknowledgements Work supported by Groupement MOMAS and ANR VFSitCom

References

1. R. Eymard, T. Gallouët, and R. Herbin. Discretisation of heterogeneous and anisotropic diffusion problems on general non-conforming meshes, SUSHI: a scheme using stabilisation and hybrid interfaces. *IMA J. Numer. Anal.*, 30(4):1009–1043, 2010. see also http://hal.archives-ouvertes.fr/.
2. R. Eymard and R. Herbin. Gradient schemes for diffusion problem. *these proceedings*, 2011.

The paper is in final form and no similar paper has been or is being submitted elsewhere.

Benchmark 3D: the VAG scheme

Robert Eymard, Cindy Guichard, and Raphaèle Herbin

1 Presentation of the scheme

Let Ω be a bounded open domain of $\mathbb{R}^3$, let $f \in L^2(\Omega)$ and let Λ be a measurable function from Ω to the set $\mathscr{M}_3(\mathbb{R})$ of 3×3 matrices, such that for a.e. $x \in \Omega$, $\Lambda(x)$ is symmetric, and such that the set of its eigenvalues is included in $[\underline{\lambda}, \overline{\lambda}]$, where $0 < \underline{\lambda} \leq \overline{\lambda}$. We wish to approximate the function u solution of

$$u \in H_0^1(\Omega) \text{ and } \forall v \in H_0^1(\Omega), \int_\Omega \Lambda(x)\nabla u(x) \cdot \nabla v(x) \mathrm{d}x = \int_\Omega f(x)v(x)\mathrm{d}x, \quad (1)$$

by the approximate gradient scheme [2,4] which reads:

$$U \in X_\mathscr{D}, \forall V \in X_\mathscr{D}, \int_\Omega \Lambda(x) \nabla_\mathscr{D} U(x) \cdot \nabla_\mathscr{D} V(x) \mathrm{d}x = \int_\Omega f(x) \Pi_\mathscr{D} V(x) \mathrm{d}x, \quad (2)$$

where $\Pi_\mathscr{D}$ is a reconstruction operator and $\nabla_\mathscr{D}$ a discrete gradient operator which act on the discrete functional space $X_\mathscr{D}$, where the index $\mathscr{D}$ denotes the discretization; these operators are defined as defined as follows:

R. Eymard
Université Paris-Est, e-mail: robert.eymard@univ-mlv.fr

C. Guichard
IFP Energies nouvelles and Université Paris-Est, e-mail: guichard@ifpenergiesnouvelles.fr

R. Herbin
Université Aix-Marseille, e-mail: Raphaele.Herbin@latp.univ-mrs.fr

1. $\mathcal{M}$ is the set of control volumes, that are disjoint open subsets of Ω such that $\bigcup_{K\in\mathcal{M}} \overline{K} = \overline{\Omega}$, $\mathcal{V} = \mathcal{V}_{int} \cup \mathcal{V}_{ext}$ is the set of vertices of the mesh; any element K of $\mathcal{M}$ is defined by its vertices $s \in \mathcal{V}_K$, its faces $\sigma \in \mathcal{F}_K$; each face is also defined by the set of its vertices $s \in \mathcal{V}_\sigma$, using a suitable geometric definition for the resulting surface in the case of non-planar faces; we assume that Λ is constant on all $K \in \mathcal{M}$, and we denote by Λ_K its value in K;
2. the set of discrete unknowns $X_\mathcal{D}$ is the finite dimensional vector space on $\mathbb{R}$, containing all real families $U = ((u_K)_{K\in\mathcal{M}}, (u_s)_{s\in\mathcal{V}})$, such that $u_s = 0$ if $s \in \mathcal{V}_{ext}$;
3. the mapping $\Pi_\mathcal{D} : X_\mathcal{D} \to L^2(\Omega)$ maps $U = ((u_K)_{K\in\mathcal{M}}, (u_s)_{s\in\mathcal{V}}) \in X_\mathcal{D}$ to the piecewise constant function $u_\mathcal{D} \in L^2(\Omega)$ equal to u_K on each cell $K \in \mathcal{M}$;
4. the mapping $\nabla_\mathcal{D} : X_\mathcal{D} \to L^2(\Omega)^3$ is the reconstruction of a gradient from the values $U = ((u_K)_{K\in\mathcal{M}}, (u_s)_{s\in\mathcal{V}}) \in X_\mathcal{D}$; different expressions for this reconstruction are proposed below, which all lead to convergent gradient schemes in the sense of [2, 4]. Their theoretical analysis is related to that of the SUSHI scheme [1, 3]. Detailed numerical results are given in this paper only using the method described in subsection 1.2; the differences obtained using the other expressions are commented in the last section.

The exterior faces are those of the form $\partial K \cap \partial \Omega$ for any boundary control volume K, and the interior faces are those of the form $\partial K \cap \partial L$ for two neighbouring control volumes K and L. For any face σ, we define a point $\boldsymbol{x}_\sigma$, which is a barycentre with non-negative weights $\beta_{\sigma,s}$ of the elements of the set $\mathcal{V}_\sigma$ including all the vertices of the face, and the value u_σ is defined by

$$u_\sigma = \sum_{s\in\mathcal{V}_\sigma} \beta_{\sigma,s} u_s \text{ with } \sum_{s\in\mathcal{V}_\sigma} \beta_{\sigma,s} = 1.$$

In the next three subsections, we describe three ways of defining a gradient operator which satisfies the VAG requirements. The first gradient is constructed from the Stokes formula on the cells of the mesh (we call it the primal cell to distinguish it from further constructed cells), and requires a stabilization. The second gradient and third gradients are constructed on tetrahedral or octahedral sub–cells of the primal mesh, and are natively stable.

1.1 Stabilised gradient on the primal mesh cells

For a face $\sigma \in \mathcal{F}$, we denote by τ any triangular sub-face with vertices $\boldsymbol{x}_\sigma, s$ and s', where s and s' are two consecutive vertices of σ. The barycentre $\boldsymbol{x}_\tau$ of each sub-face τ may thus be expressed by the following barycentric combination:

$$\boldsymbol{x}_\tau = \sum_{s\in\mathcal{V}_\sigma} \beta_{\tau,s} s \text{ with } \sum_{s\in\mathcal{V}_\sigma} \beta_{\tau,s} = 1,$$

Benchmark 3D: the VAG scheme

where $\beta_{\tau,s} \geq 0$ for all $s \in \mathcal{V}_\sigma$. We then define $\beta_{\tau,s} = 0$ for all $s \in \mathcal{V} \setminus \mathcal{V}_\sigma$. Next, we reconstruct a value u_τ at the point x_τ, by $u_\tau = \sum_{s \in \mathcal{V}_\sigma} \beta_{\tau,s} u_s$. Let $K \in \mathcal{M}$ be an element of the mesh. We denote by $\mathcal{T}_K$ the set of all sub-faces of the faces of K.

We first define, for $U = ((u_K)_{K \in \mathcal{M}}, (u_s)_{s \in \mathcal{V}})$, an approximation of the gradient on cell K:

$$\nabla_K U = \frac{1}{|K|} \sum_{\tau \in \mathcal{T}_K} |\tau| (u_\tau - u_K) n_{K,\tau} = \sum_{s \in \mathcal{V}_K} (u_s - u_K) v_{K,s}, \quad (3)$$

where we denote by

$$v_{K,s} = \frac{1}{|K|} \sum_{\tau \in \mathcal{T}_K} \beta_{\tau,s} |\tau| n_{K,\tau},$$

where $n_{K,\tau}$ is the unit normal vector to τ, outward to K, and $|\tau|, |K|$ are respectively the area and the volume of τ and K. We then define a partition $M_{K,s}$ of K (there is no need to define this partition precisely), such that $|M_{K,s}| = |K|/N_K$, where N_K is the number of vertices of K and we introduce

$$R_{K,s} U = u_s - u_K - \nabla_K U \cdot (s - x_K).$$

We then define, for a given $\gamma > 0$, the constant value $\nabla_{K,s} U$ in $M_{K,s}$:

$$\nabla_{K,s} U = \nabla_K u + \gamma R_{K,s} U v_{K,s}.$$

We finally define a piecewise constant gradient by $\nabla_{\mathcal{D}} U(x) = \nabla_{K,s} U$ for a.e. $x \in M_{K,s}$. This scheme is denoted by "VAG" in [5].

1.2 Piecewise constant gradient on octahedral sub-cells

For a given face σ of a control volume K and for any vertex s of σ, we respectively denote by s^- and s^+ the preceding and the following vertices of s in the face σ (defining any orientation on σ), and we consider the (degenerate) octahedron, denoted by $V_{K,\sigma,s}$ and depicted in Fig. 1, whose vertices are $A_1 = x_K$, $A_2 = x_\sigma$, $A_3 = \frac{1}{2}(s^- + s)$, $A_5 = s$, $A_6 = \frac{1}{2}(s^+ + s)$ and $A_4 = \frac{1}{2}(x_\sigma + s)$ (note that all these octahedra are disjoint, and that the union of their closure is $\overline{\Omega}$). The approximate values of U at the vertices of $V_{K,\sigma,s}$ are respectively $u_1 = u_K$, $u_2 = u_\sigma$, $u_3 = \frac{1}{2}(u_{s^-} + u_s)$, $u_5 = u_s$, $u_6 = \frac{1}{2}(u_{s^+} + u_s)$ and $u_4 = \frac{1}{2}(u_\sigma + u_s)$ (the main diagonals of $V_{K,\sigma,s}$ are therefore (A_1, A_4), (A_2, A_5) and (A_3, A_6)). We then define the following approximate gradient:

$$\nabla_{K,\sigma,s} U = \sum_{i=1}^{3} (u_{i+3} - u_i) \frac{\overrightarrow{A_{i+1} A_{i+4}} \wedge \overrightarrow{A_{i+2} A_{i+5}}}{\text{Det}(\overrightarrow{A_{i+1} A_{i+4}}, \overrightarrow{A_{i+2} A_{i+5}}, \overrightarrow{A_i A_{i+3}})}, \quad (4)$$

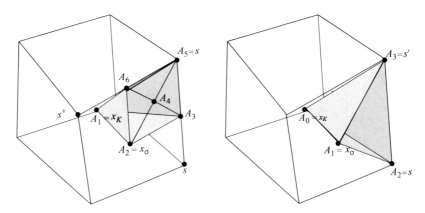

Fig. 1 The octahedral (left) and tetrahedral (right) cells for the definition of the gradient

setting $A_7 = A_1$ and $A_8 = A_2$. We finally define a piecewise constant gradient by $\nabla_\mathcal{D} U(x) = \nabla_{K,\sigma,s} U$ for a.e. $x \in V_{K,\sigma,s}$. Remark that, denoting for simplicity V instead of $V_{K,\sigma,s}$, defining $\mathcal{F}_V$ as the set of the 8 triangular faces of V and $\mathcal{V}_\tau$ as the set of the 3 vertices of each triangular face τ of V, one may check that

$$\nabla_{K,\sigma,s} U = \frac{1}{|V|} \sum_{\tau \in \mathcal{F}_V} |\tau| n_{V,\tau} \Big(\frac{1}{3} \sum_{s \in \mathcal{V}_\tau} u_s\Big). \tag{5}$$

We then set define $\nabla_K U$, used in the tables below, by

$$|K| \nabla_K U = \sum_{\sigma \in \mathcal{F}_K} \sum_{s \in \mathcal{V}_\sigma} |V_{K,\sigma,s}| \nabla_{K,\sigma,s} U.$$

This scheme is denoted by "VAGR" in [5].

1.3 Piecewise constant gradient on tetrahedral sub–cells

For a given face σ of a control volume K and for any pair of consecutive vertices (s, s') of σ, we consider the tetrahedron, denoted by $V_{K,\sigma,s,s'}$ and depicted in Fig. 1, whose vertices are $A_0 = x_K$, $A_1 = x_\sigma$, $A_2 = s$ and $A_3 = s'$ (note that all these tetrahedra are disjoint, and that the union of their closure is $\overline{\Omega}$). The approximate values of U at the vertices of $V_{K,\sigma,s,s'}$ are respectively $u_0 = u_K$, $u_1 = u_\sigma$, $u_2 = u_s$ and $u_3 = u_{s'}$. We then define the following approximate gradient:

$$\nabla_{K,\sigma,s,s'} U = \sum_{i=1}^{3} (u_i - u_0) \frac{\overrightarrow{A_0 A_{i+1}} \wedge \overrightarrow{A_0 A_{i+2}}}{\mathrm{Det}(\overrightarrow{A_0 A_{i+1}}, \overrightarrow{A_0 A_{i+2}}, \overrightarrow{A_0 A_i})}, \tag{6}$$

where $A_4 = A_1$ and $A_5 = A_2$. We finally define a piecewise constant gradient by $\nabla_\mathcal{D} U(\mathbf{x}) = \nabla_{K,\sigma,s,s'} U$ for a.e. $\mathbf{x} \in V_{K,\sigma,s,s'}$. Remark that (5) also holds in this case, denoting V instead of $V_{K,\sigma,s,s'}$.

2 Numerical results

We provide the detailed numerical results obtained, using the scheme VAGR for computing the discrete gradient. In the numerical implementation, the values u_K are locally eliminated, and the unknowns of the linear solver are the values u_s. Denoting by $|\cdot|$ denotes the Euclidean norm, the norms used in the bench tables have been computed using the following formulae:

$$\text{normg} = \sum_{K \in \mathcal{M}} |K| \, |\nabla_K U|,$$

$$\text{erl2} = \left(\left(\sum_{K \in \mathcal{M}} |K| \, (u_K - u(\mathbf{x}_K))^2 \right) / \left(\sum_{K \in \mathcal{M}} |K| \, u(\mathbf{x}_K)^2 \right) \right)^{1/2},$$

$$\text{ergrad} = \left(\left(\sum_{K \in \mathcal{M}} |K| \, |\nabla_K U - \nabla u(\mathbf{x}_K)|^2 \right) / \left(\sum_{K \in \mathcal{M}} |K| \, |\nabla u(\mathbf{x}_K)|^2 \right) \right)^{1/2},$$

$$\text{ener} = \left(\left(\sum_{K \in \mathcal{M}} |K| \, |\nabla_K U - \nabla u(\mathbf{x}_K)|_\Lambda^2 \right) / \left(\sum_{K \in \mathcal{M}} |K| \, |\nabla u(\mathbf{x}_K)|_\Lambda^2 \right) \right)^{1/2},$$

setting, for any $K \in \mathcal{M}$, $|\xi|_\Lambda^2 = \Lambda_K \xi \cdot \xi$ for all $\xi \in \mathbb{R}^3$.

- **Test 1 Mild anisotropy,** $u(x, y, z) = 1 + \sin(\pi x) \sin\left(\pi \left(y + \frac{1}{2}\right)\right) \sin\left(\pi \left(z + \frac{1}{3}\right)\right)$ min = 0, max = 2, **Tetrahedral meshes**

i	nu	nmat	umin	uemin	umax	uemax	normg
1	488	6072	5.77E-02	2.03E-02	1.95E+00	1.99E+00	1.77E+00
2	857	11269	1.88E-02	6.84E-03	1.97E+00	1.99E+00	1.78E+00
3	1601	21675	2.19E-02	9.13E-03	1.98E+00	1.99E+00	1.79E+00
4	2997	41839	1.13E-02	5.52E-03	1.99E+00	2.00E+00	1.79E+00
5	5692	81688	8.73E-03	1.49E-03	1.99E+00	2.00E+00	1.79E+00
6	10994	160852	3.63E-03	1.83E-03	1.99E+00	2.00E+00	1.80E+00

i	nu	erl2	ratiol2	ergrad	ratiograd	ener	ratioener
1	488	1.76E-02	-	2.30E-01	-	2.28E-01	-
2	857	1.02E-02	2.93E+00	1.79E-01	1.35E+00	1.77E-01	1.35E+00
3	1601	6.79E-03	1.94E+00	1.44E-01	1.05E+00	1.42E-01	1.08E+00
4	2997	4.44E-03	2.03E+00	1.13E-01	1.14E+00	1.11E-01	1.17E+00
5	5692	2.79E-03	2.18E+00	9.02E-02	1.06E+00	8.89E-02	1.03E+00
6	10994	1.75E-03	2.13E+00	7.04E-02	1.13E+00	6.92E-02	1.15E+00

- **Test 1 Mild anisotropy,** $u(x, y, z) = 1 + \sin(\pi x) \sin\left(\pi \left(y + \frac{1}{2}\right)\right) \sin\left(\pi \left(z + \frac{1}{3}\right)\right)$
min = 0, max = 2, **Voronoi meshes**

i	nu	nmat	umin	uemin	umax	uemax	normg
1	146	5936	7.54E-02	1.56E-01	2.15E+00	1.86E+00	1.14E+00
2	339	16267	-3.42E-01	1.79E-01	1.95E+00	1.81E+00	1.43E+00
3	684	37194	8.40E-04	2.67E-02	2.02E+00	1.93E+00	1.60E+00
4	1227	71069	-9.54E-02	1.20E-02	2.06E+00	1.91E+00	1.66E+00
5	2023	127883	-1.78E-02	3.85E-03	2.06E+00	1.97E+00	1.70E+00

i	nu	erl2	ratiol2	ergrad	ratiograd	ener	ratioener
1	146	1.82E-01	-	3.96E-01	-	4.05E-01	-
2	339	1.87E-01	-8.43E-02	2.49E-01	1.65E+00	2.53E-01	1.68E+00
3	684	9.92E-02	2.70E+00	1.55E-01	2.02E+00	1.62E-01	1.90E+00
4	1227	7.15E-02	1.68E+00	1.19E-01	1.35E+00	1.23E-01	1.42E+00
5	2023	4.74E-02	2.47E+00	9.56E-02	1.33E+00	9.92E-02	1.29E+00

- **Test 1 Mild anisotropy,** $u(x, y, z) = 1 + \sin(\pi x) \sin\left(\pi \left(y + \frac{1}{2}\right)\right) \sin\left(\pi \left(z + \frac{1}{3}\right)\right)$
min = 0, max = 2, **Kershaw meshes**

i	nu	nmat	umin	uemin	umax	uemax	normg
1	729	15625	7.80E-02	3.03E-02	1.96E+00	1.96E+00	1.56E+00
2	4913	117649	1.72E-02	1.06E-02	1.98E+00	1.99E+00	1.68E+00
3	35937	912673	-2.58E-04	1.75E-03	1.99E+00	2.00E+00	1.74E+00
4	274625	7189057	-2.64E-04	7.14E-04	2.00E+00	2.00E+00	1.78E+00

i	nu	erl2	ratiol2	ergrad	ratiograd	ener	ratioener
1	729	9.17E-02	-	4.91E-01	-	4.84E-01	-
2	4913	5.53E-02	7.96E-01	3.09E-01	7.28E-01	2.84E-01	8.40E-01
3	35937	2.97E-02	9.38E-01	1.74E-01	8.70E-01	1.54E-01	9.22E-01
4	274625	1.22E-02	1.31E+00	7.40E-02	1.26E+00	6.44E-02	1.29E+00

Benchmark 3D: the VAG scheme

- **Test 1 Mild anisotropy,** $u(x, y, z) = 1 + \sin(\pi x) \sin\left(\pi \left(y + \frac{1}{2}\right)\right) \sin\left(\pi \left(z + \frac{1}{3}\right)\right)$
min = 0, max = 2, **Checkerboard meshes**

i	nu	nmat	umin	uemin	umax	uemax	normg
1	97	2413	-9.81E-02	1.54E-01	2.08E+00	1.85E+00	1.34E+00
2	625	22585	-1.90E-01	4.01E-02	2.19E+00	1.96E+00	1.70E+00
3	4417	188641	-6.12E-02	1.01E-02	2.06E+00	1.99E+00	1.78E+00
4	33025	1529617	-1.70E-02	2.54E-03	2.02E+00	2.00E+00	1.79E+00
5	254977	12295153	-4.33E-03	6.36E-04	2.00E+00	2.00E+00	1.80E+00

i	nu	erl2	ratiol2	ergrad	ratiograd	ener	ratioener
1	97	3.25E-01	-	4.37E-01	-	3.97E-01	-
2	625	1.11E-01	1.73E+00	1.50E-01	1.72E+00	1.52E-01	1.54E+00
3	4417	3.01E-02	2.00E+00	5.73E-02	1.47E+00	6.09E-02	1.41E+00
4	33025	7.92E-03	1.99E+00	2.51E-02	1.23E+00	2.77E-02	1.18E+00
5	254977	2.03E-03	2.00E+00	1.18E-02	1.11E+00	1.32E-02	1.08E+00

- **Test 2 Heterogeneous anisotropy,** $u(x, y, z) = x^3 y^2 z + x \sin(2\pi xz) \sin(2\pi xy) \sin(2\pi z)$, min = −0.862, max = 1.048, **Prism meshes**

i	nu	nmat	umin	uemin	umax	uemax	normg
1	3080	99634	-8.73E-01	-8.41E-01	1.10E+00	9.84E-01	1.53E+00
2	20160	710894	-8.25E-01	-8.39E-01	1.04E+00	1.01E+00	1.66E+00
3	63240	2301754	-8.52E-01	-8.59E-01	1.05E+00	1.03E+00	1.69E+00
4	144320	5340214	-8.53E-01	-8.57E-01	1.04E+00	1.03E+00	1.70E+00

i	nu	erl2	ratiol2	ergrad	ratiograd	ener	ratioener
1	3080	1.66E-01	-	1.40E-01	-	1.38E-01	-
2	20160	4.26E-02	2.17E+00	3.71E-02	2.13E+00	3.64E-02	2.13E+00
3	63240	1.93E-02	2.08E+00	1.67E-02	2.10E+00	1.63E-02	2.10E+00
4	144320	1.10E-02	2.05E+00	9.44E-03	2.06E+00	9.25E-03	2.07E+00

- **Test 3 Flow on random meshes,** $u(x, y, z) = \sin(2\pi x) \sin(2\pi y) \sin(2\pi z)$, min = −1, max = 1, **Random meshes**

i	nu	nmat	umin	uemin	umax	uemax	normg
1	125	2197	-1.51E+00	-7.55E-01	1.68E+00	6.98E-01	1.53E+00
2	729	15625	-1.13E+00	-9.39E-01	1.21E+00	9.24E-01	2.99E+00
3	4913	117649	-1.08E+00	-9.85E-01	1.06E+00	9.82E-01	3.44E+00
4	35937	912673	-1.01E+00	-9.96E-01	1.01E+00	9.96E-01	3.56E+00

i	nu	erl2	ratiol2	ergrad	ratiograd	ener	ratioener
1	125	1.15E+00	-	6.19E-01	-	6.26E-01	-
2	729	2.56E-01	2.56E+00	2.02E-01	1.90E+00	1.81E-01	2.11E+00
3	4913	5.93E-02	2.30E+00	8.04E-02	1.45E+00	5.30E-02	1.93E+00
4	35937	1.49E-02	2.09E+00	3.45E-02	1.28E+00	1.74E-02	1.68E+00

- **Test 4 Flow around a well, Well meshes**

i	nu	nmat	umin	uemin	umax	uemax	normg
1	1248	27072	3.89E-01	4.29E-01	5.32E+00	5.32E+00	1.68E+03
2	2800	65184	2.41E-01	2.50E-01	5.33E+00	5.33E+00	1.65E+03
3	5889	143079	1.55E-01	1.57E-01	5.33E+00	5.33E+00	1.64E+03
4	12582	314964	1.18E-01	1.20E-01	5.33E+00	5.33E+00	1.63E+03
5	25300	645210	9.03E-02	9.09E-02	5.34E+00	5.34E+00	1.63E+03
6	45668	1178094	7.27E-02	7.30E-02	5.34E+00	5.35E+00	1.63E+03
7	79084	2055600	5.69E-02	5.68E-02	5.36E+00	5.36E+00	1.63E+03

i	nu	erl2	ratiol2	ergrad	ratiograd	ener	ratioener
1	1248	6.47E-03	-	5.78E-02	-	5.35E-02	-
2	2800	2.71E-03	3.23E+00	2.54E-02	3.05E+00	2.34E-02	3.08E+00
3	5889	1.19E-03	3.31E+00	1.23E-02	2.93E+00	1.15E-02	2.85E+00
4	12582	8.42E-04	1.37E+00	7.59E-03	1.91E+00	7.31E-03	1.79E+00
5	25300	4.47E-04	2.72E+00	5.10E-03	1.71E+00	4.95E-03	1.68E+00
6	45668	2.02E-04	4.03E+00	3.55E-03	1.83E+00	3.47E-03	1.80E+00
7	79084	1.75E-04	7.84E-01	3.26E-03	4.76E-01	3.19E-03	4.56E-01

- **Test 5 Discontinuous permeability,** $u(x, y, z) = \alpha_i \sin(2\pi x) \sin(2\pi y) \sin(2\pi z)$, min = 0, max = 1, **Locally refined meshes**

i	nu	nmat	umin	uemin	umax	uemax	normg
1	60	1148	-7.39E+02	-1.00E+02	7.39E+02	1.00E+02	1.24E+01
2	305	6825	-7.82E+01	-3.54E+01	7.82E+01	3.54E+01	5.20E+01
3	1881	46025	-9.90E+01	-7.89E+01	9.90E+01	7.89E+01	8.60E+01
4	13073	335601	-9.99E+01	-9.43E+01	9.99E+01	9.43E+01	9.56E+01
5	97185	2557793	-1.00E+02	-9.86E+01	1.00E+02	9.86E+01	9.80E+01

i	nu	erl2	ratiol2	ergrad	ratiograd	ener	ratioener
1	60	6.39E+00	-	1.60E+00	-	8.27E+00	-
2	305	1.19E+00	3.10E+00	5.97E-01	1.82E+00	6.01E-01	4.84E+00
3	1881	2.55E-01	2.55E+00	1.86E-01	1.92E+00	1.80E-01	1.99E+00
4	13073	6.10E-02	2.21E+00	5.96E-02	1.76E+00	4.78E-02	2.05E+00
5	97185	1.52E-02	2.08E+00	2.24E-02	1.46E+00	1.26E-02	2.00E+00

Benchmark 3D: the VAG scheme

3 Comments on the results

The results obtained using (3) (VAG) instead of (4) (VAGR) are systematically less precise, except in the test5 case, where we obtained the following tables:

i	nu	nmat	umin	uemin	umax	uemax	normg
1	60	1148	-7.65E+02	-1.00E+02	7.65E+02	1.00E+02	6.76E+01
2	305	6825	-7.73E+01	-3.54E+01	7.73E+01	3.54E+01	4.65E+01
3	1881	46025	-9.02E+01	-7.89E+01	9.02E+01	7.89E+01	8.19E+01
4	13073	335601	-9.72E+01	-9.43E+01	9.72E+01	9.43E+01	9.43E+01
5	97185	2557793	-9.93E+01	-9.86E+01	9.93E+01	9.86E+01	9.77E+01

i	nu	erl2	ratiol2	ergrad	ratiograd	ener	ratioener
1	60	6.71E+00	-	7.32E+00	-	2.90E+01	-
2	305	9.53E-01	3.60E+00	6.91E-01	4.36E+00	6.76E-01	6.93E+00
3	1881	1.49E-01	3.05E+00	2.24E-01	1.85E+00	2.20E-01	1.85E+00
4	13073	3.27E-02	2.35E+00	6.17E-02	2.00E+00	5.95E-02	2.03E+00
5	97185	7.98E-03	2.11E+00	1.73E-02	1.90E+00	1.54E-02	2.02E+00

The results using (6) are very similar to those obtained using (4) (VAGR). For both (3) (VAG) and (4) (VAGR), we have chosen the conjugate gradient solver of the ISTL library with ILU(0) preconditioning with tolerance (or reduction factor) set to 10^{-10}. The following observations have been made on the computing times, using (3) (VAG) (we may expect that similar observations could be done with VAGR).

1. On the fourth Kershaw mesh and test 1, we obtain the following CPU times using the conjugate gradient solver of the PETSC library: with ILU(2), 33s, with ILU(1), 17s, with ILU(0), 10s, and with Jacobi, 11s, which shows that the ILU(0) preconditioning seems the fastest one on this case. Note that this computing time is depending on the unknown orderings. For the bench computations, we used the recursive domain decomposition ordering, which is the most efficient for direct solvers, and the respective computing times with PETSC CG+ILU(0) and with ISTL CG+ILU(0) are 10.3 and 11.2 s. Using the reverse Cuthill - McKee ordering, we respectively obtain 4.4 s and 15.3 s with PETSC CG+ILU(0) and ISTL CG+ILU(0).
2. The computing times, for the conjugate gradient solver of the PETSC library with ILU(1) preconditioning, in the test 1 case on tetrahedral meshes 2 to 5, have been approximately equal to 0.01, 0.03, 0.04, 0.08, and 0.16 s, showing the possibility to apply this method on much larger meshes.

Finally, we may not exclude that the systematic choice of computing the L^2 error with respect to the values in the control volumes instead of the vertex values, makes all these results somewhat pessimistic.

Acknowledgements Work supported by Groupement MOMAS and ANR VFSitCom.

References

1. R. Eymard, T. Gallouët, and R. Herbin. Discretisation of heterogeneous and anisotropic diffusion problems on general non-conforming meshes, sushi: a scheme using stabilisation and hybrid interfaces. *IMA J. Numer. Anal.*, 30(4):1009–1043, 2010. see also http://hal.archives-ouvertes.fr/.
2. R. Eymard, C. Guichard, and R. Herbin. Small-stencil 3D schemes for diffusive flows in porous media. *submitted*, 2010. see also http://hal.archives-ouvertes.fr/.
3. R. Eymard, T. Gallouët and R. Herbin. Benchmark 3D: the SUSHI scheme *these proceedings*, 2011.
4. R. Eymard and R. Herbin. Gradient schemes for diffusion problem. *these proceedings*, 2011.
5. R. Eymard, G. Henry, R. Herbin, F. Hubert, R. Klöfkorn and G. Manzini. 3D Benchmark on Discretization Schemes for Anisotropic Diffusion Problems on General Grids. *these proceedings*, 2011.

The paper is in final form and no similar paper has been or is being submitted elsewhere.

Benchmark 3D: The Compact Discontinuous Galerkin 2 Scheme

Robert Klöfkorn

1 The Compact Discontinuous Galerkin 2 Method

In this paper we provide results for the *3d Benchmark on Anisotropic Diffusion Problems*. We consider the Compact Discontinuous Galerkin 2 (CDG2) method first presented in [3]. In [3] a detailed stability analysis as well as a numerical investigation showing that the CDG2 method outperforms other DG methods (e.g. Bassi–Rebay 2, symmetric Interior Penalty, or the original Compact Discontinuous Galerkin Method, see [1, 3] and references therein) in terms of L^2–accuracy versus computational time. Furthermore, the CDG2 method is a parameter free method in the sense that all tests have been calculated with the same set of parameters without specific test case tuning.

In this section we derive the flux and primal formulation of the CDG2 method for a scalar diffusion equation in $\mathbb{R}^d$, $d = 1, 2, 3$ of the form

$$-\nabla \cdot (\mathbf{K} \nabla u) = f \quad \text{in } \Omega, \tag{1}$$

$$u = g \quad \text{on } \partial\Omega,$$

where $\Omega \subset \mathbb{R}^d$ is a bounded polygonal area, $\mathbf{K} \in L^\infty(\Omega, \mathbb{R}^{d \times d})$ a positive definite diffusion matrix, and $f \in L^2(\Omega)$.

For a given partition $\mathcal{T}_h = \{E\}$ of Ω into polygons E, we look for an approximation u_h of u such that $u_h \in V_h^l$ and

$$V_h^l = \{\mathbf{v} \in L^\infty(\Omega, \mathbb{R}^l) \ : \ \mathbf{v}|_E \in [\mathbb{P}_k(E)]^l \} \quad \text{for some } l \in \mathbb{N}.$$

Robert Klöfkorn
Section of Applied Mathematics, University of Freiburg, Hermann-Herder-Strasse 10, D-79104 Freiburg, Germany, e-mail: robertk@mathematik.uni-freiburg.de

J. Fořt et al. (eds.), *Finite Volumes for Complex Applications VI – Problems & Perspectives*, Springer Proceedings in Mathematics 4,
DOI 10.1007/978-3-642-20671-9_100, © Springer-Verlag Berlin Heidelberg 2011

In the following we use the abbreviations $V_h = V_h^1$ and $\Sigma_h = V_h^d$. Let Γ_i be the family of all interior intersections of elements $E_e^+, E_e^- \in \mathcal{T}_h$ with $e = E_e^- \cap E_e^+$ and Hausdorff measure $\mathcal{H}_{d-1}(e) > 0$. Similarly, we define Γ_D to be the family of all intersections of elements E with $\partial\Omega$. We denote $\Gamma = \Gamma_i \cup \Gamma_D$. For $e \in \Gamma_i$, $\varphi \in V_h$, and $\tau \in \Sigma_h$ we introduce operators $[[\varphi]]_e = \varphi_{|E_e^-}\mathbf{n}_{E_e^-} + \varphi_{|E_e^+}\mathbf{n}_{E_e^+}$, $\{\varphi\}_e = \frac{1}{2}(\varphi_{|E_e^-} + \varphi_{|E_e^+})$, $[\tau]_e = \tau_{|E_e^-} \cdot \mathbf{n}_{E_e^-} + \tau_{|E_e^+} \cdot \mathbf{n}_{E_e^+}$, and $\{\!\{\tau\}\!\}_e = \frac{1}{2}(\tau_{|E_e^-} + \tau_{|E_e^+})$ and for $e \subset \partial\Omega$ we set $[[\varphi]]_e = \varphi\mathbf{n}$, $\{\varphi\}_e = \varphi$, $[\tau]_e = \tau \cdot \mathbf{n}$, and $\{\!\{\tau\}\!\}_e = \tau$.

The DG method in flux formulation is derived by introducing an auxiliary variable σ such that

$$-\nabla \cdot (\mathbf{K}\sigma) = f \quad \text{in } \Omega \quad \text{and} \quad \sigma = \nabla u \quad \text{in } \Omega. \tag{2}$$

Multiplying (2) by arbitrary $\varphi \in V_h$ and $\tau \in \Sigma_h$, respectively, integrating over E, and summing up over all $E \in \mathcal{T}_h$ we arrive at the *flux formulation* of the DG method on the whole domain $\Omega = \bigcup_{E \in \mathcal{T}_h} E$:

$$-\int_\Omega \mathbf{K}\sigma_h \cdot \nabla_h \varphi = \int_\Omega \varphi f - \sum_{e \in \Gamma} \int_e \widehat{\sigma}(u_h) \cdot [[\varphi]]_e \quad \forall\, \varphi \in V_h, \tag{3a}$$

$$\int_\Omega \tau \cdot \sigma_h = -\int_\Omega \nabla_h \cdot \tau u_h + \sum_{e \in \Gamma} \int_e [\tau]_e \widehat{u}(u_h) \quad \forall\, \tau \in \Sigma_h, \tag{3b}$$

where $\nabla_h v \in \Sigma_h$ is a function whose restriction to each element $E \in \mathcal{T}_h$ is equal to ∇v. Furthermore, $\widehat{u}(u_h)$ and $\widehat{\sigma}(u_h)$ are numerical fluxes, where the second is an approximation of the diffusive flux $\mathbf{K}\sigma$, over the boundaries of E where we allow the solution to be discontinuous. The method is completely described once the physical parameters f and $\mathbf{K}$ are known and appropriate numerical fluxes have been chosen. Here, we present the numerical fluxes for the CDG2 method. Other possible choices can be found in [1, 3]. To describe the numerical fluxes we introduce two kinds of *lifting operators* $r_e : [L^2(\Gamma)]^d \to \Sigma_h$ and $l_e : L^2(\Gamma_i) \to \Sigma_h$ with

$$\int_\Omega r_e(\xi) \cdot \tau = -\int_e \xi \cdot \{\!\{\tau\}\!\}_e, \quad \int_\Omega l_e(\phi) \cdot \tau = -\int_e \phi[\tau]_e, \tag{4}$$

for all $\tau \in \Sigma_h$. We extend l_e for $e \subset \partial\Omega$ by setting $l_e(\phi) \equiv 0$ on $\partial\Omega$ for all $\phi \in L^2(\Gamma)$. For convenience we define $L_e(u) := r_e([[u]]_e) + l_e(\boldsymbol{\beta}_e \cdot [[u]]_e)$ on Γ. The parameter $\boldsymbol{\beta}_e$ (frequently denoted by $\mathbf{C}_{12}$ in the literature) is called the *switch function* and is defined by $\boldsymbol{\beta}_e = \frac{1}{2}\mathbf{n}_{E_e^*}$, where $E_e^* \in \{E_e^+, E_e^-\}$ is the element adjacent to e with the smaller volume. For this switch one can show that L_e only has support on either E_e^- or E_e^+ (see [3] for details). The fluxes for the CDG2 method are

$$\widehat{u}(u) = \{u\}_e, \quad \widehat{\sigma}(u) = \{\!\{\mathbf{K}\nabla_h u\}\!\}_e + \chi_e\big(\{\!\{\mathbf{K}L_e(u)\}\!\}_e + \boldsymbol{\beta}_e[\mathbf{K}L_e(u)]_e\big) - \delta_e(u), \tag{5}$$

Benchmark 3D: The Compact Discontinuous Galerkin 2 Scheme

with $\delta_e(u) = \eta_e(\mathbf{K})[[u]]_e \geq 0$, $\chi_e > 0$, and the switch function β_e as described above. Using this switch the method is proven to be stable for any $\chi_e \geq N_{\mathscr{T}_h}/2$ (cf. [3]), where $N_{\mathscr{T}_h}$ is the maximal number of intersections an element $E \in \mathscr{T}_h$ can have.

Since the computation of σ_h might be expensive in terms of computation time as well as memory consumptions one might be interested in deriving a *primal formulation* of the form

$$B(u_h, \varphi) = L(\varphi) \quad \forall \varphi \in V_h, \tag{6}$$

where any solution $u_h \in V_h$ of equation (6) also solves equation (3) and vice versa. The derivation of the primal formulation is, for example, described in [1, 3]. The basic idea is to express σ through u via

$$\sigma = \sigma(u) := \nabla u + \sum_{e \in \Gamma} r_e([[u]]_e) + l_e(\{u - \widehat{u}(u)\}_e) \stackrel{(5)}{=} \nabla u + \sum_{e \in \Gamma} r_e([[u]]_e).$$

Using the numerical fluxes and $\sigma(u)$ (for details see [1, 3]) we arrive at the primal form

$$B(u_h, \varphi) := \int_\Omega \mathbf{K} \nabla u_h \cdot \nabla \varphi + \sum_{e \in \Gamma} \chi_e \int_\Omega \mathbf{K} L_e(u_h) \cdot L_e(\varphi)$$

$$- \sum_{e \in \Gamma} \int_e \left(\{\!\{\mathbf{K} \nabla u_h\}\!\}_e \cdot [[\varphi]]_e + \{\!\{\mathbf{K} \nabla \varphi\}\!\}_e \cdot [[u_h]]_e \right) \tag{7a}$$

$$- \sum_{e \in \Gamma} \int_e \delta_e(u_h) \cdot [[\varphi]]_e \quad \forall \varphi \in V_h,$$

$$L(\varphi) := \int_\Omega f\varphi + \sum_{e \in \Gamma_D} \int_e \mathbf{n}_e \cdot \left(\delta_e(g)\varphi - \mathbf{K}(g\nabla\varphi + \chi_e L_e(g)\varphi) \right) \forall \varphi \in V_h$$

$$\tag{7b}$$

Note that by choosing $\chi_e = 0 \ \forall e \in \Gamma$ in (7a) and (7b) we obtain the well known symmetric Interior Penalty Galerkin (SIPG) method. Finally, we need to specify $\eta_e(\mathbf{K})$. The proof of stability of the CDG2 method given in [3] shows that for simple $\mathbf{K}$ one can choose $\eta_e(\mathbf{K}) = 0$. However, in case $\mathbf{K}$ jumps across element interfaces, this parameter can be used to increase the accuracy of the method. For all numerical tests presented in this paper we used

$$\eta_e(\mathbf{K}) := \begin{cases} \Lambda(\{\!\{\mathbf{K}\}\!\}_e)/|e| & \text{if } e \in \Gamma_i \text{ and } |\Lambda(\mathbf{K}_{|e}^+) - \Lambda(\mathbf{K}_{|e}^-)| > 0, \\ 0 & \text{otherwise.} \end{cases}$$

with $\Lambda(\mathbf{K}) = (\lambda_\mathbf{K}^{max})^2/\lambda_\mathbf{K}^{min}$, where $\lambda_\mathbf{K}^{max}$, $\lambda_\mathbf{K}^{min}$ denote the maximal and minimal eigenvalues of $\mathbf{K}$, respectively.

2 Numerical results

The implementation of the CDG2 scheme is based on the discretization framework DUNE-FEM (see [4]) which is a module of DUNE [2]. Among other things DUNE provides an interface for implementations of discretization grids.

In the following we present results for Test 1, 3, 4, and 5 of the benchmark. Since there is no grid implementation of the DUNE grid interface that can handle Voronoi cells, Test 1 with Voronoi meshes and Test 2 could not be computed.

We use three different versions of the CDG2 method differing in the choice of the basis functions used to build V_h and Σ_h. The first, and most commonly used, is the CDG2–$\mathbb{P}_k$, k being the polynomial order. On hexahedral meshes we also provide results of two alternative schemes. First, there is the possibility to use tensor product Legendre polynomials as basis functions which is denoted by CDG2–$\mathbb{Q}_k$. Another possibility is to split each hexahedron into 6 tetrahedra resulting in a conforming tetrahedral mesh, if possible. This scheme is denoted CDG2–$\mathbb{P}_k$(tetra). This feature is implemented in DUNE and could be used in this case without a complicated mesh generation procedure. An advantage of this approach is that reference mappings will be linear. A disadvantage is that the number of cells and unknowns are increased by a factor of 6. We use this scheme only for the tests with complicated meshes, i.e. Test 1 (Kershaw), Test 3, and Test 4.

Although basis functions up to order 8 are implemented we restrict ourself to $k = 1, 2$. While $k = 1$ allows a direct comparison with Finite Volume schemes, $k = 2$ shows the potential of a higher order approach. The integrals in (7a) and (7b) have been calculated using quadratures of order $2k$, $2k + 1$ for element and face integrals, respectively. If the reference mapping of the element is nonlinear then the quadrature orders have been increased by 2. The calculation of the relative L^2–error as well as the energy norm is straightforward in the DG context. For the calculation of the H^1–error the auxiliary variable σ_h given by equation (3b) has been used. The quadrature order for evaluation of the errors is $2k + 4$ ($+2$ in case of nonlinear reference mappings). The convergence order of the schemes is $k + 1$ in the L^2–norm and k in the H^1–norm.

In the following computations we used $\chi_e = 2$ for tetrahedral grids and $\chi_e = 3$ for hexahedral grids, also in the non-conforming cases where χ_e would be larger when using the theoretical values presented in Section 1.

- **Test 1 Mild anisotropy,** $u(x, y, z) = 1 + \sin(\pi x) \sin\left(\pi \left(y + \frac{1}{2}\right)\right) \sin\left(\pi \left(z + \frac{1}{3}\right)\right)$ min = 0, max = 2, **Tetrahedral meshes**

Table 1, CDG2–$\mathbb{P}_1$

i	nu	nmat	umin	uemin	umax	uemax	normg
1	8012	150576	-1.54E-02	0.00E+00	2.017	1.989	1.783
2	15592	297376	-1.22E-02	0.00E+00	2.016	1.988	1.789
3	30844	593648	-7.41E-03	0.00E+00	2.005	1.993	1.792
4	61064	1184192	-3.12E-03	0.00E+00	1.999	1.997	1.794
5	121920	2379936	0.00E+00	0.00E+00	2.002	1.997	1.795
6	244208	4791872	-6.63E-04	0.00E+00	2.002	1.997	1.796

Benchmark 3D: The Compact Discontinuous Galerkin 2 Scheme   1027

Table 1,
CDG2–$\mathbb{P}_2$

i	nu	nmat	umin	uemin	umax	uemax	normg
1	20030	941100	0.00E+00	0.00E+00	1.999	1.989	1.800
2	38980	1858600	0.00E+00	0.00E+00	1.999	1.988	1.799
3	77110	3710300	0.00E+00	0.00E+00	1.999	1.993	1.799
4	152660	7401200	0.00E+00	0.00E+00	1.999	1.997	1.798
5	304800	14874600	0.00E+00	0.00E+00	1.999	1.997	1.798
6	610520	29949200	0.00E+00	0.00E+00	1.999	1.997	1.798

Table 2,
CDG2–$\mathbb{P}_1$

i	nu	erl2	ratiol2	ergrad	ratiograd	ener	ratioener
1	8012	9.01E-03	—	1.75E-01	—	1.94E-01	—
2	15592	5.78E-03	2.001	1.40E-01	0.999	1.56E-01	0.979
3	30844	3.68E-03	1.981	1.12E-01	0.968	1.24E-01	0.998
4	61064	2.38E-03	1.925	8.94E-02	1.004	9.87E-02	1.009
5	121920	1.52E-03	1.943	7.09E-02	1.005	7.86E-02	0.989
6	244208	9.45E-04	2.049	5.61E-02	1.007	6.20E-02	1.021

Table 2,
CDG2–$\mathbb{P}_2$

i	nu	erl2	ratiol2	ergrad	ratiograd	ener	ratioener
1	20030	6.32E-04	—	1.77E-02	—	1.95E-02	—
2	38980	3.24E-04	3.012	1.13E-02	1.996	1.24E-02	2.039
3	77110	1.60E-04	3.095	6.94E-03	2.165	7.71E-03	2.097
4	152660	8.03E-05	3.035	4.40E-03	2.001	4.90E-03	1.994
5	304800	4.10E-05	2.913	2.80E-03	1.958	3.12E-03	1.961
6	610520	2.04E-05	3.022	1.75E-03	2.026	1.95E-03	2.023

- **Test 1 Mild anisotropy,** $u(x, y, z) = 1 + \sin(\pi x) \sin\left(\pi \left(y + \frac{1}{2}\right)\right) \sin\left(\pi \left(z + \frac{1}{3}\right)\right)$
  min = 0, max = 2, **Kershaw meshes**

Table 1,
CDG2–
$\mathbb{P}_1$(tetra)

i	nu	nmat	umin	uemin	umax	uemax	normg
1	12288	233472	-2.81E-02	0.00E+00	2.012	1.951	1.745
2	98304	1916928	-7.22E-03	0.00E+00	1.996	1.998	1.747
3	786432	15532032	-1.17E-03	0.00E+00	2.000	1.999	1.762
4	6291456	125042688	-4.65E-04	0.00E+00	2.000	1.999	1.780

Table 1,
CDG2–
$\mathbb{P}_2$(tetra)

i	nu	nmat	umin	uemin	umax	uemax	normg
1	30720	1459200	0.00E+00	0.00E+00	1.995	1.951	1.779
2	245760	11980800	0.00E+00	0.00E+00	1.998	1.998	1.790
3	1966080	97075200	0.00E+00	0.00E+00	1.999	1.999	1.797
4	15728640	781516800	0.00E+00	0.00E+00	1.999	1.999	1.798

Table 1,
CDG2–$\mathbb{Q}_1$

i	nu	nmat	umin	uemin	umax	uemax	normg
1	4096	204800	-2.95E-02	0.00E+00	2.016	1.958	1.781
2	32768	1736704	-8.49E-03	0.00E+00	1.999	1.992	1.778
3	262144	14286848	-2.86E-04	0.00E+00	2.001	1.996	1.783
4	2097152	115867648	-5.37E-04	0.00E+00	2.000	1.999	1.790

Table 1,
CDG2–$\mathbb{Q}_2$

i	nu	nmat	umin	uemin	umax	uemax	normg
1	13824	2332800	0.00E+00	0.00E+00	1.997	1.958	1.793
2	110592	19782144	0.00E+00	0.00E+00	1.999	1.992	1.795
3	884736	162736128	0.00E+00	0.00E+00	1.999	1.996	1.798

Table 2,
CDG2–
$\mathbb{P}_1$(tetra)

i	nu	erl2	ratiol2	ergrad	ratiograd	ener	ratioener
1	12288	8.29E-02	—	6.25E-01	—	6.12E-01	—
2	98304	5.47E-02	0.600	4.28E-01	0.547	4.04E-01	0.601
3	786432	3.07E-02	0.831	2.74E-01	0.643	2.48E-01	0.702
4	6291456	1.34E-02	1.194	1.56E-01	0.809	1.39E-01	0.841

Table 2,
CDG2–
$\mathbb{P}_2$**(tetra)**

i	nu	erl2	ratiol2	ergrad	ratiograd	ener	ratioener
1	30720	3.72E-02	—	2.74E-01	—	2.53E-01	—
2	245760	9.29E-03	2.002	1.06E-01	1.365	1.02E-01	1.310
3	1966080	1.31E-03	2.831	3.12E-02	1.771	3.15E-02	1.695
4	15728640	1.07E-04	3.611	7.98E-03	1.966	8.32E-03	1.919

Table 2,
CDG2–$\mathbb{Q}_1$

i	nu	erl2	ratiol2	ergrad	ratiograd	ener	ratioener
1	4096	7.09E-02	—	6.08E-01	—	5.88E-01	—
2	32768	4.77E-02	0.574	4.25E-01	0.516	3.97E-01	0.565
3	262144	2.61E-02	0.870	2.74E-01	0.636	2.57E-01	0.630
4	2097152	1.09E-02	1.262	1.56E-01	0.812	1.50E-01	0.774

Table 2,
CDG2–$\mathbb{Q}_2$

i	nu	erl2	ratiol2	ergrad	ratiograd	ener	ratioener
1	13824	3.06E-02	—	2.44E-01	—	2.15E-01	—
2	110592	7.57E-03	2.012	1.03E-01	1.240	9.73E-02	1.141
3	884736	1.04E-03	2.865	3.22E-02	1.682	3.18E-02	1.615

- **Test 1 Mild anisotropy,** $u(x, y, z) = 1 + \sin(\pi x) \sin\left(\pi \left(y + \frac{1}{2}\right)\right) \sin\left(\pi \left(z + \frac{1}{3}\right)\right)$ min = 0, max = 2, **Checkerboard meshes**

Table 1,
CDG2–$\mathbb{P}_1$

i	nu	nmat	umin	uemin	umax	uemax	normg
1	144	3648	0.00E+00	0.00E+00	1.901	1.846	1.538
2	1152	35328	-8.80E-03	0.00E+00	2.006	1.959	1.698
3	9216	307200	-6.73E-03	0.00E+00	2.005	1.989	1.765
4	73728	2555904	-1.86E-03	0.00E+00	2.001	1.997	1.788
5	589824	20840448	-5.50E-04	0.00E+00	2.000	1.999	1.795

Table 1,
CDG2–$\mathbb{P}_2$

i	nu	nmat	umin	uemin	umax	uemax	normg
1	360	22800	-3.34E-02	0.00E+00	2.050	1.846	1.752
2	2880	220800	-7.34E-04	0.00E+00	2.000	1.959	1.794
3	23040	1920000	0.00E+00	0.00E+00	1.999	1.989	1.798
4	184320	15974400	0.00E+00	0.00E+00	1.999	1.997	1.798
5	1474560	130252800	0.00E+00	0.00E+00	1.999	1.999	1.798

Table 1,
CDG2–$\mathbb{Q}_1$

i	nu	nmat	umin	uemin	umax	uemax	normg
1	288	14592	-7.94E-02	0.00E+00	2.081	1.846	1.581
2	2304	141312	-1.77E-02	0.00E+00	2.017	1.959	1.744
3	18432	1228800	-4.59E-03	0.00E+00	2.004	1.989	1.785
4	147456	10223616	-1.27E-03	0.00E+00	2.001	1.997	1.795
5	1179648	83361792	-3.06E-04	0.00E+00	2.000	1.999	1.797

Table 1,
CDG2–$\mathbb{Q}_2$

i	nu	nmat	umin	uemin	umax	uemax	normg
1	972	166212	0.00E+00	0.00E+00	1.998	1.846	1.807
2	7776	1609632	0.00E+00	0.00E+00	1.999	1.959	1.800
3	62208	13996800	0.00E+00	0.00E+00	1.999	1.989	1.798
4	497664	116453376	0.00E+00	0.00E+00	1.999	1.997	1.798

Table 2,
CDG2–$\mathbb{P}_1$

i	nu	erl2	ratiol2	ergrad	ratiograd	ener	ratioener
1	144	1.08E-01	—	5.66E-01	—	5.70E-01	—
2	1152	3.70E-02	1.549	3.32E-01	0.770	3.46E-01	0.720
3	9216	1.12E-02	1.719	1.76E-01	0.917	1.79E-01	0.946
4	73728	3.09E-03	1.860	8.95E-02	0.974	9.05E-02	0.987
5	589824	8.01E-04	1.950	4.51E-02	0.990	4.54E-02	0.995

Table 2, CDG2–$\mathbb{P}_2$

i	nu	erl2	ratiol2	ergrad	ratiograd	ener	ratioener
1	360	3.93E-02	—	2.88E-01	—	3.07E-01	—
2	2880	7.29E-03	2.432	8.86E-02	1.699	8.56E-02	1.845
3	23040	8.62E-04	3.080	2.24E-02	1.982	2.18E-02	1.974
4	184320	1.05E-04	3.033	5.59E-03	2.004	5.46E-03	1.997
5	1474560	1.31E-05	3.007	1.40E-03	2.001	1.37E-03	1.999

Table 2, CDG2–$\mathbb{Q}_1$

i	nu	erl2	ratiol2	ergrad	ratiograd	ener	ratioener
1	288	6.31E-02	—	3.81E-01	—	3.77E-01	—
2	2304	1.83E-02	1.789	1.96E-01	0.960	1.94E-01	0.959
3	18432	4.63E-03	1.980	9.86E-02	0.989	9.82E-02	0.983
4	147456	1.16E-03	1.998	4.96E-02	0.993	4.94E-02	0.991
5	1179648	2.89E-04	2.001	2.49E-02	0.996	2.48E-02	0.995

Table 2, CDG2–$\mathbb{Q}_2$

i	nu	erl2	ratiol2	ergrad	ratiograd	ener	ratioener
1	972	7.05E-03	—	7.17E-02	—	7.03E-02	—
2	7776	1.02E-03	2.782	1.87E-02	1.936	1.84E-02	1.935
3	62208	1.36E-04	2.910	4.75E-03	1.981	4.68E-03	1.974
4	497664	1.95E-05	2.803	1.19E-03	1.993	1.18E-03	1.989

- **Test 3 Flow on random meshes,** $u(x, y, z) = \sin(2\pi x) \sin(2\pi y) \sin(2\pi z)$, min $= -1$, max $= 1$, **Random meshes**

Table 1, CDG2–$\mathbb{P}_1$(tetra)

i	nu	nmat	umin	uemin	umax	uemax	normg
1	1536	27648	-1.261	-7.43E-01	1.167	8.36E-01	2.757
2	12288	233472	-1.009	-9.35E-01	1.033	9.33E-01	3.198
3	98304	1916928	-1.016	-9.82E-01	1.017	9.84E-01	3.484
4	786432	15532032	-1.008	-9.95E-01	1.002	9.96E-01	3.568

Table 1, CDG2–$\mathbb{P}_2$(tetra)

i	nu	nmat	umin	uemin	umax	uemax	normg
1	3840	172800	-1.238	-7.43E-01	1.295	8.36E-01	3.637
2	30720	1459200	-1.042	-9.35E-01	1.028	9.33E-01	3.577
3	245760	11980800	-1.000	-9.82E-01	1.000	9.84E-01	3.598
4	1966080	97075200	-1.000	-9.95E-01	1.000	9.96E-01	3.596

Table 1, CDG2–$\mathbb{Q}_1$

i	nu	nmat	umin	uemin	umax	uemax	normg
1	512	22528	-1.143	-7.59E-01	1.244	6.91E-01	3.016
2	4096	204800	-1.076	-9.39E-01	1.074	9.23E-01	3.432
3	32768	1736704	-1.026	-9.85E-01	1.021	9.82E-01	3.564
4	262144	14286848	-1.009	-9.96E-01	1.000	9.96E-01	3.587

Table 1, CDG2–$\mathbb{Q}_2$

i	nu	nmat	umin	uemin	umax	uemax	normg
1	1728	256608	-1.015	-7.59E-01	1.034	6.91E-01	3.635
2	13824	2332800	-1.002	-9.39E-01	9.95E-01	9.23E-01	3.568
3	110592	19782144	-1.000	-9.85E-01	9.99E-01	9.82E-01	3.596
4	884736	162736128	-1.000	-9.96E-01	1.000	9.96E-01	3.595

Table 2, CDG2–$\mathbb{P}_1$(tetra)

i	nu	erl2	ratiol2	ergrad	ratiograd	ener	ratioener
1	1536	3.49E-01	—	5.82E-01	—	5.80E-01	—
2	12288	1.17E-01	1.572	3.43E-01	0.764	3.07E-01	0.920
3	98304	3.48E-02	1.751	2.00E-01	0.777	1.60E-01	0.935
4	786432	9.59E-03	1.861	1.08E-01	0.892	8.12E-02	0.981

Table 2,
CDG2–
$\mathbb{P}_2$**(tetra)**

i	nu	erl2	ratiol2	ergrad	ratiograd	ener	ratioener
1	3840	1.05E-01	—	3.08E-01	—	2.21E-01	—
2	30720	1.66E-02	2.659	9.93E-02	1.632	5.82E-02	1.925
3	245760	2.48E-03	2.739	2.80E-02	1.829	1.54E-02	1.917
4	1966080	3.06E-04	3.021	6.50E-03	2.106	3.90E-03	1.980

Table 2,
CDG2–$\mathbb{Q}_1$

i	nu	erl2	ratiol2	ergrad	ratiograd	ener	ratioener
1	512	3.05E-01	—	4.98E-01	—	5.00E-01	—
2	4096	8.38E-02	1.862	2.58E-01	0.947	2.53E-01	0.982
3	32768	2.17E-02	1.952	1.30E-01	0.988	1.25E-01	1.015
4	262144	5.86E-03	1.885	6.72E-02	0.954	6.31E-02	0.988

Table 2,
CDG2–$\mathbb{Q}_2$

i	nu	erl2	ratiol2	ergrad	ratiograd	ener	ratioener
1	1728	4.41E-02	—	1.25E-01	—	1.14E-01	—
2	13824	6.02E-03	2.875	2.97E-02	2.067	2.82E-02	2.012
3	110592	8.07E-04	2.897	7.58E-03	1.971	7.16E-03	1.980
4	884736	1.03E-04	2.968	1.91E-03	1.993	1.78E-03	2.009

- **Test 4 Flow around a well, Well meshes,** min $= 0$, max $= 5.415$

Table 1,
CDG2–
$\mathbb{P}_1$**(tetra)**

i	nu	nmat	umin	uemin	umax	uemax	normg
1	21360	369664	0.00E+00	0.00E+00	5.406	5.358	1633.939
2	53568	941888	0.00E+00	0.00E+00	5.408	5.366	1628.212
3	120384	2168128	0.00E+00	0.00E+00	5.408	5.368	1626.468
4	269280	4963456	0.00E+00	0.00E+00	5.408	5.371	1626.554
5	557040	10433536	0.00E+00	0.00E+00	5.409	5.376	1625.759
6	1023192	19371936	0.00E+00	0.00E+00	5.409	5.379	1624.751
7	1792296	34218592	0.00E+00	0.00E+00	5.410	5.387	1624.488

Table 1,
CDG2–
$\mathbb{P}_2$**(tetra)**

i	nu	nmat	umin	uemin	umax	uemax	normg
1	53400	2310400	-5.92E-03	0.00E+00	5.414	5.358	1623.061
2	133920	5886800	-1.61E-03	0.00E+00	5.414	5.366	1622.951
3	300960	13550800	-1.42E-04	0.00E+00	5.414	5.368	1623.127
4	673200	31021600	0.00E+00	0.00E+00	5.414	5.371	1623.224
5	1392600	65209600	0.00E+00	0.00E+00	5.414	5.376	1623.329
6	2557980	121074600	0.00E+00	0.00E+00	5.414	5.379	1623.423
7	4480740	213866200	0.00E+00	0.00E+00	5.414	5.387	1623.472

Table 1,
CDG2–$\mathbb{Q}_1$

i	nu	nmat	umin	uemin	umax	uemax	normg
1	7120	356736	0.00E+00	0.00E+00	5.406	5.316	1685.638
2	17856	931328	0.00E+00	0.00E+00	5.407	5.328	1653.938
3	40128	2139904	0.00E+00	0.00E+00	5.407	5.328	1639.060
4	89760	4857216	0.00E+00	0.00E+00	5.407	5.330	1633.474
5	185680	10136320	0.00E+00	0.00E+00	5.408	5.339	1629.453
6	341064	18717760	0.00E+00	0.00E+00	5.409	5.345	1627.258
7	597432	32900416	0.00E+00	0.00E+00	5.410	5.360	1626.055

Table 1,
CDG2–$\mathbb{Q}_2$

i	nu	nmat	umin	uemin	umax	uemax	normg
1	24030	4063446	0.00E+00	0.00E+00	5.408	5.316	1623.988
2	60264	10608408	-5.25E-04	0.00E+00	5.409	5.328	1623.197
3	135432	24374844	0.00E+00	0.00E+00	5.409	5.328	1623.222
4	302940	55326726	0.00E+00	0.00E+00	5.409	5.330	1623.226
5	626670	115459020	0.00E+00	0.00E+00	5.410	5.339	1623.337
6	1151091	213206985	0.00E+00	0.00E+00	5.410	5.345	1623.423
7	2016333	374756301	0.00E+00	0.00E+00	5.411	5.360	1623.446

Benchmark 3D: The Compact Discontinuous Galerkin 2 Scheme

Table 2, CDG2–$\mathbb{P}_1$(tetra)

i	nu	erl2	ratiol2	ergrad	ratiograd	ener	ratioener
1	21360	1.83E-03	—	1.69E-01	—	2.07E-01	—
2	53568	1.03E-03	1.878	1.13E-01	1.316	1.41E-01	1.259
3	120384	6.75E-04	1.570	7.87E-02	1.328	9.92E-02	1.302
4	269280	5.63E-04	0.675	6.12E-02	0.938	7.65E-02	0.968
5	557040	4.16E-04	1.253	4.83E-02	0.980	6.02E-02	0.991
6	1023192	2.70E-04	2.119	3.93E-02	1.015	4.91E-02	1.005
7	1792296	2.24E-04	1.014	3.34E-02	0.862	4.15E-02	0.905

Table 2, CDG2–$\mathbb{P}_2$(tetra)

i	nu	erl2	ratiol2	ergrad	ratiograd	ener	ratioener
1	53400	2.46E-04	—	3.53E-02	—	4.46E-02	—
2	133920	9.83E-05	2.993	1.60E-02	2.569	2.06E-02	2.530
3	300960	4.37E-05	3.005	7.94E-03	2.607	1.02E-02	2.606
4	673200	2.91E-05	1.517	4.68E-03	1.973	5.95E-03	1.997
5	1392600	1.72E-05	2.162	2.92E-03	1.944	3.69E-03	1.978
6	2557980	9.30E-06	3.037	1.92E-03	2.075	2.43E-03	2.050
7	4480740	6.79E-06	1.684	1.42E-03	1.598	1.77E-03	1.710

Table 2, CDG2–$\mathbb{Q}_1$

i	nu	erl2	ratiol2	ergrad	ratiograd	ener	ratioener
1	7120	6.22E-03	—	2.47E-01	—	2.46E-01	—
2	17856	2.92E-03	2.461	1.70E-01	1.232	1.69E-01	1.220
3	40128	1.45E-03	2.593	1.19E-01	1.312	1.19E-01	1.304
4	89760	9.46E-04	1.595	9.08E-02	1.008	9.09E-02	1.006
5	185680	5.92E-04	1.937	7.07E-02	1.032	7.08E-02	1.030
6	341064	3.63E-04	2.419	5.71E-02	1.055	5.72E-02	1.056
7	597432	2.68E-04	1.610	4.77E-02	0.962	4.78E-02	0.960

Table 2, CDG2–$\mathbb{Q}_2$

i	nu	erl2	ratiol2	ergrad	ratiograd	ener	ratioener
1	24030	3.48E-04	—	3.93E-02	—	3.93E-02	—
2	60264	1.07E-04	3.843	1.61E-02	2.912	1.61E-02	2.910
3	135432	4.55E-05	3.175	7.48E-03	2.838	7.49E-03	2.834
4	302940	3.03E-05	1.511	4.40E-03	1.978	4.40E-03	1.977
5	626670	1.70E-05	2.383	2.64E-03	2.101	2.65E-03	2.098
6	1151091	8.46E-06	3.451	1.69E-03	2.194	1.70E-03	2.194
7	2016333	5.65E-06	2.159	1.17E-03	1.968	1.18E-03	1.965

- **Test 5 Discontinuous permeability,**
$u(x, y, z) = a_i \sin(2\pi x) \sin(2\pi y) \sin(2\pi z)$, min $= -100$, max $= 100$, **Locally refined meshes**

Table 1, CDG2–$\mathbb{P}_1$

i	nu	nmat	umin	uemin	umax	uemax	normg
1	88	1984	-8.502	-100.000	8.502	100.000	10.584
2	704	17792	-52.177	-35.355	52.177	35.355	46.202
3	5632	150016	-83.767	-78.858	83.767	78.858	68.596
4	45056	1230848	-95.672	-94.345	95.672	94.345	80.340
5	360448	9969664	-98.927	-98.562	98.927	98.562	88.092

Table 1, CDG2–$\mathbb{P}_2$

i	nu	nmat	umin	uemin	umax	uemax	normg
1	220	12400	-18.447	-100.000	18.447	100.000	48.277
2	1760	111200	-99.735	-35.355	99.735	35.355	92.615
3	14080	937600	-102.184	-78.858	102.184	78.858	99.167
4	112640	7692800	-100.544	-94.345	100.544	94.345	99.221
5	901120	62310400	-100.098	-98.562	100.098	98.562	99.100

Table 1, CDG2–$\mathbb{Q}_1$

i	nu	nmat	umin	uemin	umax	uemax	normg
1	176	7936	-12.747	-100.000	12.747	100.000	8.513
2	1408	71168	-118.320	-35.355	118.320	35.355	79.760
3	11264	600064	-103.899	-78.858	103.899	78.858	94.412
4	90112	4923392	-100.970	-94.345	100.970	94.345	97.877
5	720896	39878656	-100.241	-98.562	100.241	98.562	98.740

Table 1, CDG2–$\mathbb{Q}_2$

i	nu	nmat	umin	uemin	umax	uemax	normg
1	594	90396	-94.815	-100.000	94.815	100.000	106.736
2	4752	810648	-100.376	-35.355	100.376	35.355	100.759
3	38016	6835104	-99.836	-78.858	99.836	78.858	99.387
4	304128	56080512	-99.951	-94.345	99.951	94.345	99.084
5	2433024	454242816	-99.987	-98.562	99.987	98.562	99.024

Table 2, CDG2–$\mathbb{P}_1$

i	nu	erl2	ratiol2	ergrad	ratiograd	ener	ratioener
1	88	9.26E-01	—	9.86E-01	—	1.003	—
2	704	6.95E-01	0.412	8.49E-01	0.216	7.90E-01	0.345
3	5632	2.86E-01	1.283	4.95E-01	0.777	4.30E-01	0.879
4	45056	9.88E-02	1.532	2.84E-01	0.802	2.08E-01	1.045
5	360448	2.99E-02	1.724	1.39E-01	1.034	1.01E-01	1.042

Table 2, CDG2–$\mathbb{P}_2$

i	nu	erl2	ratiol2	ergrad	ratiograd	ener	ratioener
1	220	7.70E-01	—	8.98E-01	—	7.83E-01	—
2	1760	3.18E-01	1.276	5.07E-01	0.825	4.17E-01	0.909
3	14080	5.14E-02	2.629	1.77E-01	1.519	1.09E-01	1.939
4	112640	5.63E-03	3.191	4.87E-02	1.862	2.56E-02	2.086
5	901120	7.55E-04	2.898	1.57E-02	1.633	6.27E-03	2.031

Table 2, CDG2–$\mathbb{Q}_1$

i	nu	erl2	ratiol2	ergrad	ratiograd	ener	ratioener
1	176	9.16E-01	—	1.001	—	1.005	—
2	1408	2.36E-01	1.960	4.57E-01	1.133	4.54E-01	1.146
3	11264	6.32E-02	1.897	2.27E-01	1.007	2.27E-01	1.001
4	90112	1.61E-02	1.970	1.13E-01	1.003	1.13E-01	1.001
5	720896	4.06E-03	1.992	5.67E-02	1.001	5.67E-02	1.000

Table 2, CDG2–$\mathbb{Q}_2$

i	nu	erl2	ratiol2	ergrad	ratiograd	ener	ratioener
1	594	6.22E-02	—	1.51E-01	—	1.42E-01	—
2	4752	2.89E-02	1.107	9.46E-02	0.670	9.23E-02	0.626
3	38016	3.80E-03	2.925	2.36E-02	2.002	2.32E-02	1.993
4	304128	4.92E-04	2.949	5.83E-03	2.019	5.78E-03	2.002
5	2433024	6.39E-05	2.945	1.45E-03	2.011	1.44E-03	2.002

3 Comments

For all tests the solver reduction tolerance was taken to be 10^{-10}. Test 1 (except Kershaw) and Test 5 were uncritical for all meshes. The solution was a matter of minutes rather than hours. Test 3 was the one that was most difficult to solve. Here, solving took a long time and the only combination of solver and preconditioning that produced results was GMRES + JACOBI. For all other tests practically any combination of solver and preconditioning from the solver-bench package produced good results. The scheme CDG2–$\mathbb{P}_k$(tetra) seems to be a good alternative to the scheme CDG2–$\mathbb{Q}_k$, since the reference mappings are linear and quadratures of lower order can be used. Also, the solution of the resulting linear system seemed to be much easier (hours rather than days). This is payed by a factor of 6 more elements and, in case of $k = 2$ by a factor of 3 more DoFs. For CDG2–$\mathbb{Q}_2$ we had to skip the last mesh of Test 1 (checkerboard) due to memory limitations.

Using CDG2–$\mathbb{P}_k$ for Test 3 and 4 we were not able to produce satisfying results. The next table shows the L^2–**projection** of the exact solution of **Test 3** onto V_h using the **random mesh** series:

Table 2, CDG2–$\mathbb{P}_2$

i	nu	erl2	ratiol2	ergrad	ratiograd	ener	ratioener
1	640	1.39E-01	—	3.09E-01	—	3.85E-01	—
2	5120	2.31E-02	2.586	9.51E-02	1.699	1.28E-01	1.591
3	40960	6.30E-03	1.875	4.65E-02	1.033	6.45E-02	0.987
4	327680	2.76E-03	1.189	3.99E-02	0.222	5.26E-02	0.294

As we can see, the space $\mathbb{P}_k$ seems not to be suitable to project the solution properly onto the discrete spaces. The convergence does not show the expected *ratiol2* of $k + 1$.

To conclude, with the CDG2 scheme all test cases (except Test 2, see Section 2 for explanation) could have been computed. The CDG2 provides a higher order alternative to solve anisotropic diffusion problems in 3d without specific parameter tuning required by the user. Finally, I would like to thank Peter Bastian for fruitful discussions about the test cases and for providing the test case implementation which saved a lot of time.

References

1. D.N. Arnold, F. Brezzi, B. Cockburn, and L.D. Marini. Unified analysis of discontinuous Galerkin methods for elliptic problems. *SIAM J. Numer. Anal.*, 39(5):1749–1779, 2002.
2. P. Bastian, M. Blatt, A. Dedner, C. Engwer, R. Klöfkorn, R. Kornhuber, M. Ohlberger, and O. Sander. A generic grid interface for parallel and adaptive scientific computing. II: Implementation and tests in dune. *Computing*, 82(2-3):121–138, 2008.
3. S. Brdar, A. Dedner, and R. Klöfkorn. Compact and stable Discontinuous Galerkin methods for convection-diffusion problems. Preprint No. 2/2010-15.11-2010, Mathematisches Institut, Universität Freiburg, 2010. submitted to SIAM J. Sci. Comput.
4. A. Dedner, R. Klöfkorn, M. Nolte, and M. Ohlberger. A generic interface for parallel and adaptive discretization schemes: abstraction principles and the DUNE–FEM; module. *Computing*, 90:165–196, 2010.

The paper is in final form and no similar paper has been or is being submitted elsewhere.

Benchmark 3D: Mimetic Finite Difference Method for Generalized Polyhedral Meshes

Konstantin Lipnikov and Gianmarco Manzini

1 Presentation of the scheme

Let Ω be a subset of $\Re^3$ with a Lipschitz continuous boundary. We consider the mixed (velocity-pressure) formulation of the diffusion problem,

$$\mathbf{u} = -\mathbb{K}\nabla p \quad \text{and} \quad \operatorname{div}\mathbf{u} = b \quad \text{in} \quad \Omega, \tag{1}$$

subject to Dirichlet boundary conditions on $\partial\Omega$. Here $\mathbb{K}$ is the diffusion tensor and b is the source function.

Let Ω_h be a conformal partition of Ω into generalized polyhedral elements E. We assume that each element E is shape-regular and satisfies assumptions (M2)–(M3) formulated in [1]. Let f denote a face of a generalized polyhedron E and $|f|$ be its area. Furthermore, let $\mathbf{n}_f(\mathbf{x})$ be a unit normal vector to face f at point $\mathbf{x}$. Direction of $\mathbf{n}_f(\mathbf{x})$ is fixed once and for all. Let $\mathbf{n}_{E,f}(\mathbf{x})$ be a unit normal vector external to E, so that $\mathbf{n}_{E,f}(\mathbf{x}) \cdot \mathbf{n}_f(\mathbf{x}) = \pm 1$. We introduce the average normal to face f as

$$\widetilde{\mathbf{n}}_f = \frac{1}{|f|} \int_f \mathbf{n}_f(\mathbf{x}) \, dA.$$

Maximal deviation of the average normal $\widetilde{\mathbf{n}}_f$ from a pointwise normal characterizes deviation of face f from a planar face. More precisely, we say that a face f is *moderately curved* if

---

Konstantin Lipnikov
Los Alamos National Laboratory, Theoretical Division, MS B284, Los Alamos, NM 87545, USA, e-mail: lipnikov@lanl.gov

Gianmarco Manzini
IMATI-CNR and CESNA-IUSS, Pavia, Italy, e-mail: marco.manzini@imati.cnr.it

$$\max_{\mathbf{x}\in f} \|\mathbf{n}_f(\mathbf{x}) - \widetilde{\mathbf{n}}_f\| \leq \sigma_* |f|^{1/2},$$

where σ_* is a positive constant independent of the mesh. Otherwise, we say that the face is *strongly curved*. For example, for a polyhedral mesh with planar faces, all faces are classified as moderately curved.

Integrating the second equation in (1) over element E and using the divergence theorem, we get

$$\sum_{f\in\partial E} u^h_{E,f} |f| = \int_E b\, dV, \qquad u^h_{E,f} = \frac{1}{|f|}\int_f \mathbf{u}\cdot \mathbf{n}_{E,f}(\mathbf{x})\, dA. \qquad (2)$$

Thus, it is natural to take average normal components of the velocity $\mathbf{u}$ on mesh faces as discrete unknowns. For a moderately curved face, this is the *sole* unknown representing the velocity $\mathbf{u}$ on this face. If face f is shared by two elements E and E', we impose the following continuity condition

$$u^h_{E,f} = -u^h_{E',f}. \qquad (3)$$

For a strongly curved face, regardless of the number of its vertices, we introduce two additional velocity degrees of freedom as

$$u^h_{E,f,i} = \frac{1}{|f|}\int_f \mathbf{u}\cdot\mathbf{a}_{f,i}\, dA, \qquad i = 2, 3,$$

where $\mathbf{a}_{f,i}$ are two arbitrary chosen unit vectors orthogonal to $\widetilde{\mathbf{n}}_f$ (see Fig. 1).

If this face is shared by two elements E and E', we impose the following continuity conditions

$$u^h_{E,f,i} = u^h_{E',f,i}, \qquad i = 2,3. \qquad (4)$$

For problems with discontinuous coefficients, it is more natural to define additional discrete unknowns as tangential components of the gradient ∇p, rather than velocity $\mathbf{u}$. Fortunately, in practical applications, material interfaces are composed of moderately curved faces; therefore, we did not investigate other definitions of degrees of freedom.

Taken into account continuity conditions, the total number of discrete velocity unknowns is equal to the number of moderately curved faces plus three times the number of strongly curved faces. The threshold σ_* affects the number of strongly curved faces. Smaller value of σ_* results in a more accurate method at a cost of solving larger system of equations.

Fig. 1 Local coordinate system associated with a curved face

Benchmark 3D: Mimetic Finite Difference Method

The scalar (pressure) variable p is represented by its average values over elements E and faces f. For a moderately curved face f, we introduce

$$p_E = \frac{1}{|E|} \int_E p\, dV \quad \text{and} \quad p_{E,f} = \frac{1}{|f|} \int_f (\mathbf{n}_f(\mathbf{x}) \cdot \widetilde{\mathbf{n}}_f) p\, dA, \quad (5)$$

where $|E|$ denote the volume of element E. For a strongly curved face, we need two additional degrees of freedom to match the number of velocity unknowns:

$$p_{E,f,i} = \frac{1}{|f|} \int_f (\mathbf{n}_f(\mathbf{x}) \cdot \mathbf{a}_{f,i}) p\, dA, \quad i = 2, 3. \quad (6)$$

The total number of discrete pressure unknowns is equal to the number of elements plus the number of moderately curved faces plus three times the number of strongly curved faces.

Let us consider an element E with m faces $f_1, \ldots, f_m$. Without loss of generality, we assume that only face f_1 is classified as strongly curved and the other faces are planar, as shown in Fig. 1. We assume that there exists a matrix $\mathbb{W}_E$ and the following linear relations between the discrete unknowns:

$$\begin{bmatrix} u_{E,f_1} \\ u_{E,f_1,2} \\ u_{E,f_1,3} \\ u_{E,f_2} \\ \vdots \\ u_{E,f_m} \end{bmatrix} = \mathbb{W}_E \begin{bmatrix} |f_1|(p_{E,f_1} - p_E) \\ |f_1|\, p_{E,f_1,2} \\ |f_1|\, p_{E,f_1,3} \\ |f_2|(p_{E,f_2} - p_E) \\ \vdots \\ |f_m|(p_{E,f_m} - p_E) \end{bmatrix}. \quad (7)$$

The key of the MFD method is in construction of a proper $(m+2) \times (m+2)$ matrix $\mathbb{W}_E$. Let $\mathbb{K}_E$ be a constant tensor approximating tensor $\mathbb{K}$ in element E. In practice, we take $\mathbb{K}_E = \mathbb{K}(\mathbf{x}_E)$, where $\mathbf{x}_E$ is the center of mass of E. We will define the matrix $\mathbb{W}_E$ such that equation (7) is exact for any linear function p and the corresponding constant vector $\mathbf{u}$.

It is trivial for $p = 1$ with $\mathbf{u} = (0, 0, 0)^T$, since the vectors on the left and the right-hand sides are zero vectors. For $p(x, y, z) = x$ with $\mathbf{u} = -\mathbb{K}_E (1, 0, 0)^T$, $p(x, y, z) = y$ with $\mathbf{u} = -\mathbb{K}_E (0, 1, 0)^T$, and $p(x, y, z) = z$ with $\mathbf{u} = -\mathbb{K}_E (0, 0, 1)^T$, we obtain three matrix equations:

$$\mathbb{N}_{E,x} = \mathbb{W}_E \mathbb{R}_{E,x}, \quad \mathbb{N}_{E,y} = \mathbb{W}_E \mathbb{R}_{E,y} \quad \text{and} \quad \mathbb{N}_{E,z} = \mathbb{W}_E \mathbb{R}_{E,z}. \quad (8)$$

The left and right hand-side vectors can be calculated using only geometric data for faces of E which results in relatively simple calculations for an arbitrary-shaped element.

We define $(m + 2) \times 3$ matrices $\mathbb{N} = [\mathbb{N}_{E,x}; \mathbb{N}_{E,y}; \mathbb{N}_{E,z}]$ and $\mathbb{R} = [\mathbb{R}_{E,x}; \mathbb{R}_{E,y}; \mathbb{R}_{E,z}]$. It has been proved in [2] that a particular solution to the matrix equations (8) is

$$\mathbb{W}_{E,0} = \frac{1}{|E|} \mathbb{N} \mathbb{K}_E^{-1} \mathbb{N}^T.$$

The rank of this matrix is 3 and therefore less than $m + 2$. To build a positive definite $(m + 2) \times (m + 2)$ matrix $\mathbb{W}_E$, we have to add a matrix $\mathbb{W}_{E,1}$ such that $\mathbb{W}_{E,1} \mathbb{R} = 0$. In practice, we take

$$\mathbb{W}_{E,1} = a_E \left(\mathbb{I} - \mathbb{R} \left(\mathbb{R}^T \mathbb{R} \right)^{-1} \mathbb{R}^T \right), \qquad a_E = \frac{\text{trace}(\mathbb{K}_E)}{|E|}.$$

It has been proved in [1] that the matrix $\mathbb{W}_E$ given by

$$\mathbb{W}_E = \mathbb{W}_{E,0} + \mathbb{W}_{E,1} \tag{9}$$

is positive definite. Moreover, its condition number depends only on the anisotropy of tensor $\mathbb{K}_E$ and the shape-regularity of element E.

The mimetic finite difference method is defined by (2), (3), (7), (9), and boundary conditions. The Dirichlet boundary conditions are incorporated in a straightforward manner by prescribing values of integrals in (5) and (6) to corresponding pressure unknowns.

Substituting (7) into (2), (3) and (4), we may easily get an algebraic problem with a sparse symmetric and positive definite matrix. In practice, we also eliminate the cell-based pressure unknowns p_E. The size of the final problem is equal to the number of moderately curved faces plus 3 times the number of strongly curved faces. This number is reported in tables in the next section.

To get a method that is exact for linear solutions, we have to classify all non-planar faces as strongly curved.

Under assumptions of the mesh shape regularity and the solution regularity, the second-order convergence estimate for the pressure p in a discrete norm and the first-order convergence estimate for the velocity **u** have been proved in [1].

2 Numerical results

The discrete energy norm is calculated using the inner product of the mass matrices $\mathbb{W}_E^{-1}$ with vectors of discrete velocities $u_{E,f}^h$. Due to the lack of accurate quadrature rules for generalized polyhedra, we used a mid-point quadrature rule. The quadrature points were located at the mass centers of elements. In practice, an error in average pressure values is of greater interest to engineers than an accurate estimate of the exact L^2 error. Instead of the L^1 norm of a discrete gradient, we provide a discrete energy norm, *norme*. Calculation of the discrete gradient is feasible via a post-processing of fluxes; however, such a capability was not available at the moment of writing this paper.

- **Test 1 Mild anisotropy,** $u(x, y, z) = 1 + \sin(\pi x) \sin\left(\pi \left(y + \frac{1}{2}\right)\right) \sin\left(\pi \left(z + \frac{1}{3}\right)\right)$
  $\min = 0$, $\max = 2$, **Tetrahedral meshes**

i	nu	nmat	umin	uemin	umax	uemax	normg	norme
1	4308	28344	2.29E-02	2.03E-02	1.987	1.989	—	1.925
2	8248	55024	2.33E-03	6.84E-03	1.994	1.989	—	1.924
3	16148	108680	7.67E-03	9.13E-03	1.995	1.994	—	1.924
4	31691	214883	3.17E-03	5.52E-03	1.997	1.997	—	1.924
5	62787	428547	2.49E-03	1.49E-03	1.996	1.997	—	1.924
6	124988	857612	1.66E-03	1.83E-03	1.998	1.997	—	1.924

i	nu	erl2	ratiol2	ergrad	ratiograd	ener	ratioener
1	4308	5.37E-03	—	—	—	1.22E-01	
2	8248	3.67E-03	1.758	—	—	9.95E-02	0.942
3	16148	2.26E-03	2.165	—	—	7.48E-02	1.274
4	31691	1.53E-03	1.736	—	—	6.02E-02	0.966
5	62787	9.55E-04	2.068	—	—	4.89E-02	0.912
6	124988	5.96E-04	2.054	—	—	3.81E-02	1.088

- **Test 1 Mild anisotropy,** $u(x, y, z) = 1 + \sin(\pi x) \sin\left(\pi \left(y + \frac{1}{2}\right)\right) \sin\left(\pi \left(z + \frac{1}{3}\right)\right)$
  $\min = 0$, $\max = 2$, **Voronoi meshes**

i	nu	nmat	umin	uemin	umax	uemax	normg	norme
1	172	2964	-1.36E-02	8.51E-02	1.936	1.870	—	1.906
2	402	7788	-9.00E-02	1.43E-01	1.911	1.854	—	2.128
3	811	17205	1.92E-02	3.85E-02	2.039	1.925	—	2.058
4	1452	32446	-2.92E-02	1.74E-02	1.975	1.914	—	2.046
5	2376	57180	-3.24E-02	2.84E-02	2.055	1.979	—	2.026

i	nu	erl2	ratiol2	ergrad	ratiograd	ener	ratioener
1	172	9.11E-02	—	—	—	6.96E-01	
2	402	1.33E-01	-1.337	—	—	6.44E-01	0.274
3	811	8.59E-02	1.869	—	—	4.46E-01	1.570
4	1452	6.77E-02	1.226	—	—	3.52E-01	1.219
5	2376	5.75E-02	0.995	—	—	2.92E-01	1.138

- **Test 1 Mild anisotropy,** $u(x, y, z) = 1 + \sin(\pi x) \sin\left(\pi \left(y + \frac{1}{2}\right)\right) \sin\left(\pi \left(z + \frac{1}{3}\right)\right)$ min = 0, max = 2, **Kershaw meshes**

i	nu	nmat	umin	uemin	umax	uemax	normg	norme
1	1728	17088	-2.52E-02	3.03E-02	1.973	1.958	—	1.920
2	13056	135936	-5.20E-03	1.06E-02	1.998	1.993	—	1.925
3	101376	1084416	-1.48E-03	1.75E-03	1.998	1.997	—	1.925
4	798720	8663040	2.71E-04	7.14E-04	1.999	1.999	—	1.924

i	nu	erl2	ratiol2	ergrad	ratiograd	ener	ratioener
1	1728	6.67E-02	—	—	—	1.45E-00	—
2	13056	3.35E-02	1.022	—	—	6.05E-01	1.297
3	101376	1.09E-02	1.643	—	—	1.71E-01	1.850
4	798720	2.79E-03	1.981	—	—	4.42E-02	1.966

- **Test 1 Mild anisotropy,** $u(x, y, z) = 1 + \sin(\pi x) \sin\left(\pi \left(y + \frac{1}{2}\right)\right) \sin\left(\pi \left(z + \frac{1}{3}\right)\right)$ min = 0, max = 2, **Checkerboard meshes**

i	nu	nmat	umin	uemin	umax	uemax	normg	norme
1	93	921	2.91E-01	5.17E-01	1.880	1.846	—	2.193
2	636	6588	1.42E-01	1.54E-01	1.968	1.960	—	1.961
3	4656	49584	3.45E-02	4.01E-02	1.992	1.990	—	1.932
4	35520	384192	8.63E-03	1.01E-02	1.998	1.997	—	1.926
5	277248	3023616	2.15E-03	2.54E-03	1.999	1.999	—	1.924

i	nu	erl2	ratiol2	ergrad	ratiograd	ener	ratioener
1	93	1.08E-01	—	—	—	4.34E-01	—
2	636	2.26E-02	2.441	—	—	1.13E-01	2.100
3	4656	5.25E-03	2.200	—	—	3.49E-02	1.771
4	35520	1.23E-03	2.143	—	—	1.14E-02	1.652
5	277248	2.89E-04	2.115	—	—	3.83E-03	1.593

- **Test 2 Heterogeneous anisotropy,** $u(x, y, z) = x^3 y^2 z + x \sin(2\pi xz) \sin(2\pi xy) \sin(2\pi z)$, min = -0.862, max = 1.0487, **Prism meshes**

i	nu	nmat	umin	uemin	umax	uemax	normg	norme
1	5331	72291	-0.873	-0.862	0.832	0.831	—	2.794
2	37261	529581	-0.861	-0.861	0.925	0.925	—	2.790
3	119791	1731871	-0.883	-0.883	0.951	0.951	—	2.790
4	276921	4039161	-0.890	-0.890	0.963	0.963	—	2.790

i	nu	erl2	ratiol2	ergrad	ratiograd	ener	ratioener
1	5331	4.07E-02	—	—	—	9.23E-02	
2	37261	1.10E-02	2.019	—	—	2.95E-02	1.760
3	119791	5.02E-03	2.015	—	—	1.49E-02	1.755
4	276921	2.85E-03	2.027	—	—	9.24E-03	1.711

- **Test 3 Flow on random meshes,** $u(x, y, z) = \sin(2\pi x) \sin(2\pi y) \sin(2\pi z)$, min $= -1$, max $= 1$, **Random meshes**

i	nu	nmat	umin	uemin	umax	uemax	normg	norme
1	240	2160	-1.268	-0.756	1.430	0.712	—	
2	1728	17088	-1.184	-0.939	1.397	0.926	—	
3	13056	135936	-1.135	-0.986	1.111	0.982	—	
4	107118	1223876	-1.027	-0.996	1.021	0.996	—	

i	nu	erl2	ratiol2	ergrad	ratiograd	ener	ratioener
1	240	9.26E-01	—	—	—	1.53E+00	
2	1728	3.34E-01	1.550	—	—	8.63E-01	0.870
3	13056	1.06E-01	1.702	—	—	4.84E-01	0.858
4	107118	3.14E-02	1.734	—	—	2.53E-01	0.925

- **Test 4 Flow around a well, Well meshes,** min $= 0$, max $= 5.415$

i	nu	nmat	umin	uemin	umax	uemax	normg	norme
1	3004	29782	5.37E-01	4.57E-01	5.317	5.317	—	2.08E+01
2	7232	74192	2.90E-01	2.62E-01	5.329	5.329	—	2.16E+01
3	15886	166366	1.74E-01	1.62E-01	5.329	5.329	—	2.18E+01
4	34983	371583	1.29E-01	1.23E-01	5.330	5.330	—	2.19E+01
5	71683	768167	9.66E-02	9.28E-02	5.339	5.339	—	2.20E+01
6	130894	1410160	7.67E-02	7.42E-02	5.345	5.345	—	2.20E+01
7	228463	2471077	5.91E-02	5.75E-02	5.361	5.361	—	2.20E+01

i	nu	erl2	ratiol2	ergrad	ratiograd	ener	ratioener
1	3004	1.12E-02	—	—	—	1.40E-01	
2	7232	4.30E-03	3.269	—	—	5.61E-02	3.123
3	15886	1.90E-03	3.114	—	—	2.76E-02	2.704
4	34983	1.04E-03	2.290	—	—	1.62E-02	2.025
5	71683	6.14E-04	2.204	—	—	1.01E-02	1.976
6	130894	4.04E-04	2.086	—	—	6.74E-03	2.015
7	228463	2.99E-04	1.621	—	—	4.95E-03	1.663

- **Test 5 Discontinuous permeability,** $u(x, y, z) = a_i \sin(2\pi x) \sin(2\pi y) \sin(2\pi z)$, min = -100, max = 100, **Locally refined meshes**

i	nu	nmat	umin	uemin	umax	uemax	normg	norme
1	93	921	-1.66E+02	-1.00E+02	1.66E+02	1.00E+02	—	1.43E+03
2	636	6588	-5.43E+01	-3.54E+01	5.43E+01	3.54E+01	—	4.81E+02
3	4656	49584	-9.06E+01	-7.89E+01	9.06E+01	7.89E+01	—	4.14E+02
4	35520	384192	-9.79E+01	-9.44E+01	9.79E+01	9.44E+01	—	3.93E+02
5	277248	3023616	-9.95E+01	-9.86E+01	9.95E+01	9.86E+01	—	3.87E+02
6	2190336	23989248	-9.99E+01	-9.96E+01	9.99E+01	9.96E+01	—	3.86E+02

i	nu	erl2	ratiol2	ergrad	ratiograd	ener	ratioener
1	93	1.94E+00	—	—	—	3.12E+00	—
2	636	5.34E-01	2.013	—	—	3.07E-01	3.618
3	4656	1.48E-01	1.934	—	—	7.56E-02	2.112
4	35520	3.78E-02	2.015	—	—	1.89E-02	2.047
5	277248	9.50E-03	2.016	—	—	4.79E-03	2.004
6	2190336	2.38E-03	2.009	—	—	1.23E-03	1.973

3 Comments

All problems have been solved with a preconditioned conjugate gradient method applied to a system for face-based pressure unknowns. We used the diagonal of the stiffness matrix as the preconditioner. Relative reduction of the residual by factor 10^{-12} required 1382 iterations for the last Kershaw mesh. The other tests require less than 679 iterations.

The method is superconvergent on all smooth meshes for both primary variables. We used $\sigma_* = 10$ in Test 3 and $\sigma_* = 0.1$ in Test 4. Our experience shows that the presented meshes are too coarse to see significant impact of this parameter on the asymptotic convergence rate.

References

1. F. Brezzi, K. Lipnikov, and M. Shashkov. Convergence of mimetic finite difference method for diffusion problems on polyhedral meshes with curved faces. *Math. Models Methods Appl. Sci.*, 16(2):275–297, 2006.
2. F. Brezzi, K. Lipnikov, M. Shashkov, and V. Simoncini. A new discretization methodology for diffusion problems on generalized polyhedral meshes. *Comput. Methods Appl. Mech. Engrg.*, 196, 3682–3692, 2007.

The paper is in final form and no similar paper has been or is being submitted elsewhere.

Benchmark 3D: CeVe-DDFV, a Discrete Duality Scheme with Cell/Vertex Unknowns

Yves Coudière and Charles Pierre

1 Presentation of the scheme

1.1 General presentation of DDFV methods

"Discrete Duality" Finite Volumes (*DDFV*) schemes have been specifically designed for anisotropic and/or heterogeneous diffusion problems working on general meshes: distorted, non-conformal and locally refined. They first were introduced in 2D independently by Hermeline [9, 10] and Domelevo and Omnès [8], though the key ideas already appear in the work of Nicolaides [14].

As originally defined in [8], a 2D *DDFV* scheme consists in associating a second mesh (the dual mesh) to the original (primal) mesh by building dual cells around each (primal) mesh vertex. Cell and vertex centered *scalar data* are associated to this *double mesh* framework (one data per primal and dual cell), whereas a *vector data* consists in one vector per (primal) mesh edge. To a scalar data is associated a discrete gradient that is a vector data. A *gradient reconstruction* method is used to define this discrete gradient: precisely using the diamond method [6]. A discrete divergence acts on vector data by averaging their normal component on the primal and dual cell boundaries, which procedure is classical for finite volume methods. The key feature is a duality property between the discrete gradient and the discrete divergence operators of Green formula type.

Yves Coudière
LMJL, Université de Nantes, France, e-mail: Yves.Coudiere@univ-nantes.fr

Charles Pierre
LMA, Université de Pau et des pays de l'Adour, France, e-mail: charles.pierre@univ-pau.fr

Extensions of *DDFV* schemes to 3D [1–3, 11, 12, 15] are of two types.

CV-DDFV. The original 2D *double mesh* framework is conserved, dual cells are built around the primal mesh vertices and scalar data consist in a double set of unknowns associated with the (primal) mesh cells and vertices.

CeVeFE-DDFV. Recently Coudière and Hubert [3, 4] modified the 2D framework by considering a third mesh (triple mesh method), with unknowns associated with cells, faces, edges and vertices of the primal mesh.

The method considered here is of *CV-DDFV* type, *CV* holding for Cell and Vertex centered. Two versions have been developed so far.

(A) A first 3D construction was introduced by Pierre in [15] for anisotropic and/or heterogeneous diffusion problems. The dual cells here do not form a mesh in the classical sense: they recover the domain twice.

(B) A second version, independently introduced by Hermeline [12] and Andreianov & al. [1, 2] differs from the previous one by the dual cell definition that here form a partition of Ω.

For both versions, in presence of heterogeneity, auxiliary (locally eliminated) data are added relatively to faces, as presented in [5, 12]. In case of complex meshes, involving face shapes other than triangles or quadrangles, this local elimination procedure is made difficult enforcing to consider auxiliary data as real unknowns inside the algorithm, which drastically increases the problem size.

We first emphasize the similarities between *(A)* and *(B)*. These two versions are based on the same definition of the discrete gradient. They also induce comparable discrete duality properties. Indeed, after a careful examination of these duality properties in [15] and in [2] it turns out that they do involve exactly the same stiffness and mass matrices. As a result, between these two versions, only the averaging of the source terms on the dual cells will differ.

In this paper, version *(A)* will be considered without auxiliary data on the mesh faces. The fifth test case, including heterogeneity and thus necessitating these auxiliary unknowns per face center, will not be treated here because of a lack of time.

1.2 CV- DDFV version (A), discrete duality

Let the domain $\Omega \subset \mathbb{R}^3$ be a connected open subset, its boundary is assumed to be polyhedral. Let $\mathcal{M}$ be a (general) mesh of Ω, possibly non conformal, and whose (primal) cells (*resp.* faces) are general polyhedral (*resp.* polygonal). The set of cells, faces and vertices of $\mathcal{M}$ are respectively denoted $\mathcal{C}$, $\mathcal{F}$ and $\mathcal{V}$. To any vertex $v \in \mathcal{V}$ is associated a dual cell v^* and to any face $f \in \mathcal{F}$ is associated a diamond cell D_f. Diamond cells form a partition of Ω, whereas dual cells intersect and recover Ω exactly twice.

A vector data is a piecewise constant vector function on the diamond cells. A scalar data is provided by one scalar per cell and per vertex of $\mathcal{M}$. The space of vector data is denoted $\mathbb{Q}_h$ and the space of scalar data $\mathbb{F}_h$. A discrete function is obtained by supplementing a scalar data with one scalar data per boundary face. The space of discrete functions is denoted $\mathbb{U}_h$. As developed in Sect. 1.3, $u_h \in \mathbb{U}_h$ will be interpreted as a function defined on the diamond cell boundaries:

$$\partial\mathscr{D} := \bigcup_{f \in \mathscr{F}} \partial D_f, \quad u_h : \partial\mathscr{D} \longrightarrow \mathbb{R}, \tag{1}$$

that moreover is continuous and piecewise affine on the diamond cell faces.

Two discrete operators will be defined, $\nabla_h : \mathbb{U}_h \longrightarrow \mathbb{Q}_h$ and $\mathrm{div}_h : \mathbb{Q}_h \longrightarrow \mathbb{F}_h$. that satisfy the *discrete duality property* (see [15])

$$\int_\Omega \nabla_h u_h \cdot \mathbf{q}_h \, dx = -\langle\!\langle u_h, \mathrm{div}_h \, \mathbf{q}_h \rangle\!\rangle + \int_{\partial\Omega} u_h \, \mathbf{q}_h \cdot \mathbf{n} \, ds \tag{2}$$

for any functions $u_h \in \mathbb{U}_h$ and $\mathbf{q}_h \in \mathbb{Q}_h$, with $\mathbf{n}$ the unit normal on $\partial\Omega$ pointing outside Ω, and the pairing:

$$\langle\!\langle u_h, \mathrm{div}_h \, \mathbf{q}_h \rangle\!\rangle = \frac{1}{3} \sum_{c \in \mathscr{C}} u_c \, \mathrm{div}_c \, \mathbf{q}_h |c| + \frac{1}{3} \sum_{v \in \mathscr{V}} u_v \, \mathrm{div}_v \, \mathbf{q}_h |v^*|. \tag{3}$$

Here, $|c|$ and $|v^*|$ are the volumes of the primal and dual cells c and v^*, u_c and $\mathrm{div}_c \, \mathbf{q}_h$ are the values associated to the cell c of the two scalar data u_h and $\mathrm{div}_h \, \mathbf{q}_h$, and similarly u_v and $\mathrm{div}_v \, \mathbf{q}_h$ are the values associated to the vertex v of the two scalar data u_h and $\mathrm{div}_h \, \mathbf{q}_h$.

In (2) the two integrals are well defined. The first integral is an L^2 product on Ω since both $\mathbf{q}_h$ and $\nabla_h u_h$ are piecewise constant vector functions on the diamond cells. The second integral is an L^2 product on $\partial\Omega$: $\mathbf{q}_h$ is piecewise constant on the boundary faces and its normal component $\mathbf{q}_h \cdot \mathbf{n}$ also, moreover $\partial\Omega \subset \partial\mathscr{D}$ defined in (1) and so u_h has a restriction to $\partial\Omega$ that is continuous.

1.3 Dual and diamond cells

A center x_c (resp. x_f) is associated to each cell $c \in \mathscr{C}$ (resp. $f \in \mathscr{F}$).

Diamond cells. Let $f \in \mathscr{F}$. In case $f \not\subset \partial\Omega$ then f is the interface between two cells $c_1, c_2 \in \mathscr{C}$: $f = \overline{c_1} \cap \overline{c_2}$. Denoting x_i the center of c_i, then D_f is the union of the two pyramids with apex x_i and with base f as depicted on Fig. 1. In case $f \subset \partial\Omega$, then $f = \partial\Omega \cap \overline{c}$ for one cell $c \in \mathscr{C}$. In this case D_f is the pyramid with apex x_c and base f. Still in this cases, f can be considered as a degenerated (flat) pyramid of apex its own center x_f and base f. Thus, in all cases, D_f is the union of two pyramids, and its boundary can be partitioned into triangles. The vertices of

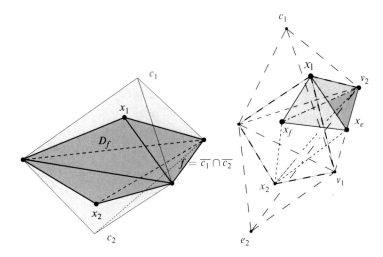

Fig. 1 Left: diamond cell for an internal triangular face f. Right: dual cell construction

these triangles either are: cell centers, vertices or boundary face centers of $\mathcal{M}$. As a result providing a scalar value to each cell, vertex and boundary face of $\mathcal{M}$ defines a unique continuous piecewise affine function $u_h : \partial\mathcal{D} \mapsto \mathbb{R}$, with $\partial\mathcal{D}$ defined in (1). This is precisely the lift from the discrete function in $\mathbb{U}_h$ presented in Sect. 1.2 into continuous piecewise affine functions on $\partial\mathcal{D}$ in (1).

Dual cells. Let $v \in \mathcal{V}$ and consider a cell $c \in \mathcal{C}$ and a face $f \in \mathcal{F}$ so that v is a vertex of f and f is a face of c. This configuration is denoted by $v \prec f \prec c$. To a triple (v, f, c) so that $v \prec f \prec c$ is associated an element $T_{v,f,c}$. The dual cell $v^\star$ then is defined as $v^\star = \bigcup_{f,c:\ v \prec f \prec c} T_{v,f,c}$. Let us eventually define the element $T_{v,f,c}$, as depicted on Fig. 1. Introduce w_1 and w_2 the two vertices of f such that $[v, w_1]$ and $[v, w_2]$ are two edges of f. Then $T_{v,f,c}$ is the union of the two tetrahedra $vx_c x_f w_i$ for $i = 1, 2$.
As one can see, for a fixed face f and a fixed cell c such that $f \subset \partial c$, considering all elements $T_{v,f,c}$ for all the vertices v of f recovers exactly twice $D_f \cap c$. As a result the dual cells recover the whole domain exactly twice: $\sum_{v \in \mathcal{V}} |v^\star| = 2|\Omega|$.

1.4 Discrete operators

The discrete divergence is classically defined by averaging the normal component of $\mathbf{q}_h \in \mathbb{Q}_h$ on the primal and dual cells, for all $c \in \mathcal{C}$ and all $v \in \mathcal{V}$:

$$\operatorname{div}_c \mathbf{q}_h = \frac{1}{|c|} \int_{\partial c} \mathbf{q}_h \cdot \mathbf{n} ds, \quad \operatorname{div}_v \mathbf{q}_h = \frac{1}{|v^\star|} \int_{\partial v^\star} \mathbf{q}_h \cdot \mathbf{n} ds, \qquad (4)$$

for **n** the unit normal on ∂c (resp. ∂v^*) pointing outside c (resp. v^*). This definition is well posed since the discontinuity set of $\mathbf{q}_h \in \mathbb{Q}_h$ has a zero 2-dimensional measure intersection with the primal and dual cell boundaries.

The discrete gradient is defined as follows. Let $u_h \in \mathbb{U}_h$, for all $f \in \mathscr{F}$:

$$\nabla_f u_h = \frac{1}{|D_f|} \int_{\partial D_f} u_h \mathbf{n} ds, \qquad (5)$$

where $\nabla_f u_h$ is the (vector) value of $\nabla_h u_h$ on D_f and for **n** the unit normal on ∂D_f pointing outside D_f.

In practice, definition (5) can always be reformulated in terms of data differences as in the 2D case where (see e.g. [7]):

$$\nabla_f u_h = (u_{c_1} - u_{c_2})\mathbf{N}_f + (u_{v_1} - u_{v_2})\mathbf{M}_f,$$

f is a mesh interface (edge), c_1 and c_2 the two cells on each side of f, v_1 and v_2 the two vertices of f and $\mathbf{N}_f, \mathbf{M}_f$ two vectors. We refer to [5, 15] for similar expansions in 3D.

1.5 The scheme

The linear diffusion problem $-\mathrm{div}(\mathbf{K}\nabla u) = f$ is considered together with a Dirichlet boundary condition $u_{|\partial\Omega} = g$. The tensor $\mathbf{K}$ is discretized into $\mathbf{K}_h$ by averaging $\mathbf{K}$ on each diamond cells and the source term f is discretized as $f_h \in \mathbb{F}_h$ by averaging f over each primal and dual cells. The problem reads: find $u_h \in \mathbb{U}_h$ such that

$$\forall c \in \mathscr{C}: \mathrm{div}_c(\mathbf{K}_h \nabla_h u_h) = f_c, \quad \forall v \in \mathscr{V}, v \notin \partial\Omega: \mathrm{div}_v(\mathbf{K}_h \nabla_h u_h) = f_v \qquad (6)$$

$$\forall v \in \mathscr{V}, v \in \partial\Omega: u_h(v) = g(v), \quad \forall f \in \mathscr{F}, f \subset \partial\Omega: u_h(x_f) = g(x_f), \qquad (7)$$

To solve (6) (7), we split $\mathbb{U}_h = \mathbb{U}_{h,0} \oplus \mathbb{B}$ where $\mathbb{U}_{h,0}$ is the subset of discrete functions equal to zero on $\partial\Omega$. Then u_h decomposes as $u_h = u_0 + \tilde{u}$, where $\tilde{u} \in \mathbb{B}$ is uniquely determined by (7). Now $u_0 \in \mathbb{U}_{h,0}$ satisfies $-\mathrm{div}_h(\mathbf{K}_h \nabla_h u_0) = f_h + \mathrm{div}_h(\mathbf{K}_h \nabla_h \tilde{u}) := \tilde{f}_h$ for all primal cells and all interior vertices. This is a square linear system equivalent with: find $u_0 \in \mathbb{U}_{h,0}$ so that for all $v \in \mathbb{U}_{h,0}$ we have:

$$-\langle\!\langle \mathrm{div}_h(\mathbf{K}_h \nabla_h u_0), v \rangle\!\rangle = \langle\!\langle \tilde{f}_h, v \rangle\!\rangle$$

With the help of the discrete duality property (2) it is also equivalent with finding $u_0 \in \mathbb{U}_{h,0}$ so that for all $v \in \mathbb{U}_{h,0}$:

$$\int_\Omega \mathbf{K}_h \nabla_h u_0 \cdot \nabla_h v \, dx = \langle\!\langle \tilde{f}_h, v \rangle\!\rangle.$$

In practice, introducing the stiffness matrix S associated to the discrete tensor $\mathbf{K}_h$, this problem is rewritten as the square positive symmetric linear system

$$S U_0 = \tilde{F}, \tag{8}$$

with U_0 (resp. $\tilde{F}$) the vector formed by the values of u_0 (resp. $\tilde{f}_h$) at the cell centers and interior vertices. The stiffness matrix S has the coefficients $S_{ij} = \int_\Omega \mathbf{K}_h \nabla_h w_i \cdot \nabla_h w_j \, dx$, with $w_i \in \mathbb{U}_{h,0}$ the base function having value 1 at one cell center or interior vertex and 0 everywhere else. This matrix is clearly symmetric and positive.

2 Numerical results

The cell centers as well as the face centers are set to their iso-barycenter.

Let us first define the data (source term f and anisotropy tensor $\mathbf{K}$) discretization. Primal, dual and diamond cells are partitioned using a single set of tetrahedra of type $E = x_c x_f v_1 v_2$, with $c \in \mathscr{C}$, f a face of c and v_1, v_2 two vertices of f forming one of its edges. To form the scalar data f_h, f is averaged on the tetrahedra E partitioning each primal and dual cells whereas the discrete tensor $\mathbf{K}_h$ is obtained by averaging $\mathbf{K}$ on the tetrahedra E partitioning the diamond cells. Averaging is made by the mean of Gaussian quadrature on each tetrahedra E using a 15 points quadrature formula of order 5, see e.g. [13]. Assembling the discrete source term f_h and tensor $\mathbf{K}_h$ requires one loop on the mesh faces.

The stiffness matrix S in (8) also is assembled using a loop on the mesh faces. Precisely two base functions w_i and w_j have a non zero interaction (i.e. $S_{ij} = \int_\Omega \mathbf{K}_h \nabla_h w_i \cdot \nabla_h w_j \, dx \neq 0$) in case they are associated to two vertices of a same diamond D_f.

Let us now define the L^2, H^1 and energy errors reported in the following tables as erl2, ergrad and ener respectively. Let u_h denote the discrete solution of one of the test case, and u the solution of the associated continuous problem. The discrete function u_h is lifted to a function $\bar{u}_h \in L^2(\Omega)$ as follows. Consider a face f, u_h provides a value at each vertex of D_f and also at the face center x_f in case of a boundary face. In case of an interior face, a supplementary value u_f is computed at x_f as $u_f = (\sum_{i=1}^n u_{v_i})/n$ where the v_i are the n vertices of f, which definition is consistent since x_f is the iso-barycenter of f. With these additional values, scalars are available for every vertices of the tetrahedra E that partition Ω: this defines a unique function $\bar{u}_h$ by P^1 interpolation, which then is continuous piecewise affine on Ω. We define:

$$\text{erl2}^2 = \frac{\int_\Omega |\bar{u}_h - u|^2 dx}{\int_\Omega |u|^2 dx}.$$

The discrete vector data $\nabla_h u_h$ is a piecewise constant vector function on the diamond cells. Therefore $\nabla_h u_h$ is an L^2 functions on Ω and the H^1 and energy errors reported in the following tables are defined as:

$$\text{ergrad}^2 = \frac{\int_\Omega |\nabla_h u_h - \nabla u|^2 dx}{\int_\Omega |\nabla u|^2 dx}, \quad \text{ener}^2 = \frac{\int_\Omega K(\nabla_h u_h - \nabla u) \cdot (\nabla_h u_h - \nabla u)}{\int_\Omega K \nabla u \cdot \nabla u dx}.$$

- **Test 1 Mild anisotropy,** $u(x, y, z) = 1 + \sin(\pi x) \sin\left(\pi \left(y + \frac{1}{2}\right)\right) \sin\left(\pi \left(z + \frac{1}{3}\right)\right)$
  min = 0, max = 2, **Tetrahedral meshes**

i	nu	nmat	umin	uemin	umax	uemax	normg
1	2187	21287	1.34E-02	1.53E-02	1.99E+00	1.99E+00	1.80E+00
2	4301	44813	3.24E-03	6.84E-03	1.99E+00	1.99E+00	1.80E+00
3	8584	94088	8.78E-03	7.44E-03	2.00E+00	1.99E+00	1.80E+00
4	17102	195074	4.74E-03	5.52E-03	2.00E+00	2.00E+00	1.80E+00
5	34343	405077	5.90E-04	1.49E-03	2.00E+00	2.00E+00	1.80E+00
6	69160	838856	1.30E-03	6.19E-04	2.00E+00	2.00E+00	1.80E+00

i	nu	erl2	ratiol2	ergrad	ratiograd	ener	ratioener
1	2187	1.39E-02	–	1.85E-01	–	1.80E-01	–
2	4301	8.80E-03	2.04E+00	1.48E-01	1.01E+00	1.44E-01	9.89E-01
3	8584	5.64E-03	1.93E+00	1.18E-01	9.73E-01	1.15E-01	9.97E-01
4	17102	3.61E-03	1.94E+00	9.36E-02	1.01E+00	9.10E-02	1.01E+00
5	34343	2.26E-03	2.01E+00	7.43E-02	9.92E-01	7.24E-02	9.81E-01
6	69160	1.42E-03	2.00E+00	5.87E-02	1.01E+00	5.70E-02	1.02E+00

- **Test 1 Mild anisotropy,** $u(x, y, z) = 1 + \sin(\pi x) \sin\left(\pi \left(y + \frac{1}{2}\right)\right) \sin\left(\pi \left(z + \frac{1}{3}\right)\right)$
  min = 0, max = 2, **Voronoï meshes**

i	nu	nmat	umin	uemin	umax	uemax	normg
1	87	1433	1.23E-01	1.79E-01	1.91E+00	1.85E+00	1.80E+00
2	235	4393	6.66E-02	2.93E-03	1.87E+00	2.00E+00	1.80E+00
3	527	10777	1.32E-02	9.56E-03	1.93E+00	1.97E+00	1.80E+00
4	1013	21793	-1.76E-03	4.97E-03	1.93E+00	2.00E+00	1.80E+00
5	1776	40998	5.42E-04	4.30E-03	1.98E+00	1.97E+00	1.80E+00

i	nu	erl2	ratiol2	ergrad	ratiograd	ener	ratioener
1	87	6.19E-02	–	4.43E-01	–	4.29E-01	–
2	235	3.36E-02	1.85E+00	3.37E-01	8.28E-01	3.29E-01	7.96E-01
3	527	2.10E-02	1.74E+00	2.55E-01	1.03E+00	2.49E-01	1.04E+00
4	1013	1.35E-02	2.03E+00	2.05E-01	1.01E+00	2.01E-01	9.85E-01
5	1776	9.99E-03	1.62E+00	1.75E-01	8.38E-01	1.71E-01	8.47E-01

- **Test 1 Mild anisotropy,** $u(x,y,z) = 1 + \sin(\pi x)\sin\left(\pi\left(y+\frac{1}{2}\right)\right)\sin\left(\pi\left(z+\frac{1}{3}\right)\right)$ min = 0, max = 2, **Kershaw meshes**

i	nu	nmat	umin	uemin	umax	uemax	normg
1	855	13819	7.16E-02	2.88E-02	1.94E+00	1.96E+00	1.80E+00
2	7471	138691	1.26E-02	6.45E-03	1.99E+00	1.99E+00	1.80E+00
3	62559	1237459	1.30E-03	1.75E-03	2.00E+00	2.00E+00	1.80E+00
4	512191	10443763	4.61E-04	5.45E-04	2.00E+00	2.00E+00	1.80E+00

i	nu	erl2	ratiol2	ergrad	ratiograd	ener	ratioener
1	855	5.64E-02	–	4.57E-01	–	4.51E-01	–
2	7471	1.71E-02	1.65E+00	1.91E-01	1.20E+00	1.89E-01	1.21E+00
3	62559	3.45E-03	2.26E+00	7.74E-02	1.28E+00	7.67E-02	1.27E+00
4	512191	7.62E-04	2.15E+00	3.47E-02	1.14E+00	3.41E-02	1.16E+00

- **Test 1 Mild anisotropy,** $u(x,y,z) = 1 + \sin(\pi x)\sin\left(\pi\left(y+\frac{1}{2}\right)\right)\sin\left(\pi\left(z+\frac{1}{3}\right)\right)$ min = 0, max = 2, **Checkerboard meshes**

i	nu	nmat	umin	uemin	umax	uemax	normg
1	59	703	1.46E-01	3.41E-02	1.86E+00	1.97E+00	1.80E+00
2	599	9835	3.87E-02	8.56E-03	1.96E+00	1.99E+00	1.80E+00
3	5423	101539	9.24E-03	2.14E-03	1.99E+00	2.00E+00	1.80E+00
4	46175	917395	2.15E-03	5.35E-04	2.00E+00	2.00E+00	1.80E+00
5	381119	7788403	5.01E-04	1.34E-04	2.00E+00	2.00E+00	1.80E+00

i	nu	erl2	ratiol2	ergrad	ratiograd	ener	ratioener
1	59	4.79E-02	–	4.01E-01	–	3.94E-01	–
2	599	1.08E-02	1.93E+00	1.95E-01	9.34E-01	1.92E-01	9.31E-01
3	5423	2.55E-03	1.96E+00	9.58E-02	9.66E-01	9.37E-02	9.73E-01
4	46175	6.27E-04	1.96E+00	4.75E-02	9.83E-01	4.63E-02	9.89E-01
5	381119	1.56E-04	1.98E+00	2.36E-02	9.92E-01	2.30E-02	9.95E-01

Benchmark 3D: CeVe-DDFV, a Discrete Duality Scheme

- **Test 2 Heterogeneous anisotropy,** $u(x, y, z) = x^3 y^2 z + x \sin(2\pi x z) \sin(2\pi x y) \sin(2\pi z)$, min $= -0.862$, max $= 1.0487$, **Prism meshes**

i	nu	nmat	umin	uemin	umax	uemax	normg
1	3010	64158	-8.54E-01	-8.41E-01	1.00E+00	1.00E+00	1.71E+00
2	24020	555528	-8.56E-01	-8.59E-01	1.02E+00	1.05E+00	1.71E+00
3	81030	1924098	-8.61E-01	-8.59E-01	1.04E+00	1.04E+00	1.71E+00
4	192040	4619868	-8.59E-01	-8.61E-01	1.04E+00	1.05E+00	1.71E+00

i	nu	erl2	ratiol2	ergrad	ratiograd	ener	ratioener
1	3010	5.06E-02	–	2.45E-01	–	2.48E-01	–
2	24020	1.85E-02	1.45E+00	1.26E-01	9.63E-01	1.27E-01	9.66E-01
3	81030	1.46E-02	5.90E-01	8.51E-02	9.63E-01	8.59E-02	9.66E-01
4	192040	1.37E-02	2.08E-01	6.49E-02	9.44E-01	6.53E-02	9.50E-01

- **Test 3 Flow on random meshes,** $u(x, y, z) = \sin(2\pi x) \sin(2\pi y) \sin(2\pi z)$, min $= -1$, max $= 1$, **Random meshes**

i	nu	nmat	umin	uemin	umax	uemax	normg
1	91	1063	-1.58E+00	-9.78E-01	1.54E+00	9.31E-01	3.65E+00
2	855	13819	-1.08E+00	-9.94E-01	1.12E+00	9.82E-01	3.57E+00
3	7471	138691	-1.04E+00	-9.95E-01	1.01E+00	9.91E-01	3.60E+00
4	62559	1237459	-1.01E+00	-9.98E-01	1.01E+00	9.98E-01	3.60E+00

i	nu	erl2	ratiol2	ergrad	ratiograd	ener	ratioener
1	91	3.06E-01	–	5.89E-01	–	5.70E-01	–
2	855	8.29E-02	1.75E+00	3.14E-01	8.56E-01	2.87E-01	9.21E-01
3	7471	2.28E-02	1.79E+00	1.65E-01	8.90E-01	1.46E-01	9.28E-01
4	62559	6.98E-03	1.67E+00	8.96E-02	8.58E-01	7.34E-02	9.68E-01

- **Test 4 Flow around a well, Well meshes,** min $= 0$, max $= 5.415$

i	nu	nmat	umin	uemin	umax	uemax	normg
1	1482	23942	4.85E-01	-6.02E-06	5.32E+00	5.42E+00	1.62E+03
2	3960	70872	2.71E-01	-5.68E-06	5.33E+00	5.42E+00	1.62E+03
3	9229	173951	1.66E-01	-5.76E-06	5.33E+00	5.42E+00	1.62E+03
4	21156	412240	1.25E-01	-7.39E-06	5.33E+00	5.42E+00	1.62E+03
5	44420	882520	9.37E-02	-6.93E-06	5.34E+00	5.42E+00	1.62E+03
6	82335	1654893	7.48E-02	-6.94E-06	5.35E+00	5.42E+00	1.62E+03
7	145079	2937937	5.80E-02	-8.05E-06	5.36E+00	5.42E+00	1.62E+03

i	nu	erl2	ratiol2	ergrad	ratiograd	ener	ratioener
1	1482	2.92E-03	–	1.79E-01	–	1.78E-01	–
2	3960	1.38E-03	2.29E+00	1.22E-01	1.18E+00	1.21E-01	1.16E+00
3	9229	7.45E-04	2.19E+00	8.57E-02	1.25E+00	8.56E-02	1.24E+00
4	21156	5.53E-04	1.08E+00	6.56E-02	9.71E-01	6.55E-02	9.72E-01
5	44420	3.77E-04	1.55E+00	5.14E-02	9.85E-01	5.13E-02	9.83E-01
6	82335	2.44E-04	2.11E+00	4.18E-02	1.01E+00	4.17E-02	1.01E+00
7	145079	1.83E-04	1.53E+00	3.51E-02	9.27E-01	3.50E-02	9.26E-01

3 Comments

The linear system (8) to be solved is symmetric and positive: a Conjugate Gradient algorithm has been applied, together with a basic Jacobi preconditioner. The sparsity pattern of the stiffness matrix is not compact, especially for matrix lines corresponding to vertex nodes. The stiffness matrix lines corresponding to cell nodes have $1 + n_f + n_s$ nonzero terms with n_f and n_v the number of faces and vertices of the considered cell; for a tetrahedra $1 + n_f + n_v = 9$. The maximum principle is not fulfilled by *DDFV* schemes. In practice it has been violated only once for test one on Voronoï meshes and more significantly on test 3. Meanwhile no oscillation phenomena are observed. Expected order 2 convergence on erl2 is observed for all tests excepted test 2. Order 1 convergence is observed for ergrad and ener on all tests.

References

1. Andreianov, B., Bendahmane, M., Karlsen, K.H.: A gradient reconstruction formula for finite volume schemes and discrete duality. Proceedings of FVCA5 (2008)
2. Andreianov, B., Bendahmane, M., Karlsen, K.H.: Discrete duality finite volume schemes for doubly nonlinear degenerate hyperbolic-parabolic equations. J. Hyperbolic Differ. Equ. **7**(1), 1–67 (2010)
3. Coudière, Y., Hubert, F.: A 3d discrete duality finite volume method for nonlinear elliptic equation. HAL Preprint URL http://hal.archives-ouvertes.fr/docs/00/45/68/37/PDF/ddfv3d.pdf
4. Coudière, Y., Hubert, F.: A 3d duality finite volume method for nonlinear elliptic equations. Proceedings of Algoritmy 2009 pp. 51–60 (2009)
5. Coudiere, Y., Pierre, C., Rousseau, O., Turpault, R.: 2D/3D DDFV scheme for anisotropic-heterogeneous elliptic equations, application to electrograms simulation from medical data. Int. J. Finite Volumes (2009)
6. Coudière, Y., Vila, J.P., Villedieu, P.: Convergence rate of a finite volume scheme for a two dimensional convection-diffusion problem. M2AN **33**(3), 493–516 (1999)
7. Delcourte, S., Domelevo, K., Omnès, P.: Discrete-duality finite volume method for second order elliptic problems. Proceedings of FVCA4 pp. 447–458 (2005)

8. Domelevo, K., Omnes, P.: A finite volume method for the Laplace equation on almost arbitrary two-dimensional grids. M2AN Math. Model. Numer. Anal. **39**(6), 1203–1249 (2005)
9. Hermeline, F.: Une méthode de volumes finis pour les équations elliptiques du second ordre. C. R. Acad. Sci. **326**(12), 1433–1436 (1998)
10. Hermeline, F.: A finite volume method for the approximation of diffusion operators on distorted meshes. J. Comput. Phys. **160**(2), 481–499 (2000)
11. Hermeline, F.: Approximation of 2-D and 3-D diffusion operators with variable full tensor coefficients on arbitrary meshes. Comput. Methods Appl. Mech. Engrg. **196**(21-24), 2497–2526 (2007)
12. Hermeline, F.: A finite volume method for approximating 3D diffusion operators on general meshes. Comput. Meth. Appl. Mech. Engrg. (2009)
13. Jinyun, Y.: Symmetric gaussian quadrature formulae for tetrahedronal regions. Computer Methods in Applied Mechanics and Engineering (1981)
14. Nicolaides, R.: Direct discretization of planar div-curl problems. SIAM J. Numer. Anal. **29**(1), 32–56 (1992)
15. Pierre, C.: Modelling and simulating the electrical activity of the heart embedded in the torso, numerical analysis and finite volumes methods. . PhD Thesis, Université de Nantes (2005)

The paper is in final form and no similar paper has been or is being submitted elsewhere.

Benchmark 3D: A multipoint flux mixed finite element method on general hexahedra

Mary F. Wheeler, Guangri Xue, and Ivan Yotov

1 Presentation of the scheme

In this paper we discuss a family of numerical schemes for modeling Darcy flow, the multipoint flux mixed finite element (MFMFE) methods. The MFMFE methods allow for an accurate and efficient treatment of irregular geometries and heterogeneities such as faults, layers, and pinchouts that require highly distorted grids and discontinuous coefficients. The methods can be reduced to cell-centered discretizations and have convergent pressures and velocities on general hexahedral and simplicial grids.

The development of the MFMFE methods has been motivated by the multipoint flux approximation (MPFA) methods [1, 2, 7, 8]. In the MPFA finite volume framework, sub-edge (sub-face) fluxes are introduced, which allows for local flux elimination and reduction to a cell-centered scheme. Similar elimination is achieved in the MFMFE variational framework, by employing appropriate finite element spaces and special quadrature rules. Our approach is based on the BDM$_1$ [5] or the BDDF$_1$ [3] spaces with a trapezoidal quadrature rule applied on the reference element. We refer to [4] for a similar approach on simplicial grids, as well as to [10, 11] for a related work on quadrilateral grids using a broken Raviart-Thomas space. Mortar MFMFE methods on non-matching grids have been developed in [14].

We describe the method for a single phase Darcy flow in a domain $\Omega \subset \mathbb{R}^3$

$$\psi = -K\nabla u, \quad \nabla \cdot \psi = f \quad \text{in } \Omega, \quad u = 0 \quad \text{on } \partial\Omega,$$

Mary F. Wheeler and Guangri Xue
The University of Texas at Austin, USA, e-mail: mfw@ices.utexas.edu, gxue@ices.utexas.edu

Ivan Yotov
University of Pittsburgh, USA, e-mail: yotov@math.pitt.edu

where ψ is the Darcy velocity, u is the pressure, and K is a symmetric, uniformly positive definite tensor representing the rock permeability divided by the fluid viscosity. Other boundary conditions can also be treated. The weak formulation of the problem reads: find $\psi \in H(\mathbf{div}; \Omega)$ and $u \in L^2(\Omega)$, such that

$$(K^{-1}\psi, \mathbf{v}) - (u, \nabla \cdot \mathbf{v}) = 0, \qquad \forall \mathbf{v} \in H(\mathbf{div}; \Omega), \tag{1}$$

$$(\nabla \cdot \psi, w) = (f, w), \qquad \forall w \in L^2(\Omega), \tag{2}$$

where $H(\mathbf{div}; \Omega) := \{ \mathbf{v} \in (L^2(\Omega))^d : \nabla \cdot \mathbf{v} \in L^2(\Omega) \}$ and $(\cdot, \cdot)$ denotes the inner product in $L^2(\Omega)$.

Multipoint flux mixed finite element (MFMFE) methods have been developed and analyzed in [9, 13–15] for simplicial, quadrilateral, and hexahedral grids. The method is defined as follows: find $\psi_h \in V_h$ and $u_h \in W_h$ such that

$$(K^{-1}\psi_h, \mathbf{v})_Q - (u_h, \nabla \cdot \mathbf{v}) = 0, \qquad \forall \mathbf{v} \in V_h, \tag{3}$$

$$(\nabla \cdot \psi_h, w) = (f, w), \qquad \forall w \in W_h \tag{4}$$

In the above V_h and W_h are suitable mixed finite element spaces and $(\cdot, \cdot)_Q$ is a special quadrature rule. Appropriate choices allow for a flux variable defined at a vertex to be expressed by cell-centered pressures surrounding the vertex. This results in a 27 point pressure stencil on logically rectangular 3D grids.

The quadrature rule (9) can be symmetric or non-symmetric. On smooth hexahedral grids, both the symmetric and non-symmetric MFMFE methods give first-order accurate velocities and pressures, as well as second order accurate face fluxes and pressures at the cell centers [9, 13, 15]. On highly distorted hexahedral grids with non-planar faces [13], the convergence of the symmetric MFMFE can deteriorate while the non-symmetric MFMFE still gives a first order accuracy under a mild assumption on the grids and permeability anisotropy. The non-symmetric quadrature rule was first proposed in [10] for quadrilateral grids.

Finite element spaces. Let Ω be a polyhedral domain partitioned into a union of hexahedral finite elements of characteristic size h. Let us denote the partition by $\mathcal{T}_h$ and assume that it is shape-regular and quasi-uniform [6]. The velocity and pressure finite element spaces on any physical grid-block E are defined, respectively, via the Piola transformation

$$\mathbf{v} \leftrightarrow \hat{\mathbf{v}} : \hat{\mathbf{v}} = \frac{1}{J_E} DF_E \hat{\mathbf{v}} \circ F_E^{-1},$$

and the scalar transformation

$$w \leftrightarrow \hat{w} : w = \hat{w} \circ F_E^{-1},$$

where $\hat{E}$ is the reference cube or tetrahedron, F_E denotes a trilinear mapping from $\hat{E}$ to E, DF_E is the Jacobian of F_E, and J_E is its determinant. The Piola transformation preserves the normal components of the vectors. The finite element spaces V_h and

Benchmark 3D: A multipoint flux mixed finite element method on general hexahedra

W_h on $\mathcal{T}_h$ are given by

$$V_h = \left\{\mathbf{v} \in H(\mathbf{div};\Omega): \quad \mathbf{v}|_E \leftrightarrow \hat{\mathbf{v}},\ \hat{\mathbf{v}} \in \hat{V}(\hat{E}),\quad \forall E \in \mathcal{T}_h\right\},$$
$$W_h = \left\{w \in L^2(\Omega): \quad w|_E \leftrightarrow \hat{w},\ \hat{w} \in \hat{W}(\hat{E}),\quad \forall E \in \mathcal{T}_h\right\}, \tag{5}$$

where $\hat{V}(\hat{E})$ and $\hat{W}(\hat{E})$ are finite element spaces on the reference element $\hat{E}$.

The spaces on the reference cube are defined by enhancing the BDDF$_1$ spaces:

$$\hat{V}(\hat{E}) = \text{BDDF}_1(\hat{E}) + r_2\text{curl}(0,0,\hat{x}^2\hat{z})^T + r_3\text{curl}(0,0,\hat{x}^2\hat{y}\hat{z})^T + s_2\text{curl}(\hat{x}\hat{y}^2,0,0)^T$$
$$+ s_3\text{curl}(\hat{x}\hat{y}^2\hat{z},0,0)^T + t_2\text{curl}(0,\hat{y}\hat{z}^2,0)^T + t_3\text{curl}(0,\hat{x}\hat{y}\hat{z}^2,0)^T,$$

$$\hat{W}(\hat{E}) = P_0(\hat{E}),$$

where the BDDF$_1(\hat{E})$ space is defined as [3]:

$$\text{BDDF}_1(\hat{E}) = P_1(\hat{E})^3 + r_0\text{curl}(0,0,\hat{x}\hat{y}\hat{z})^T + r_1\text{curl}(0,0,\hat{x}\hat{y}^2)^T + s_0\text{curl}(\hat{x}\hat{y}\hat{z},0,0)^T,$$
$$+ s_1\text{curl}(\hat{y}\hat{z}^2,0,0)^T + t_0\text{curl}(0,\hat{x}\hat{y}\hat{z},0)^T + t_1\text{curl}(0,\hat{x}^2\hat{z},0)^T.$$

In above equations, r_i, s_i, t_i ($i = 0, \ldots, 3$) are real constants, P_k denotes polynomials of degree at most k, and $(\hat{x}, \hat{y}, \hat{z})^T$ denotes a point in the reference element. The enhancement of the BDDF$_1$ space is needed to obtain a space with four degrees of freedom per face, rather than three in the original formulation. This allows to associate a degree of freedom with each vertex of the face, which is needed in the reduction to a cell-centered pressure stencil as described later in this section.

There are four degrees of freedom (DOF) per reference face. The DOF are chosen to be the normal components at the vertices. This choice of DOF guarantees continuity of the normal component of the velocity vector across element faces, which is needed for an $H(\mathbf{div};\Omega)$-conforming velocity space as required by (5).

A quadrature rule. The integration on a physical element is performed by mapping to the reference element and choosing a quadrature rule on $\hat{E}$. Using the Piola transformation, we write $(K^{-1}\cdot,\cdot)$ in (1) as

$$(K^{-1}\mathbf{q},\mathbf{v})_E = \left(\frac{1}{J_E}DF_E^T K^{-1}(F_E(\hat{x}))DF_E\hat{\mathbf{q}},\hat{\mathbf{v}}\right)_{\hat{E}} \equiv (\mathcal{M}_E\hat{\mathbf{q}},\hat{\mathbf{v}})_{\hat{E}},$$

where

$$\mathcal{M}_E = \frac{1}{J_E}DF_E^T K^{-1}(F_E(\hat{x}))DF_E. \tag{6}$$

Define a perturbed $\widetilde{\mathcal{M}}_E$ as

$$\widetilde{\mathcal{M}}_E = \frac{1}{J_E}DF_E^T(\hat{\mathbf{r}}_{c,\hat{E}})\overline{K}_E^{-1}DF_E, \tag{7}$$

where $\hat{\mathbf{r}}_{c,\hat{E}}$ is the centroid of $\hat{E}$ and $\overline{K}_E$ denotes the mean of K on E. In addition, denote the trapezoidal rule on $\hat{E}$ by $\text{Trap}(\cdot,\cdot)_{\hat{E}}$:

$$\text{Trap}(\hat{\mathbf{q}},\hat{\mathbf{v}})_{\hat{E}} \equiv \frac{|\hat{E}|}{k}\sum_{i=1}^{k}\hat{\mathbf{q}}(\hat{\mathbf{r}}_i)\cdot\hat{\mathbf{v}}(\hat{\mathbf{r}}_i), \tag{8}$$

where $\{\hat{\mathbf{r}}_i\}_{i=1}^{k}$ are the vertices of $\hat{E}$.

The symmetric quadrature rule is based on the original $\mathcal{M}_E$ while the non-symmetric one is based on the perturbed $\widetilde{\mathcal{M}}_E$. The quadrature rule on an element E is defined as

$$(K^{-1}\mathbf{q},\mathbf{v})_{Q,E} \equiv \begin{cases} \text{Trap}(\mathcal{M}_E\hat{\mathbf{q}},\hat{\mathbf{v}})_{\hat{E}} = \frac{|\hat{E}|}{k}\sum_{i=1}^{k}\mathcal{M}_E(\hat{\mathbf{r}}_i)\hat{\mathbf{q}}(\hat{\mathbf{r}}_i)\cdot\hat{\mathbf{v}}(\hat{\mathbf{r}}_i), & \text{symmetric}, \\ \text{Trap}(\widetilde{\mathcal{M}}_E\hat{\mathbf{q}},\hat{\mathbf{v}})_{\hat{E}} = \frac{|\hat{E}|}{k}\sum_{i=1}^{k}\widetilde{\mathcal{M}}_E(\hat{\mathbf{r}}_i)\hat{\mathbf{q}}(\hat{\mathbf{r}}_i)\cdot\hat{\mathbf{v}}(\hat{\mathbf{r}}_i), & \text{non-symmetric}. \end{cases} \tag{9}$$

The non-symmetric quadrature rule has certain critical properties on the physical elements that lead to a convergent method on rough hexahedra [13].

Reduction to a cell-centered pressure system. The choice of trapezoidal quadrature rule implies that on each element, the velocity degrees of freedom associated with a vertex become decoupled from the rest of the degrees of freedom. As a result, the assembled velocity mass matrix in (3) has a block-diagonal structure with one block per grid vertex. The dimension of each block equals the number of velocity DOF associated with the vertex. Inverting each local block in the mass matrix in (3) allows for expressing the velocity DOF associated with a vertex in terms of the pressures at the centers of the elements that share the vertex. Substituting these expressions into the mass conservation equation (4) leads to a cell-centered system for the pressures. The stencil is 27 points on logically rectangular hexahedral grids. The local linear systems and the resulting global pressure system are positive definite and therefore invertible for the symmetric MFMFE method and, under a mild restriction on the shape regularity of the grids and/or the anisotropy of the permeability, for the non-symmetric MFMFE method; see (11) below. The reader is referred to [9, 13, 15] for further details on the reduction to a cell-centered pressure system.

Theoretical convergence results. Let $W^{k,\infty}_{\mathcal{T}_h}$ consist of functions ϕ such that $\phi|_E \in W^{k,\infty}(E)$ for all $E \in \mathcal{T}_h$. Here k is a multi-index with integer components and $W^{k,\infty}(E)$ denotes the Sobolev space of functions whose derivatives of order k belong to $L^\infty(E)$. Let $\|\cdot\|_k$ be the norm in the Hilbert space $H^k(\Omega)$ with functions whose derivatives of order k belong to $L^2(\Omega)$. The norm in $L^2(\Omega)$ is denoted by $\|\cdot\|$. Let $X \lesssim (\gtrsim) Y$ denote that there exists a constant C, independent of the mesh size h, such that $X \leq (\geq) CY$. The notation $X \approx Y$ means that both $X \lesssim Y$ and $X \gtrsim Y$ hold.

The following convergence results have been established for the symmetric MFMFE method on h^2-perturbed parallelepipeds.

Theorem 1 ([9, 15]). *If $K^{-1} \in W^{1,\infty}_{\mathcal{T}_h}$, then, the velocity ψ_h and pressure u_h of the symmetric MFMFE method (3)–(4) satisfy*

$$\|\psi - \psi_h\| \lesssim h\|\psi\|_1, \quad \|\nabla \cdot (\psi - \psi_h)\| \lesssim h\|\nabla \cdot \psi\|_1, \quad \|u - u_h\| \lesssim h(\|\psi\|_1 + \|u\|_1).$$

On h^2-perturbed parallelepipeds, the non-symmetric MFMFE method has same order of accuracy as the symmetric method. In addition, the non-symmetric method has first order convergence for the velocity and pressure on general quadrilaterals and for the face flux and pressure on general hexahedra with non-planar faces.

For the analysis of the non-symmetric MFMFE method, we require some properties of the bilinear form $(K^{-1} \cdot, \cdot)_Q$ defined on the space V_h. Note that

$$(K^{-1}\mathbf{q}, \mathbf{v})_Q = \sum_{E \in \mathcal{T}_h} (K^{-1}\mathbf{q}, \mathbf{v})_{Q,E} = \sum_{c \in \mathcal{C}_h} \mathbf{v}_c^T \mathbf{M}_c \mathbf{q}_c, \qquad (10)$$

where $\mathcal{C}_h$ denotes the set of corner or vertex points in $\mathcal{T}_h$, $\mathbf{v}_c := \{(\mathbf{v} \cdot \mathbf{n}_e)(\mathbf{x}_c)\}_{e=1}^{n_c}$, $\mathbf{x}_c$ is the coordinate vector of point c, and n_c is the number of faces (or edges in 2D) that share the vertex point c.

Lemma 1 ([13]). *Assume that $\mathbf{M}_c$ is uniformly positive definite for all $c \in \mathcal{C}_h$:*

$$h^d \boldsymbol{\xi}^T \boldsymbol{\xi} \lesssim \boldsymbol{\xi}^T \mathbf{M}_c \boldsymbol{\xi}, \quad \forall \boldsymbol{\xi} \in \mathbb{R}^{n_c}. \qquad (11)$$

Then the bilinear form $(K^{-1} \cdot, \cdot)_Q$ is coercive in $\mathbf{V}_h$ and induces a norm in V_h equivalent to the L^2-norm:

$$(K^{-1}\mathbf{v}, \mathbf{v})_Q \simeq \|\mathbf{v}\|^2, \quad \forall \mathbf{v} \in V_h. \qquad (12)$$

If in addition

$$\boldsymbol{\xi}^T \mathbf{M}_c^T \mathbf{M}_c \boldsymbol{\xi} \lesssim h^{2d} \boldsymbol{\xi}^T \boldsymbol{\xi}, \quad \forall \boldsymbol{\xi} \in \mathbb{R}^{n_c}, \qquad (13)$$

then the following Cauchy-Schwarz type inequality holds:

$$(K^{-1}\mathbf{q}, \mathbf{v})_Q \lesssim \|\mathbf{q}\|\|\mathbf{v}\| \quad \forall \mathbf{q}, \mathbf{v} \in V_h, \qquad (14)$$

Conditions (11) and (13) impose mild restrictions on the element geometry and the anisotropy of the permeability tensor K, see [10, 12].

Theorem 2 ([13]). *Let $K \in W^{1,\infty}_{\mathcal{T}_h}(\Omega)$ and $K^{-1} \in W^{0,\infty}(\Omega)$. If (11) and (13) hold, then the velocity ψ_h and the pressure u_h of the non-symmetric MFMFE method (3)—(4) satisfy*

$$\|\Pi\psi - \psi_h\| + \|Q_h u - u_h\| \lesssim h(|\psi|_1 + \|u\|_2), \qquad (15)$$

where Π is the canonical interpolation operator onto V_h and Q_h is the L^2-orthogonal projection onto W_h.

This result further implies convergence of the computed normal velocity to the true normal velocity on the element faces. First, define a norm for vectors in Ω based on the normal components on the faces of $\mathcal{T}_h$:

$$\|\mathbf{v}\|_{\mathcal{F}_h}^2 := \sum_{E \in \mathcal{T}_h} \sum_{e \in \partial E} \frac{|E|}{|e|} \|\mathbf{v} \cdot \mathbf{n}_e\|_e^2, \tag{16}$$

where $|E|$ is the volume of E and $|e|$ is the area of e. This norm gives an appropriate scaling of $|\Omega|^{1/2}$ for a unit vector.

Theorem 3 ([13]). *Let $K \in W_{\mathcal{T}_h}^{1,\infty}(\Omega)$ and $K^{-1} \in W_{\mathcal{T}_h}^{0,\infty}(\Omega)$. If (11) and (13) hold, then the velocity ψ_h of the non-symmetric MFMFE method (3)–(4) satisfies*

$$\|\psi - \psi_h\|_{\mathcal{F}_h} \lesssim h(\|\psi\|_1 + \|u\|_2). \tag{17}$$

2 Numerical results

We note that in all tests we report absolute errors. Both the pressure error $\|u - u_h\|$ and the velocity error $\|\Pi\psi - \psi_h\|$ are approximated by the trapezoidal quadrature rule on the reference unit cube. For the velocity face error $\|\psi - \psi_h\|_{\mathcal{F}_h}$ and the mean velocity face error

$$\|\psi - \psi_h\|_{\mathcal{F}_h}^2 \equiv \sum_{E \in \mathcal{T}_h} \sum_{e \in \partial E} |E| \left(\frac{1}{|e|} \int_e \psi \cdot \mathbf{n}_e - \frac{1}{|e|} \int_e \psi_h \cdot \mathbf{n}_e \right)^2,$$

the face integrals are approximated by the 9-point Gaussian quadrature rule on the reference face.

- **Test 1 Mild anisotropy,** $u(x, y, z) = 1 + \sin(\pi x) \sin\left(\pi \left(y + \frac{1}{2}\right)\right) \sin\left(\pi \left(z + \frac{1}{3}\right)\right)$, min $= 0$, max $= 2$, **Kershaw meshes**

Symmetric MFMFE

i	nu	nmat	umin	uemin	umax	uemax
1	512	10648	4.66E-03	3.03E-02	1.97E+00	1.96E+00
2	4096	97336	4.23E-03	1.06E-02	1.99E+00	1.99E+00
3	32768	830584	-2.42E-03	1.75E-03	2.00E+00	2.00E+00
4	262144	6859000	7.49E-05	7.14E-04	2.00E+00	2.00E+00

Benchmark 3D: A multipoint flux mixed finite element method on general hexahedra

i	nu	$\|u-u_h\|$	rate	$\|\Pi\psi-\psi_h\|$	rate	$\|\psi-\psi_h\|_{\mathscr{F}_h}$	rate	$\|\psi-\psi_h\|_{\mathscr{F}_h}$	rate	Iters
1	512	2.08E-01	–	3.01E+00	–	4.74E+00	–	4.30E+00	–	9
2	4096	1.17E-01	0.83	1.11E+00	1.44	2.17E+00	1.13	1.94E+00	1.15	17
3	32768	5.96E-02	0.97	3.95E-01	1.45	7.44E-01	1.54	6.62E-01	1.55	32
4	262144	2.95E-02	1.01	1.54E-01	1.36	2.43E-01	1.61	2.01E-01	1.72	65

Non-symmetric MFMFE

i	nu	nmat	umin	uemin	umax	uemax
1	512	10648	-1.25E-03	3.03E-02	2.01E+00	1.96E+00
2	4096	97336	-3.35E-03	1.06E-02	2.00E+00	1.99E+00
3	32768	830584	-2.08E-03	1.75E-03	2.00E+00	2.00E+00
4	262144	6859000	5.02E-05	7.14E-04	2.00E+00	2.00E+00

i	nu	$\|u-u_h\|$	rate	$\|\Pi\psi-\psi_h\|$	rate	$\|\psi-\psi_h\|_{\mathscr{F}_h}$	rate	$\|\psi-\psi_h\|_{\mathscr{F}_h}$	rate	Iters
1	512	2.20E-01	–	2.81E+00	–	2.52E+00	–	2.14E+00	–	8
2	4096	1.19E-01	0.89	9.15E-01	1.62	1.23E+00	1.03	1.07E+00	1.00	16
3	32768	5.95E-02	1.00	3.06E-01	1.58	4.27E-01	1.53	3.66E-01	1.55	33
4	262144	2.94E-02	1.02	1.18E-01	1.37	1.40E-01	1.61	1.00E-01	1.87	73

- **Test 1 Flow on random meshes,** $u(x,y,z) = 1 + \sin(\pi x)\sin\left(\pi\left(y+\frac{1}{2}\right)\right)\sin\left(\pi\left(z+\frac{1}{3}\right)\right)$, min = 0, max = 2, **Random meshes**

Symmetric MFMFE

i	nu	nmat	umin	uemin	umax	uemax
1	64	1000	-1.43E-02	4.46E-02	1.90E+00	1.82E+00
2	512	10648	2.03E-02	3.17E-02	1.96E+00	1.95E+00
3	4096	97336	-1.07E-03	2.69E-03	1.99E+00	1.99E+00
4	32768	830584	1.14E-03	1.23E-03	2.00E+00	2.00E+00

i	nu	$\|u-u_h\|$	rate	$\|\Pi\psi-\psi_h\|$	rate	$\|\psi-\psi_h\|_{\mathscr{F}_h}$	rate	$\|\psi-\psi_h\|_{\mathscr{F}_h}$	rate	Iters
1	64	2.54E-01	–	1.15E+00	–	1.01E+00	–	4.26E-01	–	5
2	512	1.25E-01	1.02	6.14E-01	0.91	4.91E-01	1.04	1.79E-01	1.25	6
3	4096	6.29E-02	0.99	3.82E-01	0.68	2.86E-01	0.78	1.00E-01	0.84	7
4	32768	3.15E-02	1.00	2.96E-01	0.37	2.34E-01	0.29	8.84E-02	0.18	8

Non-symmetric MFMFE

i	nu	nmat	umin	uemin	umax	uemax
1	64	1000	-2.17E-02	4.46E-02	1.90E+00	1.82E+00
2	512	10648	1.52E-02	3.17E-02	1.96E+00	1.95E+00
3	4096	97336	-1.42E-03	2.69E-03	1.99E+00	1.99E+00
4	32768	830584	5.59E-04	1.23E-03	2.00E+00	2.00E+00

i	nu	$\|u-u_h\|$	rate	$\|\Pi\psi-\psi_h\|$	rate	$\|\psi-\psi_h\|_{\mathcal{F}_h}$	rate	$\|\!|\psi-\psi_h|\!\|_{\mathcal{F}_h}$	rate	Iters
1	64	2.54E-01	–	1.19E+00	–	1.01E+00	–	3.83E-01	–	5
2	512	1.25E-01	1.02	5.54E-01	1.10	4.32E-01	1.23	1.23E-01	1.64	7
3	4096	6.30E-02	0.99	2.78E-01	0.99	2.05E-01	1.08	4.63E-02	1.41	7
4	32768	3.15E-02	1.00	1.39E-01	1.00	1.09E-01	0.91	2.32E-02	1.00	8

- **Test 3 Flow on random meshes**, $u(x, y, z) = \sin(2\pi x)\sin(2\pi y)\sin(2\pi z)$, min $= -1$, max $= 1$, **Random meshes**

Symmetric MFMFE

i	nu	nmat	umin	uemin	umax	uemax
1	64	1000	-6.20E+00	-7.59E-01	5.75E+00	6.91E-01
2	512	10648	-1.93E+00	-9.39E-01	2.05E+00	9.23E-01
3	4096	97336	-1.20E+00	-9.85E-01	1.19E+00	9.82E-01
4	32768	830584	-1.06E+00	-9.96E-01	1.04E+00	9.96E-01

i	nu	$\|u-u_h\|$	rate	$\|\Pi\psi-\psi_h\|$	rate	$\|\psi-\psi_h\|_{\mathcal{F}_h}$	rate	$\|\!|\psi-\psi_h|\!\|_{\mathcal{F}_h}$	rate	Iters
1	64	1.88E+00	–	1.67E+03	–	1.76E+03	–	1.22E+03	–	17
2	512	4.27E-01	2.14	5.84E+02	1.52	5.31E+02	1.73	3.34E+02	1.87	36
3	4096	1.48E-01	1.53	2.97E+02	0.98	2.32E+02	1.19	1.19E+02	1.49	59
4	32768	6.71E-02	1.14	2.02E+02	0.56	1.59E+02	0.55	7.57E+01	0.65	77

Non-symmetric MFMFE

i	nu	nmat	umin	uemin	umax	uemax
1	64	1000	-1.20E+02	-7.59E-01	3.76E+01	6.91E-01
2	512	10648	-5.01E+02	-9.39E-01	6.34E+02	9.23E-01
3	4096	97336	-3.34E+01	-9.85E-01	4.97E+01	9.82E-01
4	32768	830584	-2.16E+03	-9.96E-01	4.12E+03	9.96E-01

i	nu	$\|u-u_h\|$	rate	$\|\Pi\psi-\psi_h\|$	rate	$\|\psi-\psi_h\|_{\mathcal{F}_h}$	rate	$\|\!|\psi-\psi_h|\!\|_{\mathcal{F}_h}$	rate
1	64	2.94E+01	–	1.07E+05	–	9.36E+04	–	3.45E+04	–
2	512	8.96E+01	< 0	3.45E+05	< 0	2.78E+05	< 0	1.41E+05	< 0
3	4096	5.84E+00		3.39E+04		2.50E+04		1.11E+04	
4	32768	3.56E+02	< 0	3.23E+06	< 0	2.53E+06	< 0	1.08E+06	< 0

- **Test 5 Discontinuous permeability**,

$u(x, y, z) = a_i \sin(2\pi x)\sin(2\pi y)\sin(2\pi z)$, min $= -100$, max $= 100$, **Locally refined meshes**

The locally refined grids are treated by introducing mortar finite elements on the subdomain interfaces to approximate the interface pressure and impose weakly continuity of flux; for details see [14]. Here we take the mortar grid to be the trace of the coarser subdomain grid and choose the mortar space to consist of discontinuous piecewise bilinear functions. It is easy to check that this results in forcing on each interface element the four fine grid normal velocities to be equal to the coarse grid normal velocity.

For this test we also report the velocity error $\|\psi - \psi_h\|$, approximated by 27-point Gaussian quadrature rule on the reference unit cube, as well as the norm $\|\psi - \Pi^{RT}\psi_h\|$ defined as follows. For a scalar function $\phi(x_1, x_2, x_3)$ in a cubic element E, let $\|\phi\|_{i,E}$ denote an approximation integral of $|\phi|^2$ using exact integration rule in x_i and midpoint rule in the other directions. Then, for $\mathbf{q} = (q_1, q_2, q_3)^T$, let

$$\|\mathbf{q}\|^2 = \sum_{E \in \mathcal{T}_h} \sum_{i=1}^{3} \|q_i\|_{i,E}^2.$$

In the reported error norm, Π^{RT} is the canonical interpolation operator in the lowest order Raviart-Thomas space.

i	nu	umin	uemin	umax	uemax
2	176	-4.36E+01	-3.54E+01	4.36E+01	3.54E+01
3	1408	-8.30E+01	-7.89E+01	8.30E+01	7.89E+01
4	11264	-9.56E+01	-9.43E+01	9.56E+01	9.43E+01
5	90112	-9.89E+01	-9.86E+01	9.89E+01	9.86E+01

i	$\|u - u_h\|$	rate	$\|\Pi\psi - \psi_h\|$	rate	$\|\psi - \psi_h\|$	rate	$\|\psi - \Pi^{RT}\psi_h\|$	rate	CGiter
2	2.28E+01	–	1.77E+03	–	9.49E+02	–	3.38E+02	–	12
3	1.19E+01	0.94	8.80E+02	1.00	4.96E+02	0.94	8.18E+01	2.05	18
4	6.02E+00	0.98	4.38E+02	1.00	2.51E+02	0.98	2.03E+01	2.01	21
5	3.02E+00	1.00	2.19E+02	1.00	1.26E+02	0.99	5.06E+00	2.00	31

3 Comments

In Test 1 and Test 3, the resulting linear algebraic system is solved using the software HYPRE (high performance preconditioners) developed by researchers at Lawrence Livermore National Laboratory[1]. Specifically, we use the generalized minimum residual (GMRES) method with one V-cycle of algebraic multigrid method as a preconditioner. The stopping criteria for GMRES is relative residual less than 10^{-9}. The number of iterations is reported in each table.

---

[1] https://computation.llnl.gov/casc/hypre/software.html

In Test 5, the problem is reduced to an interface problem in terms of mortar variables. We use the conjugate gradient (CG) method and the stopping criteria is the relative residual less than 10^{-9}. The number of CG iterations is given in the table.

In Test 1 Mild anisotropy, both the symmetric and non-symmetric methods are first order accurate for the pressure and the velocity, as well as superconvergent of order approaching $O(h^2)$ for the face velocities.

In Test 1 Flow on random meshes, the pressure is first order for both methods. However, the velocity convergence of the symmetric method deteriorates due to the element distortion, while the non-symmetric method maintains first order accuracy in the velocity. These results are consistent with Theorems 1, 2, and 3.

In Test 3 Flow on random meshes, the symmetric method is first order convergent for the pressure and approximately $O(h^{1/2})$ convergent for the velocity, as expected by the theory. For the non-symmetric method, the severe anisotropy in the permeability combined with element distortion leads to near violation of conditions (11) and (13). As a result, the algebraic system is very ill-conditioned and the method fails to converge.

In Test 5 Discontinuous permeability, the two methods are identical, since the elements are cuboids. We observe first order convergence for the pressure and velocity, as well as second order superconvergence for the error $\|\psi - \Pi^{RT}\psi_h\|$, as predicted by the theory from [14].

The paper is in final form and no similar paper has been or is being submitted elsewhere.

References

1. I. Aavatsmark. An introduction to multipoint flux approximations for quadrilateral grids. *Comput. Geosci.*, 6:405–432, 2002.
2. I. Aavatsmark, T. Barkve, O. Boe, and T. Mannseth. Discretization on unstructured grids for inhomogeneous, anisotropic media, part ii: Discussion and numerical results. *SIAM J. Sci. Comput.*, 19(5):1717–1736, 1998.
3. F. Brezzi, J. Douglas, R. Duran, and M. Fortin. Mixed finite elements for second order elliptic problems in three variables. *Numer. Math.*, 51:237–250, 1987.
4. F. Brezzi, M. Fortin, and L. D. Marini. Error analysis of piecewise constant pressure approximations of Darcy's law. *Comput. Methods Appl. Mech. Eng.*, 195:1547–1559, 2006.
5. Franco Brezzi, Jim Douglas, and L. D. Marini. Two families of mixed finite elements for second order elliptic problems. *Numer. Math.*, 47(2):217–235, 1985.
6. P. G. Ciarlet. *The Finite Element Method for Elliptic Problems*. Stud. Math. Appl. 4, North-Holland, Amsterdam, 1978; reprinted, SIAM, Philadelphia, 2002.
7. M. G. Edwards. Unstructured control-volume distributed, full-tensor finite-volume schemes with flow based grids. *Comput. Geosci.*, 6:433–452, 2002.
8. M. G. Edwards and C. F. Rogers. Finite volume discretization with imposed flux continuity for the general tensor pressure equation. *Comput. Geosci.*, 2:259–290, 1998.
9. R. Ingram, M. F. Wheeler, and I. Yotov. A multipoint flux mixed finite element method on hexahedra. *SIAM J. Numer. Anal.*, 48:1281–1312, 2010.

10. R. A. Klausen and R. Winther. Robust convergence of multi point flux approximation on rough grids. *Numer. Math.*, 104:317–337, 2006.
11. Runhild A. Klausen and Ragnar Winther. Convergence of multipoint flux approximations on quadrilateral grids. *Numer. Methods Partial Differential Equations*, 22(6):1438–1454, 2006.
12. K. Lipnikov, M. Shashkov, and I. Yotov. Local flux mimetic finite difference methods. *Numer. Math.*, 112(1):115–152, 2009.
13. M. F. Wheeler, G. Xue, and I. Yotov. A multipoint flux mixed finite element method on distorted quadrilaterals and hexahedra. *ICES REPORT 10-34, The Institute for Computational Engineering and Sciences, The University of Texas at Austin, Submitted*, 2010.
14. M. F. Wheeler, G. Xue, and I. Yotov. A multiscale mortar multipoint flux mixed finite element method. *ICES REPORT 10-33, The Institute for Computational Engineering and Sciences, The University of Texas at Austin, Submitted*, 2010.
15. M. F. Wheeler and I. Yotov. A multipoint flux mixed finite element method. *SIAM. J. Numer. Anal.*, 44(5):2082–2106, 2006.